CONTROL SYSTEM DESIGN AND SIMULATION

CONTROL SYSTEM DESIGN AND SIMULATION

Jack Golten

Principal Lecturer
Department of Mechanical Engineering
Manchester Polytechnic

and

Andy Verwer

Senior Lecturer
Department of Mechanical Engineering
Manchester Polytechnic

McGRAW-HILL BOOK COMPANY

London · New York · St Louis · San Francisco · Auckland
Bogotá · Caracas · Hamburg · Lisbon · Madrid · Mexico · Milan
Montreal · New Delhi · Panama · Paris · San Juan · São Paulo
Singapore · Sydney · Tokyo · Toronto

Published by
McGRAW-HILL Book Company Europe
Shoppenhangers Road, Maidenhead, Berkshire, SL6 2QL, England
TEL: 0628 23432; FAX: 0628 770224

British Library Cataloguing in Publication Data
Golten, Jack
  Control system design and simulation.
  1. Control systems. Design
  I. Title II. Verwer, Andy
  629.8312

  ISBN 0-07-707412-2

Library of Congress Cataloging-in-Publication Data
Golten, Jack
  Control system design and simulation/Jack Golten and Andy Verwer.
    p.   cm.
  Includes bibliographical references and index.
  ISBN 0-07-707412-2:
  1. Feedback control systems—Design and construction—Data processing. I. Verwer, Andy. II.
Title.
TJ216.G64   1991   90-24724
629.8'3—dc20   CIP

234 CL 94321

Typeset by Mid-County Press, London
and printed and bound in Great Britain by Clays Ltd, St Ives plc

In fond memory of Marcel and Jean

CONTENTS

This book and accompanying computer software form a unit which is suitable as a course in feedback control systems on a wide range of degree level studies. The approach is based on student centred learning, where readers can try out various ideas and exercises on their own. This book is therefore also suitable as a 'teach yourself' text for individuals not pursuing a formal course.

The book emphasizes the sound grasp of principles, and has many examples of how to apply these principles to produce working designs that meet specifications. The mathematical content has been kept down to a minimum and where possible heuristic argument has been used rather than rigorous and over-facing analysis. Many of the exercises and problems are designed to encourage the reader to explore the subject area and in conjunction with the accompanying software will allow imaginative students to test and develop their knowledge. The case studies show how the design methods can be applied to practical problems.

The software supplied with the book (CODAS) is a special issue of a professional package (CODAS-II) with restricted features. The accompanying disk has all the capabilities of the full package except for the discrete time and nonlinear facilities, also the overall system is limited to fifth order. It allows most of the exercises and problems in the book on continuous time linear systems to be tackled and validated. The exercises and problems on discrete time and nonlinear systems in Chapters 8, 9 and 10 can be done with the full version of CODAS-II. A comprehensive manual for CODAS is supplied on the disk. The file README.DOC contains information on the supplied software and any updates.

CODAS-II (Control System Design and Simulation) is a fully integrated graphics-based package for designing and simulating feedback control systems. PCS is a package concerned exclusively with process control. CODAS-II (and to a lesser extent PCS) is the computational vehicle that is used to promote the ideas developed in this book. In addition to its use as a teaching aid, CODAS-II is a powerful tool for the practising control engineer and technician. All packages are written in efficient C programming language and run on fully IBM-compatible machines which are fitted with one of the standard graphics adaptors such as Hercules, CGA, EGA, VGA and others.

CODAS-II provides time domain, frequency domain and root locus environments for the design and simulation of single-input/single-output control systems. The plant and controller dynamics are defined in terms of transfer functions and optional transport delay can be included. CODAS-II can cater for continuous and discrete time systems and can include nonlinear elements which are defined using an interactive editor. CODAS-II is designed to facilitate rapid user interaction and promote 'what if?' experimentation. The techniques described in this book are not exclusive to CODAS-II; certainly other packages do exist on which many or all of the exercises can be done and many of the techniques can be tried.

Often in the past, control engineering text books developed designs that were not or could not easily be tested. CODAS-II is a hard task master because for every design and design technique there is instant verification and there is no room for hopeful intent. Using CODAS-II often shatters preconceived views and challenges former fondly-held beliefs.

Even though the book is heavily based on computer-aided design, the design methods used are mainly classical. There is an introduction to state space but on the whole the classical methods of s-plane, z-plane and frequency domain have been used because they have an intuitive appeal which is generally lost in state space representations. State space methods come into their own when dealing with multi-variable systems and in optimal control problems; both of which are outside the scope of this book.

The reader should be familiar with complex numbers and their use for vector manipulation. A knowledge of elementary calculus is required and some familiarity with simple differential equations is useful but not essential. It is expected that the reader is sufficiently familiar with the MS–DOS operating system to run programs and manipulate and examine files.

The book begins with an introduction to feedback control systems and modelling in Chapters 1 and 2. Chapter 3 deals with the transient and steady-state responses of open-loop systems. The correlation between transient response and s-plane diagram is explored. First- and second-order dynamics are covered in detail as well as the effects of additional poles and zeros. Closed-loop systems are introduced towards the end of Chapter 3 where steady-state performance is covered. In Chapters 4 and 5 frequency domain methods are developed. Chapter 4 concentrates primarily on open-loop systems whereas Chapter 5 deals with the design of closed-loop systems using frequency domain techniques. Some new, or at least revised design techniques using the 'D' contour are covered here.

Chapter 6 returns to the s-plane and examines root locus and root contours as design techniques. Two full case studies are included which look at the design of control systems for two very different applications. Chapter 7 deals with process control and the tuning of three-term controllers. A thorough evaluation of popular tuning techniques is included. Chapters 8 and 9 deal with discrete time and computer control systems. Chapter 8 examines industrial control computer hardware and software and covers the analytical techniques required for discrete time system analysis. Chapter 9 develops design methods for discrete time controllers and filters. Practical aspects, including sample rate selection, anti-aliasing filters and quantization are also dealt with. A comprehensive case study concludes this chapter. The final chapter looks at nonlinear control systems using perturbation methods and describing functions.

We would like to acknowledge the support given to us by the Department of Mechanical Engineering, Design and Manufacture; Manchester Polytechnic. We also thank all the students over the years who have acted as guinea pigs in helping to develop our computer-aided design and simulation approach to teaching control engineering.

Jack Golten
Andy Verwer

ONE

INTRODUCTION TO CONTROL SYSTEMS

1.1 INTRODUCTION

The automatic control of machines and processes is fundamental to the successful operation of modern industry. Modern manufacturing, processing and transportation systems are heavily dependent on automatic control systems. The benefits of automatic control include more consistent operation, greater safety for the process or machine and operating personnel and reduced operating costs due to improved utilization and reduction in manpower requirements.

The need for automatic control continues to grow in terms of the range of applications and performance requirements. The development of high performance civil and military aircraft, missile technology and space vehicles has placed great demands on the speed and accuracy of attitude control systems. In manufacturing, the developing use of robots and automated production has further increased the need for reliable, high performance control systems. In the process industry, stricter requirements for product quality, energy efficiency and pollution levels place tighter limits on process control systems.

The reducing cost and increased performance of digital computers has had a significant impact on the way control systems are designed and implemented. Powerful mini- and microcomputers with superb graphics capabilities are readily available for designing and simulating control systems. Computers are also widely used for implementing automatic control in an increasing variety of industrial and domestic applications. Even mass-produced consumer products can include powerful microprocessor systems to monitor and control the system. For example, the average music centre or video camera contains more computing power and automatic control than a typical engineering laboratory of the mid 1960s.

1.2 BASIC CONTROL SYSTEM TERMINOLOGY

A *control system* consists of a *controller* and a *plant*. We use the general term plant to describe the machine, vehicle or process which is being controlled. The controller can

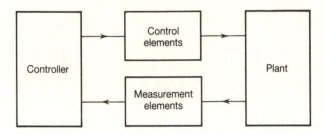

Figure 1.1 A general control system

be a person, in which case we have a *manual control system*. Alternatively, in an *automatic control system* the controller is a device, electronic circuit, computer, or mechanical linkage, etc. Figure 1.1 shows the general arrangement.

The interface between the plant and the controller requires actuators (*control elements*) to provide the control action. In addition instrumentation, detectors and sensors (*measurement elements*) are needed to provide information about the plant status to the controller. The information passing between the controller and the plant is in the form of *signals*. These signals can be very diverse, for example electrical, pneumatic or mechanical, etc. The term 'transmitter' is commonly used to describe the measurement element in a process control system because the transmitter sends an electrical or pneumatic signal representing the measured value to the controller.

Controllers are usually implemented electronically, either using analogue circuits or a digital computer (microprocessor). Pneumatic and hydraulic controllers are also to be found. Actuators are commonly pneumatic, electric or hydraulic depending on the application and power level required.

The behaviour and performance of a control system depends on the interaction of all the elements. The individual components cannot generally be considered in isolation. The plant itself is probably the most important element in any control system; the best controller in the world cannot make an inadequate plant operate well.

Feedback

In everyday life, feedback occurs when we are made aware of the consequences of our actions. Feedback is so natural that we take it for granted. Imagine trying to accomplish the simplest of tasks without feedback, for example, trying to walk without visual feedback. Feedback not only gives verification of our actions: it allows us to cope with a changing environment by adjusting our actions in the presence of unforeseen events and changing conditions.

Feedback has similar advantages when applied to automatic control. Feedback occurs in automatic control systems when the control action depends upon the measured state of the machine or process being controlled. Feedback gives an automatic control system the ability to deal with unexpected disturbances and changes in the plant behaviour.

Sequential and Quantitative Control Systems

A *sequential control system* involves *logic control functions*. The sensors monitoring the plant provide switched outputs which produce only on/off signals. For example an automatic door may be fitted with limit switches to detect the position of the door and an infrared detector with a switched output to sense an approaching person. The control function involves the use of logical rules so that the actuators operate in the correct sequence and at the correct time. Sequential control systems are common in factory automation, automatic warehouses and the control of batch operations. The design of sequential control systems involves problems in logic and is not covered in this book.

The objectives of a *quantitative control system* are different. This type of control system is concerned with controlling the actual value of some plant quantity. Measurement elements provide quantitative information to the controller rather than just on/off signals. The division between sequential control and quantitative control can be vague and some systems can be considered in both ways.

As an example, a modern automatic washing machine clearly involves sequential control to switch the various solenoid valves and pumps on and off in the required progression for the selected wash program. Quantitative control is used for the wash drum rotational speed. Here the actual drum speed is measured and controlled by altering the power delivered to the motor. A less clear example is the thermostatic control of wash temperature. The thermostat switches the heater element on and off depending on whether the wash temperature is too low or too high. The control signal is clearly on/off but since it is the actual value of the temperature which is important the system can also be considered as quantitative.

The behaviour of a quantitative control system depends fundamentally on the rate and extent to which the plant responds to the control action. Such *dynamic* behaviour is difficult to predict and the design of quantitative control systems to achieve acceptable response is no trivial matter. This book is concerned with the behaviour and design of quantitative control systems.

1.3 OBJECTIVES OF AUTOMATIC CONTROL SYSTEMS

Regulation

A control system for maintaining the plant output constant at the desired value in the presence of external disturbances is called a *regulator*. Disturbances will cause the plant output to deviate and the regulator must apply control action or *control effort* to attempt to maintain the plant output at the reference value with the minimum of error. Feedback is fundamental to regulation because only feedback can provide information about the actual plant output. A good regulator will minimize the effects of disturbances on the plant output.

Trajectory Following

Quite often a control system is required to make the plant output follow a certain profile or trajectory. A *servo system* is a control system specifically designed to follow a changing reference value. The servo problem, as it is called, is of major concern in transportation, defence and manufacturing systems. The servo must apply control effort to make the plant output follow the desired path with the minimum of error.

It is clear that the regulation and servo problems are very similar and indeed many control systems give good regulation against disturbances and close following of a changing reference.

1.4 CONTROL STRATEGIES

In order to examine some different control strategies let us consider a simple level control problem. Figure 1.2 shows a tank holding liquid for feeding some process. The process being supplied requires a constant head feed and so a control system is required to keep the tank level constant at some reference level. A valve is located in the tank inlet to alter or modulate the flow rate.

Open-loop Control

The simplest strategy is to have a dial on the inlet valve. By experiment the valve can be moved to different positions and a note made of the dial position and the corresponding level in the vessel. The dial can be calibrated in 'metres'. Thus if it is decided to operate at a different level, the valve can be moved to the corresponding position on the dial. This strategy is termed open-loop control. Open-loop control is simple and will work well provided there is no change in the flow of liquid from the vessel and all other parameters affecting the level in the vessel remain constant.

Feedforward Control

The major cause of disturbances affecting the tank level is likely to be changes in the tank outflow rate. An increased outflow will cause the level to drop. A more sophisticated strategy is to use a set of calibrations over a number of outflow rates. By monitoring the outflow rate when the plant is in operation, the correct position of the valve can

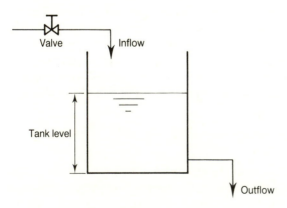

Figure 1.2   Example plant for level control

be determined by examining the calibration curve for the new flow and then opening or closing the inlet valve accordingly. This strategy is termed feedforward control.

Feedforward control requires a set of calibration curves or a *model* of the relationship between the valve position, outflow and level. The relationship can be obtained experimentally by measuring the level for various outflow rates and inlet valve positions. Alternatively the model can be formulated from a theoretical analysis of the tank. Another consequence of this strategy is that a measurement of the outlet flow rate is required to calculate the necessary change in the position of the inlet valve.

Feedforward control is an improvement over simple open-loop control. But it only caters for the one variable which is being monitored (in this case the outflow) and relies on a good model of the plant. If the model is inaccurate or the behaviour of the system varies with time then the feedforward strategy may not work too well. Disturbances can originate from many causes which may not be included in the model, or are not monitored. For example, the supply pressure upstream of the inlet valve may change or the density of the liquid could alter. These variations will cause the relationship between inlet valve position and tank inflow to change and so the tank level will be incorrect. The feedforward strategy will not correct for these factors.

Feedback Control

Rather than adding more feedforward measurements to compensate for these other factors, the obvious solution to maintain the level in the vessel is to monitor the level itself and adjust the inlet valve if the level deviates from the desired value. Such a *feedback* strategy is *error driven* in that the control effort is a function of the difference between the required level and actual level. The relationship between the error and the control effort is called the *control law*. Feedback control, unlike feedforward, can give regulation against unmeasured or unmodelled disturbances.

On/Off Feedback Control

The simplest method of monitoring the level is by means of a level switch (float switch). The level switch is mounted in the tank at the desired level. The switch produces a binary (on/off) signal that indicates whether the level is above or below the required value. The signal can be used to operate the inlet valve directly. When the level is above the reference the inlet valve is closed and when below it is opened. The control law in on/off control therefore switches the control effort between extremes depending on the sign of the error.

On/off control certainly overcomes the criticisms of the open-loop and feedforward strategies. Whatever the cause of the change in level, if the deviation in the level is large enough to activate the switch then control action will be applied to correct the situation. The required level (reference) in this simple scheme is determined by the position of the level switch on the tank. On/off control requires only very simple equipment in the form of level switches and a simple solenoid type actuator to open or shut the valve.

There are several problems with on/off control. One problem concerns the violent fluctuations in inlet flow as the valve switches between fully open and fully shut. These flow changes may appear as a significant disturbance to any process feeding the tank.

Another problem concerns plants which do not respond immediately to the control effort applied. Such delays are common in more complex plants and particularly in temperature control systems. Any delay in the plant response means that the process output will continue to rise even after the upper switching limit is reached. Eventually the output will respond to the control effort and start to fall, but of course the process output will once again continue to drop past the lower switching limit. Thus the precision of on/off control depends heavily on any plant delay.

Even with negligible delay, the tank level will be constantly fluctuating. If the switch is very sensitive it only requires a small change in level to change state. A sensitive switch will cause the inlet valve to switch between fully open and fully closed very frequently as the level cycles around the required setting. Also any waves or ripples on the liquid surface could cause a very sensitive switch to be activated very rapidly. The frequency of switching can be traded off against a loss of precision in the level control by using a less sensitive switch or even by using two switches, one at high level to close the inlet valve and another at low level to open it.

Modulating Feedback Control

Rather than switch the inlet valve open and shut, a more subtle approach is to inch the valve by an amount which depends on the difference between the actual level and the desired level. This strategy can be termed *modulating feedback control*.

Modulating control implies a more elaborate level measurement and valve actuator. In the first place it requires a signal related to the actual level (i.e. a level transmitter). Secondly the valve actuator must be able to open and close the inlet valve gradually (modulate the valve opening). Furthermore the valve itself must have a smooth characteristic so that its resistance to flow is infinitely variable.

The control law can profoundly effect the way in which a feedback control system behaves. Simple on/off control can sometimes give acceptable performance; most domestic heating systems are controlled this way. The important characteristic of modulating control is that it is capable of providing a range of control effort and can produce small, as well as large, corrections. With a well designed control law, feedback control can provide good regulation and trajectory following. One of the primary functions of a control engineer is to design or select an appropriate control law for the plant which gives acceptable performance.

Figure 1.3 shows a general block diagram of a control system with both feedforward and feedback control. The box labelled 'feedback controller' is where the feedback contribution to the control effort is produced. If there is a feedforward component, it is added to the feedback contribution to make up the total control effort signal.

1.5 SAFETY

One overriding consideration in control system design and implementation is safety. This may relate to the reliability and robustness of the control strategy, but more often deals with the monitoring of exceptional conditions and the subsequent alarm and safety systems associated with the detection of a malfunction or of a variable that has reached

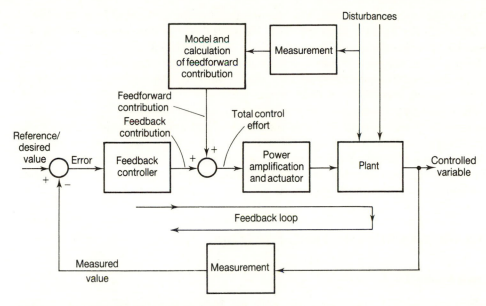

Figure 1.3 General block diagram of a control system

dangerous levels. Safety systems are usually separate from the normal control system and can be treated independently in terms of strategy and implementation.

Safety systems usually involve detection of unsafe or potentially unsafe conditions using on/off or switching sensors. Standby equipment may need to be started if there is a failure or the plant may need to be shut down safely. Safety systems involve the same problems of binary logic as sequential control systems.

Quite often control systems can be designed to *fail safe*. With the correct design, failures of instrumentation and control equipment can result in safe rather than unsafe situations. For example consider the level control problem examined above. It may be that the liquid held in the tank is corrosive or otherwise dangerous. It would therefore be unsafe if the tank were to overflow due to a control system failure. The fail safe philosophy is to design the control system and associated instrumentation such that a failure in any one element causes the control effort to act in a safe way (i.e. the tank inlet valve should close).

For a fail safe measurement, a failure in the measuring element should produce the same signal as exists in the dangerous condition. In the case of the level control system a level transmitter failure should look like a high level in the tank. This can be achieved by using a low signal to represent a high level (a reverse acting transmitter). Loss of the measurement signal would thus look like a high level in the tank and so the controller would shut the valve preventing tank overflow.

The fail safe philosophy can also be applied to the control system actuator. The argument here is simpler, the actuator should move to the safe position in the event of an actuator power failure. For the tank inlet control valve this means that the valve should close if the actuator signal is lost. This can be accomplished by making the actuator push against a spring to open the valve (i.e. a spring return actuator). The spring would then close the valve in the absence of actuator power.

The above arguments are based on the premise that the tank should not be allowed to overflow. If the liquid held in the tank is essential to the downstream process (i.e. lubricating oil or coolant) then the tank should not be allowed to run dry in the event of a failure. The situation is now reversed and the fail safe argument implies that a direct acting level transmitter should be used. Similarly the control valve must be designed to open, should the actuator power fail.

Safety considerations are very dependent on the process or system, and notwithstanding their importance, can only be touched upon in a general text of this nature. Experience and a thorough knowledge of the system operation are required before sensible decisions on safety aspects can be made. It must be emphasized that most control engineers spend far more time considering safety aspects than in designing the control system for normal operation.

1.6 EXAMPLES OF CONTROL SYSTEMS

Furnace Control

Furnaces are used in the process industry for heating feedstocks prior to further treatment. For example, in an oil refinery, crude oil is heated before it enters the crude distillation column where it is split up into fractions which eventually produce marketable products such as aviation fuel or road bitumen. The initial heating is carefully controlled as the subsequent fractionation is quite dependent on the degree of vaporization of the crude oil feed.

Figure 1.4 shows a simplified diagram of a process furnace. The temperature of the process fluid is measured at the outlet of the furnace by a temperature transmitter which sends a signal to the temperature controller. The controller has a dial so that the desired

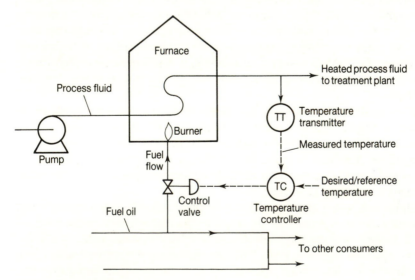

Figure 1.4 Process furnace

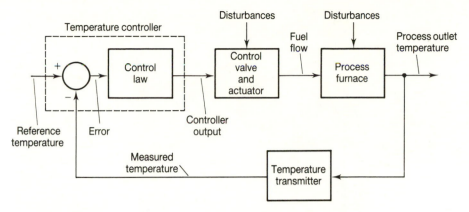

Figure 1.5   Block diagram for temperature control of a process furnace

value of the process outlet temperature (the reference temperature) can be set. The controller compares the measured temperature with the reference temperature and the fuel flow is increased or decreased accordingly. The exact relationship between the error in temperature and the fuel valve movement is determined by the control law.

A block diagram of the control system is shown in Fig. 1.5. The main source of disturbances are changes in the process fluid flowrate. Changes in throughput are inevitable for numerous reasons. For example, if there is unsufficient feedstock or if there is a sudden increase in demand because of a weather change or even fluctuations in the stock market. Feedforward control would be relatively easy to implement by monitoring the flow of process fluid and adding in a component to the feedback control effort.

Turning to the safety of this furnace control system, fail safe precautions can be taken to ensure that a control equipment failure results in the burner shutting down. However there are many other safety considerations to be considered in fired equipment such as this. For example, during normal operation disturbances in the process fluid could cause the furnace outlet temperature to rise. If the correcting control effort is excessive, the fuel valve may be shut off causing a burner flame failure. When the fuel flow is restored by the controller opening the valve, the potential for an explosion will exist because of the unignited fuel entering the hot furnace. Solutions to this problem can include placing limits on the fuel control valve travel, use of flame detectors and automatic re-ignition of the extinguished flame.

Control of a Robotic Arm

Robots are increasingly used for materials handling, automatic assembly and fabrication. Robotic arms normally have several joints or axes each fitted with an actuator enabling the arm to move in a variety of ways to position and orientate the gripper. The actuators are controlled by a dedicated computer so that the correct sequence of motions is carried out. The computer normally does this by replaying a stored sequence of desired motions. Sometimes more sophisticated systems use television cameras or touch sensors to help decide on the required gripper motion.

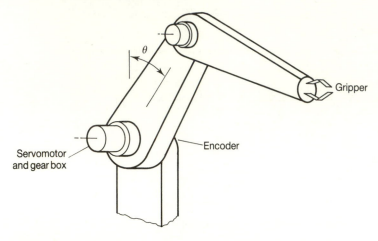

Figure 1.6 Robotic arm

The control computer must control all the robot joints simultaneously. Sometimes each axis has its own separate microprocessor which receives commands from the main control computer. However implemented, the scheme for the positioning of each axis is similar, so the control of just one arm joint will be examined as shown in Fig. 1.6.

In this example, the joint is driven by a dc electric motor (servomotor) through a gear box. A dc motor is a flexible actuator which can be driven in either direction at various speeds by altering the direction and magnitude of the motor current. The current is supplied to the motor by a power amplifier which in turn receives its input from the computer or microprocessor. The computer deals with digital data and so the interface to the power amplifier requires a digital to analogue converter. The angle of the arm joint is monitored by an encoder. An encoder is a transducer which produces a digital output representing the measured angle which can be directly interfaced to the computer.

As the computer retrieves the sequence of desired robot motions, the required positions for the axis are passed to a part of the program which implements the control law. The control calculation involves subtracting the measured arm angle from the required arm angle to find the angular position error. The computer then uses this error to determine the magnitude and direction of the control effort. This calculated control effort is converted to an analogue signal which is applied to the input of the power amplifier to drive the servomotor and hence reduce the error. Figure 1.7 shows the arrangement in block diagram form.

The main requirement in this control system is that of fast and accurate trajectory following. The controller can use the rate of change of measured position to calculate the velocity of the joint and so reduce the motor current if the arm is moving too fast. The control law in a high-performance robot may be quite complex and will be tailored to the way in which the arm is expected to respond.

A difficulty arises because the arm may respond differently when carrying a load than when empty-handed. The motor will require more current to accelerate and decelerate with the load than without. The control computer may know whether the gripper is holding a load and in theory could then compensate by using more current.

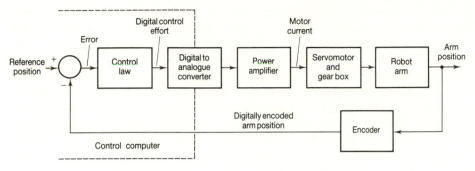

Figure 1.7 Block diagram of robotic arm control system (one axis only)

In practice however, the control law is usually chosen to give adequate performance with a range of loads.

1.7 CONTROL SYSTEM DYNAMICS, MODELLING, SIMULATION AND DESIGN

The response of any natural system to stimuli is not instantaneous, it takes time for the system to respond. We are all aware that before we can make a cup of instant coffee, it takes time for the kettle to boil after we have switched it on. When a car hits a bump in the road the suspension bounces or oscillates for a while before it settles back to an equilibrium position. When we hit the accelerator pedal of our car, it takes time for it to accelerate to a new speed, even if we happen to own the latest Lamborghini! The way the response of a system evolves as function of time to a particular stimulus is termed the dynamic response.

The dynamic behaviour of systems can be modelled by differential equations. Models can be obtained by applying the basic equations of physics and mechanics to the system. This approach can be applied to systems where the underlying principles are clear and where the system is sufficiently simple, or can be broken down into simple subsystems, to use a 'first principles' approach. Another approach is to observe the behaviour of the system in its normal working environment or introduce test signals. A model can then be proposed from the observations of the input/output behaviour.

The solution of differential equations by hand methods is difficult and so a range of numerical methods have been developed which can be implemented on digital computers. Many computer techniques and packages exist for solving differential equations and simulating the behaviour of dynamic systems.

Control system design involves simulating the dynamic behaviour of feedback systems, but it also involves other techniques which require the use of computers. Dynamic simulation alone is not enough. Design methods for control systems have evolved since the 1930s. At first methods were developed which were based on short-cut hand calculations. Analogue computers were used to do the final simulation and verify the design. This state of affairs did not change significantly until the early 1950s when the first digital computers came into commercial and scientific use. Computer programs were written to solve sets of differential equations and results produced on listing paper. Graphic displays were virtually nonexistent and the interaction between the designer and the computer was very poor. Even with the development of minicomputers in the mid 1960s the situation was not significantly better.

Graphics display terminals first became widely available in the early 1970s. Their use for control system design was generally restricted to larger universities and industrial companies. User interaction was still poor, and often the design suites were not well integrated. The 1980s saw the arrival of the 'personal computer' (PC) and the availability of graphics terminals of high quality. Many software houses adapted existing programs to the PC, but really did not take advantage of its potential as a design tool. The ancestry of the programs was clearly visible.

In addition to the improvements and cost reduction of the hardware, the second half of the 1980s saw an increasing awareness of well designed (user friendly) software. Business sodftware with programs such as word processors, spreadsheets and databases was the first area in which this development occurred. Later, as graphics capabilities grew, computer-aided design packages for draughting, three-dimensional modelling, printed circuit design, etc., appeared. Computer-Aided Control System Design (CACSD) software was slow to respond to this trend since a fairly large investment was required with a more restricted market. Today there are several good quality, low cost packages available.

PROBLEMS

1.1 Classify the following into sequential or quantitative control systems and identify whether feedback is present:

      (a) The thermostatic temperature control in a central heating boiler.
      (b) The programmer controlling a typical central heating system.
      (c) A cruise control system, as fitted to a modern car.
      (d) The over-speed cutout fitted to a car engine.

1.2 Examine the fail safe requirements for the instrumentation and actuators in the furnace temperature control system shown in Fig. 1.4. You may assume that it is unsafe for the process fluid to overheat.

1.3 Consider the oil cooling system in Fig. P1.3. The oil temperature is controlled by throttling the cooler bypass valve. Safety considerations indicate that control equipment failure should not cause loss of the cooling function. Determine the required action for the temperature transducer and the control valve. What action will the controller take if the signal from the temperature transducer reduces in value?

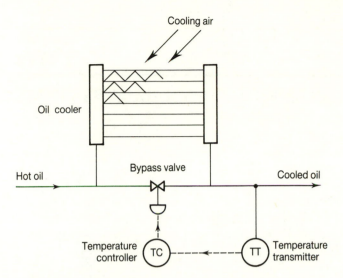

Figure P1.3   Oil cooler temperature control system

MODELLING OF DYNAMIC SYSTEMS

2.1 INTRODUCTION

In dealing with everyday tasks, we expect familiar devices and mechanisms to react to our actions in familiar ways. Our forecast of behaviour is based upon a subconscious simplified *model* of how the physical world operates. For example, when estimating the degree to which the steering wheel must be turned when cornering in a car, the experienced driver will have a mental picture of how the car will respond to the steering wheel motion, road camber and car speed, etc. He uses this mental picture to estimate the required steering effort as his journey proceeds. The driver's mental model will probably not involve the intricate details of the rack and pinion mechanism or any power-assisted servo which may be fitted, even though these details make a considerable difference to the way in which a car steers.

Engineers use models which are based upon mathematical relationships or equations. These equations model the behaviour of machines or processes and allow predictions of performance to be made. Furthermore, by rearranging the model equations it is possible to design the system components to achieve a required performance. In order to be useful, mathematical models invariably involve simplification. Assumptions concerning operation are made, small effects are neglected and idealized relationships are assumed.

When applying the ideas of modelling to engineering problems it is helpful to use the concept of a *system* which responds to stimuli or *inputs* to produce responses or *outputs*. Figure 2.1 shows the basic idea. The system is the collection of elements or components which make up the mechanism, process or device. The mathematical model will involve *parameters* which are attributable to these elements.

The common use of the term 'static' applies to objects which are not moving or stationary. When used to describe the way systems behave, it signifies that the system response can be described without reference to time. A *static system* is one in which the response is independent of time. In this context, the opposite of a static system is a dynamic one where the response is a function not only of the input but also of time. *Dynamic systems* take time to respond to changes at their inputs. The response of a dynamic system evolves with time.

14

Figure 2.1 Concept of a system

Modelling of dynamic systems is concerned with cause and effect relationships between input and output. The model equations will involve time as a variable, primarily in the form of derivatives or integrals with respect to time. These *differential equations* form the basis for analysing and designing control systems.

Example 2.1 Mathematical model of a liquid level system Obtain a mathematical model relating the flow of liquid into the tank shown in Fig. 2.2 to the tank level.

SOLUTION Assuming that the liquid density is constant, the principle of continuity of volume can be applied to the tank:

Net volumetric inflow rate = Rate of change of volume

$$q_i = \frac{dv}{dt}$$

The volume of liquid in the tank, v, is given by the product of tank area and liquid level

$$v = Ah$$

Since the tank area is independent of the level we obtain

$$q_i = A \frac{dh}{dt} \tag{2.1}$$

This differential equation (2.1) is a mathematical model of the tank system. The equation is time dependent in that the *rate of change* of level, dh/dt, is involved. The

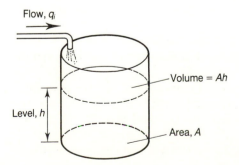

Figure 2.2 Open tank example

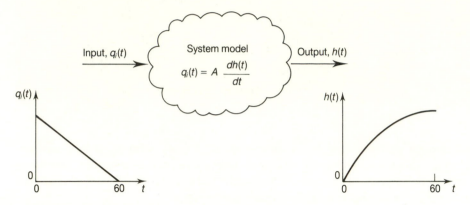

Figure 2.3 Model gives the relationship between input and output

model is therefore dynamic. The input to the system is the inlet flow rate and the output is the tank level. The tank area, A, is a parameter of the model.

It is important to realize that this mathematical model does not explicitly tell us the level at a particular time, only the relationship that exists between input, output and time. In order to determine the actual level we need to find a solution to the model equation for the specific input which is of interest.

As an illustration, suppose the inlet flow is started suddenly at 10 litre/s ($0.01\ \mathrm{m^3/s}$) and then is throttled back linearly to zero in 60 seconds as shown in Fig. 2.3. The inlet flow is in this case

$$q_i(t) = 0.01(1 - t/60) \qquad \mathrm{m^3/s} \tag{2.2}$$

Substituting this into the model equation we obtain

$$A\frac{dh}{dt} = 0.01(1 - t/60) \qquad \mathrm{m^3/s}$$

Dividing by the area, A, and assuming an initial level of zero, integration with respect to time produces

$$h(t) = \frac{0.01}{A}(t - t^2/120) \qquad \mathrm{m} \tag{2.3}$$

which represents the way in which the level in the tank will behave for this particular input.

The majority of design and analysis techniques are based upon models which are *linear ordinary differential equations*. A linear differential equation is one in which the variables and their derivatives have a proportional contribution. Linear relationships are therefore normally assumed when deriving model equations. Ordinary differential equations (as opposed to partial differential equations) occur when the elements of a system are assumed to act as concentrated or lumped entities. In other words the physical effect we are modelling is assumed to act at a point rather than throughout the element.

You should not expect a linear model to produce a straight line response. Equation

(2.3), which is the response of a linear model to a linear input, is not a linear function of time. All physical systems are nonlinear to some extent. For example, the level in the tank in Fig. 2.2 will not be proportional to the inflow when the tank is full and overflowing! Nonlinearities such as this are called limiting or saturation (no pun intended) and are always present in real systems. Since many systems are designed not to reach their physical limits during normal operation, ignoring limiting is often justified.

Example 2.2 A liquid-in-glass thermometer A mercury-filled thermometer is shown in Fig. 2.4. Obtain a linear ordinary differential equation relating the thermometer indicated temperature, T_2, to the measured temperature, T_1.

SOLUTION Providing the thermometer is correctly calibrated, the indicated temperature should correspond to the average temperature of the mercury. The mercury can be considered to behave as a lumped body at a uniform temperature of T_2. Heat must flow through the glass bulb to change the temperature of the mercury. This heat flow will depend upon the temperature difference across the bulb. According to Fourier's law, the heat flow rate through the glass bulb is

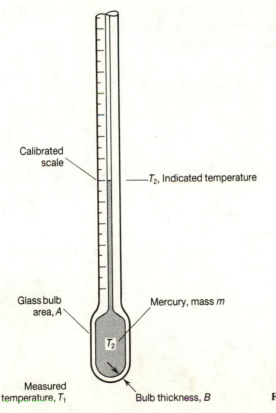

Calibrated
scale

T_2, Indicated temperature

Glass bulb
area, A

Mercury, mass m

T_2

Measured
temperature, T_1

Bulb thickness, B

Figure 2.4 Liquid-in-glass thermometer

proportional to the temperature gradient within the bulb,

$$q = kA\frac{dT}{dx} \qquad (2.4)$$

where k is the thermal conductivity of the glass bulb and A its surface area. In the previous example the symbol 'q' was used to represent **volumetric** flow rate but here it represents **heat** flow rate.

Assume that the measured medium is at a uniform temperature, T_1. The temperature difference across the glass bulb will be $(T_1 - T_2)$. Assuming a constant bulb thickness, B, the temperature gradient through the bulb can be expressed as

$$\frac{dT}{dx} = \left(\frac{T_1 - T_2}{B}\right)$$

so that

$$q = kA\left(\frac{T_1 - T_2}{B}\right) \qquad (2.5)$$

Equation (2.5) neglects any effects due to the nature of the measured fluid medium, i.e. its thermal conductivity and whether or not it is agitated. These effects, however, will not alter the **form** of the relationship and can be accommodated by increasing the effective bulb thickness to include any boundary effect.

The principle of conservation of energy means that

$$\begin{array}{c} \text{Net heat flow} \\ \text{into bulb} \end{array} = \begin{array}{c} \text{Rate of change of} \\ \text{internal heat} \end{array}$$

or

$$q = \frac{dH}{dt} \qquad (2.6)$$

The internal heat, H, of the mass, m, of mercury at uniform temperature, T_2, is given by

$$H = mC_pT_2 \qquad (2.7)$$

where C_p is the specific heat of mercury.

The three equations (2.5) to (2.7) can now be manipulated to remove the intermediate variables, H and q, so that

$$\frac{kA}{B}(T_1 - T_2) = \frac{d}{dt}(mC_pT_2)$$

Since m and C_p are constants we can write

$$kAT_2 + mC_pB\frac{dT_2}{dt} = kAT_1 \qquad (2.8)$$

2.2 TRANSFER FUNCTIONS AND TRANSFER OPERATORS

Differential equations are not an ideal way to represent dynamic systems because the system input, output and their derivatives are spread through the equation and the cause/effect relationship between input and output is not at all obvious. In addition when model equations are combined into more complex systems the mathematics involved in obtaining the overall model equation can be considerable. *Transfer functions* or *transfer operators* offer a simple way of overcoming these problems.

As an example of the way in which we would like to represent dynamic systems we will first look at a simple static system. The lever shown in Fig. 2.5 can be described by the equations

$$x = b \sin(\theta) \qquad \text{and} \qquad y = a \sin(\theta)$$

giving
$$y = \frac{a}{b} x$$

The lever can be considered to possess a *sensitivity* or *gain* of a/b. The *block diagram* representation shown in Fig. 2.6 gives a clear picture of the operation of our model equation. The input, x, is multiplied by the block gain in order to give the output, y. Obviously the direction of signal flow indicated by the arrows is important and we would obtain the wrong answer if the arrows were reversed.

The D operator

What about dynamic systems? We developed a linear model for the thermometer, but unfortunately Eq. (2.8) cannot be rearranged into the same simple form as the above static system since both the output temperature, T_2, and its derivative are present. The problem can be overcome by defining a mathematical operator, D, to represent the process of differentiation with respect to time,

i.e.
$$Dx \equiv \frac{dx}{dt} \qquad (2.9)$$

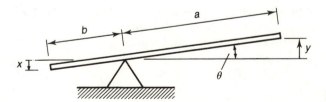

Figure 2.5 Simple static system

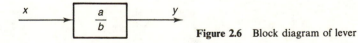

Figure 2.6 Block diagram of lever

Because the 'dt' in dx/dt has units of time, the D operator has units of $(time)^{-1}$.

Using the D operator, Eq. (2.8) can be written as

$$kAT_2 + mC_pBDT_2 = kAT_1$$

or

$$T_2(kA + mC_pBD) = kAT_1$$

giving

$$T_2 = \frac{kA}{kA + mC_pBD} T_1 \qquad (2.10)$$

We now have something with which to multiply the input, T_1, in order to obtain the output, T_2. Remember that this equation is simply a representation of the model differential equation and the multiplication cannot be done explicitly since 'D' doesn't actually possess a value, it is simply a mathematical opeerator.

The ratio of output to input as a function of D

$$\frac{T_2}{T_1}(D) = \frac{kA}{kA + mC_pBD} \qquad (2.11)$$

is called a *transfer operator*, it represents a *dynamic* sensitivity or gain. In other words it represents the fact that the system will take time to respond and carries information about the form of the response and how fast it will take to evolve.

The Laplace Transform

Just when you might be happy with transfer operators, we will introduce yet another representation of the model. Don't worry though, we will eventually come full circle and return to the simple ideas behind transfer operators.

The Laplace transform is a mathematical transformation which allows functions of time, t, to be represented in a new variable, s, the *Laplace operator*. The Laplace transform of a function of time, $f(t)$, is written as the integral

$$\mathscr{L}\{f(t)\} = F(s) = \int_0^\infty f(t)\,e^{-st}\,dt \qquad (2.12)$$

Conventionally, functions of time are written in lower case (for example $x(t)$) and functions of the Laplace operator are written in upper case (for example $X(s)$). We describe the transformation as mapping from one *domain* to another. Here the transformation is from the *time domain* to the Laplace or *s-domain*.

As an example, the Laplace transform of the exponential function

$$x(t) = e^{-at}$$

is given by

$$\mathscr{L}\{x(t)\} = X(s) = \int_0^\infty e^{-at}\,e^{-st}\,dt$$

$$= \int_0^\infty e^{-(s+a)t}\,dt$$

$$= \left[\frac{-1}{s+a} \, \mathrm{e}^{-(s+a)t} \right]_0^\infty$$

$$= \frac{-1}{s+a} \, [0-1]$$

$$\mathscr{L}\{\mathrm{e}^{-at}\} = \frac{1}{s+a} \tag{2.13}$$

Another example is the Laplace transform of the derivative of a function of time, dx/dt.

$$\mathscr{L}\left\{ \frac{dx}{dt}(t) \right\} = \int_0^\infty \frac{dx}{dt}(t) \, \mathrm{e}^{-st} \, dt$$

Integrating by parts using

$$u = \mathrm{e}^{-st} \qquad \text{so that} \qquad du = -s\,\mathrm{e}^{-st} \, dt$$

and

$$dv = \frac{dx}{dt}(t) \, dt \qquad \text{so that} \qquad v = x(t)$$

$$\int u \, dv = uv - \int v \, du$$

thus

$$\mathscr{L}\left\{ \frac{dx}{dt}(t) \right\} = [\mathrm{e}^{-st} \, x(t)]_0^\infty + \int_0^\infty x(t)s \, \mathrm{e}^{-st} \, dt$$

giving

$$\mathscr{L}\left\{ \frac{dx}{dt}(t) \right\} = -x(0) + sX(s) \tag{2.14}$$

$x(0)$ is the value of $x(t)$ at $t = 0$, which is an initial condition, and $X(s)$ is of course the Laplace transform of $x(t)$. Providing the initial conditions are zero, differentiating a function of time is equivalent to multiplying its Laplace transform by 's'.

$$\frac{dx}{dt}(t) \Rightarrow Dx(t) \Rightarrow sX(s)$$

That is, there is an equivalence between the D operator and the Laplace operator **providing the initial conditions are zero**. We must not be tempted to extend the equivalence into an equality since 's' is a variable and 'D' is simply an operator. This concept is a very useful one and can give insight into the significance of Laplace transform equations. The idea behind the equivalence between s and D can be extended to cover higher derivatives and even integrals as shown in Table 2.1.

The Laplace transform is a **linear** transformation so that if

$$z(t) = \alpha x(t) + \beta y(t)$$

then

$$Z(s) = \alpha X(s) + \beta Y(s)$$

Table 2.1

Function of time	Function of s (zero initial conditions)
$x(t)$	$X(s)$
$\dfrac{dx}{dt}(t)$	$sX(s)$
$\dfrac{d^2x}{dt^2}(t)$	$s^2 X(s)$
$\displaystyle\int x(t)\,dt$	$\dfrac{X(s)}{s}$

Therefore, going back to the thermometer example, Eq. (2.8) can be transformed into the s-domain (assuming zero initial conditions) term by term, so that

$$\mathscr{L}\{kAT_2(t)\} = kAT_2(s)$$

and

$$\mathscr{L}\left\{mC_pB\frac{dT_2}{dt}(t)\right\} = mC_pBsT_2(s)$$

etc., giving the transformed equation as

$$kAT_2(s) + mC_pBsT_2(s) = kAT_1(s)$$

or

$$\frac{T_2}{T_1}(s) = \frac{kA}{kA + mC_pBs} \tag{2.15}$$

This function is called the system *transfer function* and you will notice that it is identical to the transfer operator derived previously (Eq. (2.11)) if D is replaced by s.

This particular model is a first-order transfer function corresponding to a first-order differential equation in which only the first derivative is involved. It is normal to express a first-order transfer function in a normalized form. A *standard first-order system* is

$$G(s) = \frac{1}{1 + \tau s} \tag{2.16}$$

The parameter τ is called the system *time constant* and has units of time. It is a measure of the speed of response of the system. Dividing both numerator and denominator of Eq. (2.15) by kA

$$\frac{T_2}{T_1}(s) = \frac{1}{1 + (mC_pB/kA)s} \tag{2.17}$$

The time constant of the thermometer is therefore given by (mC_pB/kA) which you may wish to verify has units of time. Notice that the Laplace operator, s, itself has the same units as D, i.e. $(\text{time})^{-1}$, so that 'τs' is dimensionless.

Transfer functions can be obtained for higher-order systems. The general linear ordinary differential equation

$$b_0 x + b_1 \frac{dx}{dt} + b_2 \frac{d^2 x}{dt^2} \cdots + b_m \frac{d^m x}{dt^m} = a_0 y + a_1 \frac{dy}{dt} + a_2 \frac{d^2 y}{dt^2} \cdots a_n \frac{d^n y}{dt^n}$$

can be represented by the transfer function

$$\frac{Y}{X}(s) = \frac{b_0 + b_1 s + \cdots b_m s^m}{a_0 + a_1 s + \cdots a_n s^n} \tag{2.18}$$

One advantage of the transfer function representation is that, like the transfer operator, it can be used in a block diagram representation of the system as in Fig. 2.6. Additionally it can be used to analytically determine the response of the system to a particular input.

The Laplace Transform Method

The transfer function of a system is defined as the Laplace transform of the output divided by the Laplace transform of the input with zero initial conditions.

$$G(s) \equiv \frac{\mathscr{L}\{y(t)\}}{\mathscr{L}\{x(t)\}} = \frac{Y}{X}(s) \tag{2.19}$$

Provided the initial conditions are zero, the transformed output is given by multiplying the transfer function by the transformed input

$$Y(s) = G(s) \cdot X(s)$$

The basis of the Laplace transform method is to transform the model and input from their time/domain representation into an s-domain representation. The resulting equations are then algebraically manipulated into a form suitable for inverse transformation into the time/domain giving the system response. You might like to think of the technique as analogous to the use of logarithms to help with multiplying numbers (widely used in the days before calculators). The numbers were transformed by taking their logs. These logs were then added (much simpler than multiplication) and finally inverse transformed by finding the anti-log to give the result. Appendix B gives some of the more useful Laplace transforms.

Example 2.3 Step response using the Laplace method As an example we will use the Laplace transform method to find the response of the thermometer to a sudden change of input temperature as might occur when plunging the thermometer into hot liquid. Such an input is commonly called a *step input* and is both a simple signal to analyse as well as being a useful test signal. Consider an input which steps from zero to 'A' at time, $t = 0$, Fig. 2.7.

$$x(t) = \begin{Bmatrix} 0 \text{ for } t < 0 \\ A \text{ for } t \geqslant 0 \end{Bmatrix}$$

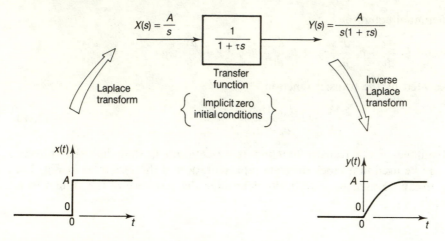

Figure 2.7 The Laplace transform method

The Laplace transform of this step input is therefore given by

$$X(s) = \int_0^\infty A\,e^{-st}\,dt = \left[-\frac{A}{s}\,e^{-st} \right]_0^\infty$$

giving

$$X(s) = \frac{A}{s}$$

Now, the Laplace transform of the output, $Y(s)$, is given by multiplying the transformed input, $X(s)$, by the transfer function

$$Y(s) = X(s) \cdot G(s)$$

$$= \frac{A}{s} \cdot \frac{1}{1+\tau s}$$

The inverse transform of $Y(s)$ will give $y(t)$. The simplest way to do this is to express the function as the sum of the two partial fractions:

$$\frac{A}{s(1+\tau s)} = \frac{B}{s} + \frac{C}{1+\tau s}$$

To find the coefficients B and C we multiply both sides of the equation by $s(1+\tau s)$ and equate corresponding powers of 's', i.e.

$$A = B(1+\tau s) + Cs$$

Equating the coefficients of s^0 leads to

$$A = B$$

and equating the coefficients of s^1 gives

$$0 = \tau B + C$$

or

$$C = -\tau B = -\tau A$$

Therefore

$$Y(s) = \frac{A}{s} + \frac{-\tau A}{1 + \tau s}$$

From the previous transforms evaluated, we can see that

$$\mathscr{L}\{A\, e^{-t/\tau}\} = \frac{\tau A}{1 + \tau s}$$

and for $t \geqslant 0$

$$\mathscr{L}\{A\} = \frac{A}{s}$$

thus

$$y(t) = A(1 - e^{-t/\tau}) \qquad\qquad (2.20)$$

Exercise Use CODAS to simulate the step response of the transfer function, Eq. (2.16) with τ entered as 1 second. Make sure that the response obtained is for the open-loop (rather than closed-loop) system. Vary the time constant, τ, and verify that the response always reaches 0.63 when $t = \tau$.

The time response of the thermometer to the step input is shown in Fig. 2.7. The assumption of zero initial conditions when using the transfer function directly in this example means that the response starts at a temperature of zero. If initial conditions are non-zero then the transfer function cannot be used directly and the differential equation must be transformed taking into account the appropriate initial values as in Eq. (2.14).

The exponential function has a value of 'A' at time zero and decays toward zero as time proceeds. The rate of decay depends on the time constant, τ. You can now see the effect of the time constant on the thermometer's step response. A large time constant means a slow response. The thermometer transfer function Eq. (2.17) indicates that its time constant is

$$\tau = \frac{mC_p B}{kA}$$

Thus to get a fast response the mercury mass should be small, the glass bulb should be thin and of large area, etc.

The response can be seen to settle down to a steady output of 'A' which is the same as the input temperature. This is to be expected for an accurate thermometer. The standard first-order system thus has a *static* or *steady-state gain* of unity. Steady-state gain is covered in greater detail in Chapter 3.

First-order dynamics such as exhibited by the thermometer are extremely common in all types of dynamic system. The response of such first-order systems always show the output lagging behind the input, trying to keep up like a reluctant dog on an elasticated lead. This type of dynamic behaviour is often called an *exponential lag* or a *first-order lag*.

The Laplace transform method can be used to predict the output of linear systems as an explicit function of time and as such, it has its place in control system analysis. However, the effort required to apply the method to all but the simplest of systems makes it time-consuming to use. Further, the Laplace transform method is not particularly suited as a *design* tool. The approach adopted in this book is to use the Laplace transform in a similar way to the *D* operator, i.e. to represent the system dynamics as a transfer function.

Exercise Use the Laplace method to show that the response of the thermometer with a step input of 100°C and an initial reading of 25°C is given by

$$y(t) = 100 - 75\,e^{-t/\tau} \qquad °C$$

Hint: You will need to transform the original differential equation taking into account both the input and initial conditions.

Simulate the situation on CODAS by using a user-defined input set up as '100' and using an initial plant output of '25'.

2.3 ELECTRICAL, MECHANICAL AND FLUID SYSTEMS

Example 2.4 Resistance/capacitance network Figure 2.8 shows a resistance/capacitance (RC) network which is often used as a simple filter in electrical circuits. Obtain a transfer function relating the output and input voltages.

SOLUTION Ohm's law can be used to obtain the current through the resistor

$$i = \frac{v}{R}$$

v is the potential difference across the resistor ($v_i - v_o$), so that

$$i = \frac{v_i - v_0}{R} \qquad (2.21)$$

A capacitance element stores a charge proportional to the voltage across the capacitor plates

$$Q = Cv_o$$

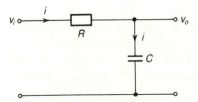

Figure 2.8 Resistance/capacitance network

But current is the rate of change of charge

$$i = \frac{dQ}{dt}$$

so that

$$i = C \frac{dv_o}{dt} \qquad (2.22)$$

Assuming that no current flows from the output terminal, the current through the resistor will flow into the capacitor and so from Eqs (2.21) and (2.22)

$$\frac{v_i - v_o}{R} = C \frac{dv_o}{dt}$$

or

$$v_i - v_0 = CR \frac{dv_o}{dt} \qquad (2.23)$$

Assuming zero initial conditions and taking Laplace transforms of each term results in

$$V_i - V_0 = CRsV_o$$

giving

$$\frac{V_o}{V_i}(s) = \frac{1}{1 + CRs} \qquad (2.24)$$

Do you recognize the transfer function in Eq. (2.24)? Compare it with Eq. (2.16), and you will see that the RC network is another standard first-order system. The time constant for this system is the product of R and C. This similarity reveals that the dynamic behaviour of the thermometer is the same as that of the RC network. Of course different time constants will result in different response times, but the shape of the response to the same input signals will be identical.

Example 2.5 Spring/mass/damper system Figure 2.9 shows a schematic representation of a spring/mass/damper system. The body of mass, m, is suspended on a spring element of stiffness, k. Friction and air drag losses can be assumed to behave linearly producing a damping force

$$F_f = C \frac{dy}{dt}$$

Assuming that the mass only moves vertically, determine the transfer function relating the mass displacement, y, to the spring displacement, x.

SOLUTION Figure 2.10 shows a *free body diagram* of the mass with all the applied forces shown. The spring, on being compressed by an amount $(x - y)$ pushes down on the mass with a force

$$F_k = k(x - y)$$

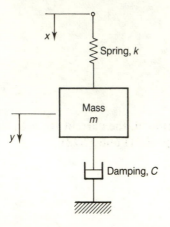

Spring, k

Mass
m

Damping, C

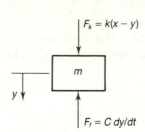

$F_k = k(x - y)$

m

$F_f = C\,dy/dt$

Figure 2.9 Spring/mass/damper system   **Figure 2.10** Free body diagram of mass

Newton's second law of motion expresses the relationship between force and acceleration for straight line motion as

$$\sum_y F = m\frac{d^2 y}{dt^2} \qquad (2.25)$$

$\sum_y$ is the sum of all the forces in the direction of y which are acting on the body. Applying Newton's second law to the body

$$\sum_y F = F_k - F_f = k(x - y) - C\frac{dy}{dt} = m\frac{d^2 y}{dt^2}$$

which upon Laplace transformation (with zero initial conditions) produces

$$k(X - Y) - CsY = ms^2 Y$$

or
$$\frac{Y}{X}(s) = \frac{k}{k + Cs + ms^2} \qquad (2.26)$$

This transfer function is second order (corresponding to a second-order differential equation). The response of a second-order system is fundamentally different to that of a first-order system and is dealt with in Chapter 3.

Exercise Simulate the step response of the spring/mass/damper system of Example 2.5 using CODAS. You may assume a mass of 1 kg, a spring stiffness of 100 N/m and no frictional loss (i.e. $C = 0$). Make sure that you obtain the open-loop response. You should change the x-axis scale to cover the time range 0 to 1.0 second. Examine the effect of introducing frictional losses by using various values for the damping coefficient, C.

With C set at 25 Ns/m, gradually reduce the mass, m, and examine the response. What happens to the system when the mass is zero?

Example 2.6 Modelling a dc electric motor Figure 2.11 shows a typical servomotor of a type often used as the actuator in position or speed control systems. The field of this small motor is provided by a permanent magnet. In larger motors the field is often produced by a separately excited field coil. With a constant field, the armature produces a torque which is proportional to the armature current. The torque drives the motor against the load and causes the armature to accelerate. Friction and drag in the armature cause the available torque to reduce with increasing motor speed.

The motor will generally be driving an external load which may itself possess inertia and friction. The load inertia and friction must be added to those of the armature in order to produce a realistic model of the motor with load. The armature angular position, θ, is a suitable choice for the model output and armature current, i, and applied load torque, L, are suitable inputs.

Figure 2.12 shows the free body diagram of the motor armature. The inertia, J_a, (armature plus load) is accelerated by the motor armature torque, T_a

$$T_a = K_a i.$$

where K_a is the motor torque constant (Nm/A).

The frictional and load torques both act in the opposite direction to the rotation. For a linear model the frictional torque will be assumed to be proportional to the armature rotational speed

$$T_f = C \frac{d\theta}{dt}$$

where C is the rotational damping coefficient due to friction and drag losses (again including those of the load).

Newton's second law of motion for rotational motion can be expressed as

$$\sum_\theta T = J \frac{d^2\theta}{dt^2} \tag{2.27}$$

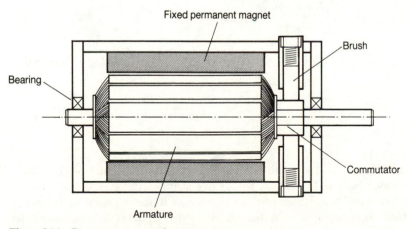

Figure 2.11 Permanent magnet dc servomotor

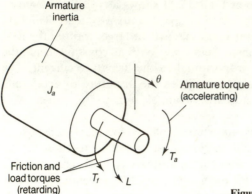

Figure 2.12   Free body diagram of servomotor armature

where $\sum_\theta T$ is the sum of all the torques in the direction of θ. J is the second moment of mass (moment of inertia) of the rotating parts.

Applying Newton's second law to the armature plus load inertia

$$\sum_\theta T = K_a i - L - C\frac{d\theta}{dt} = J_a \frac{d^2\theta}{dt^2}$$

Laplace transformation gives

$$K_a I(s) - L(s) - Cs\theta(s) = J_a s^2 \theta(s)$$

or $$(Cs + J_a s^2)\theta(s) = K_a I(s) - L(s)$$

There are three independent variables in this equation, i.e. two inputs, I and L, and the output θ. A transfer function can only represent the relationship between one input/output pair. So in order to find the dynamic relationship between say θ and I, the load torque can be assumed to be zero.

$$\frac{\theta}{I}(s) = \frac{K_a}{Cs + J_a s^2}$$

or $$\frac{\theta}{I}(s) = \frac{K_a/C}{s(1 + (J_a/C)s)} \qquad (2.28)$$

The transfer function relating θ to L is similarly obtained by putting I to zero

$$\frac{\theta}{L}(s) = \frac{-1/C}{s(1 + (J_a/C)s)} \qquad (2.29)$$

Before leaving this problem it is worth considering these transfer functions in a little more detail. Equations (2.28) and (2.29) are rather like a first-order system but with an extra factor $1/s$. What is the significance of this extra term? Referring back to Table 2.1 you will recall that integrating a variable in the time domain is equivalent to multiplication by $1/s$ in the s-domain. The factor $1/s$ is a *free integrator*. So where is the integration taking place? The natural operation of the motor is to rotate continuously when a steady armature current is applied.

Consider the relationship between motor speed, Ω, and position, θ

$$\Omega = \frac{d\theta}{dt} \tag{2.30}$$

In other words our chosen output, θ, is the integral of speed, Ω. Taking Laplace transforms of Eq. (2.30) gives

$$\Omega = s\theta$$

and using Eq. (2.28) we have

$$\frac{\Omega}{I}(s) = \frac{\Omega}{\theta}(s) \cdot \frac{\theta}{I}(s) = s \cdot \frac{K_a/C}{s(1 + (J_a/C)s)}$$

i.e.
$$\frac{\Omega}{I}(s) = \frac{K_a/C}{(1 + (J_a/C)s)} \tag{2.31}$$

Comparing Eq. (2.31) with the standard first-order transfer function we can see that the system time constant is (J_a/C). What about the term (K_a/C) in the numerator? This represents the fact that this transfer function has a different gain to the standard first-order system.

Example 2.7 Fluid resistance/capacitance system Figure 2.13 shows a holdup tank used to reduce fluctuations in a liquid supply stream. Assuming that the flow through the valve is proportional to the head across the valve, obtain a transfer function relating the outlet flow, q_o, to the inlet flow, q_i.

SOLUTION The volume balance for the tank gives

$$\text{Net inflow} = \text{Rate of accumulation} \tag{2.32}$$

$$q_i - q_o = \frac{dV}{dt} = A\frac{dh}{dt} \tag{2.33}$$

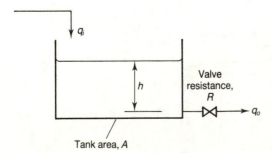

Figure 2.13 Fluid resistance/capacitance system

Since we are assuming that the flow through the valve is linearly related to the head across the valve we can write

$$q_o = \frac{\delta h}{R}$$

The constant of proportionality, R, is the valve resistance.

Since the head across the valve, δh, is just the tank head we can write

$$q_o = \frac{h}{R} \qquad\qquad (2.34)$$

Equations (2.33) and (2.34) can be transformed to develop a transfer function

$$Q_i - Q_o = AsH \qquad \text{and} \qquad Q_o = \frac{H}{R}$$

which gives

$$Q_i - Q_o = ARsQ_o$$

or

$$\frac{Q_o}{Q_i}(s) = \frac{1}{1 + ARs} \qquad\qquad (2.35)$$

2.4 TRANSPORT DELAYS

Transport delay is a common dynamic phenomenon especially where flow or transport occur. An example is illustrated in Fig. 2.14. The conveyor transports material from

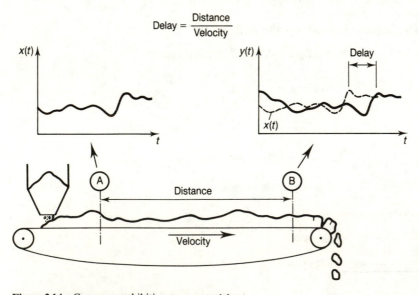

Figure 2.14 Conveyor exhibiting transport delay

one point A to another B. The time delay between the points depends upon the conveyor speed and the distance travelled. Any material property measured at point A will appear at B after this delay time. The measurement at B lags behind A by the transportation time. The terms *dead-time*, *distance/velocity lag* and *pure time delay* are also used to describe the same phenomenon.

Given the delay time between the two points is T_D, the relationship between the two signals, x and y can be expressed as

$$y(t) = x(t - T_D) \tag{2.36}$$

The transfer function of a pure transport delay is derived in Section 4.10 where it is shown that

$$\frac{Y}{X}(s) = e^{-sT_D} \tag{2.37}$$

The transfer function of a pure transport delay is an irrational function of 's' and so does not fall into the same category as the other transfer functions covered. Equation (2.36) is however linear and so linear operations can be applied to transport delay systems.

Exercise Examine the open-loop response of a pure transport delay of 1 second to various inputs using CODAS. Note that transport delay is entered using the 'T' option, not as part of the transfer function. Examine the step and ramp response. Also try a sinusoidal input of frequency 1 rad/s by setting up a user defined input as 'sin(t)'.

2.5 BLOCK DIAGRAM REPRESENTATION

The concept of block diagram representation of a transfer function was introduced in Section 2.2. The basic block diagram element, called a *block*, is a box with one input and one output. The direction of flow from input to output is indicated by arrows (Fig. 2.15(a)). The box contains the gain, or more generally the transfer function, $G(s)$, expressing the relationship between an input and an output. The block represents a **multiplication** i.e. the input is multiplied by the factor in the block to produce the output. The direction of signal flow from input to output is most important since signal flow in the opposite direction would suggest that the input is obtained by multiplying the transfer function by the output. The arrows shown in a block diagram are therefore most essential.

Block diagrams have the great advantage that they clearly show the cause and effect relationship of system components. They also show how the elements in a system are interconnected so that the interactions that take place in more involved systems can be plainly seen. Furthermore, by manipulation of the diagram, complex systems can be simplified to obtain transfer functions relating overall system inputs and outputs.

One other block diagram element is required to represent more complex linear systems. A *summing junction* is used to represent the addition or subtraction of signals, Fig. 2.15(b). Again arrows are necessary to show the direction of signal flow. Each

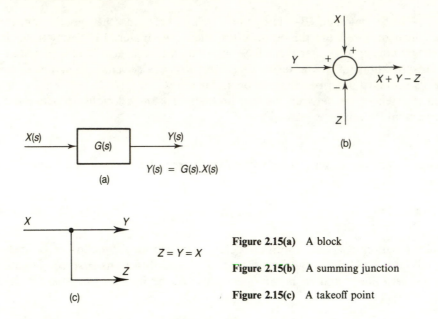

Figure 2.15(a)   A block

Figure 2.15(b)   A summing junction

Figure 2.15(c)   A takeoff point

incoming arrow has a sign associated with it indicating whether that signal has a positive or negative contribution to the summing junction output.

When a signal goes to two destinations we can show this as indicated in Fig. 2.15(c). Such a construction is called a *takeoff point*.

Block Diagram Reduction

The block diagram of a complicated linear system can be reduced to a simpler system by using a set of intuitively derived rules. The transfer function relating the system output to an input can consequently be obtained from the block diagram.

One of the most common configurations is that of blocks in *cascade* as shown in Fig. 2.16(a). The overall system transfer function will be

$$\frac{Z}{X} = \frac{Y}{X} \cdot \frac{Z}{Y} = G_1 \cdot G_2 = G_2 \cdot G_1$$

In other words we can replace the blocks in cascade with one block containing the product of the two transfer functions. There is an underlying assumption in this operation that the two individual transfer functions do not interact. In practice the second block can cause a loading effect on the first, which will cause the overall transfer function to be different from the product of the individual transfer functions. It is most important to look out for loading effects before blindly interconnecting elements in a block diagram.

Moving Takeoff Points and Summing Junctions

In order to simplify block diagrams it is often necessary to move takeoff points or summing junctions past blocks. Figures 2.16(b) to 2.16(e) show the basic rules for

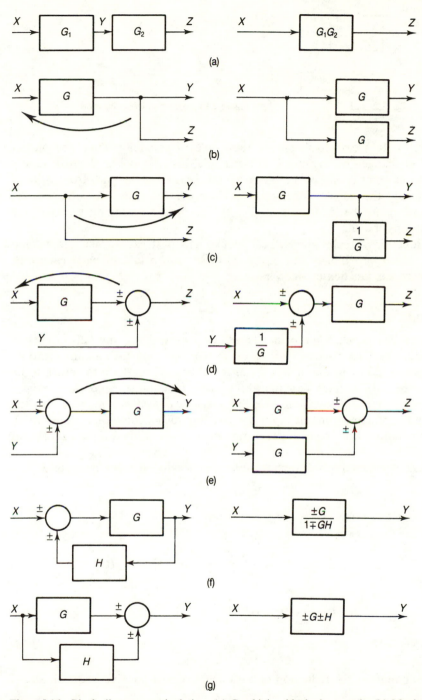

Figure 2.16 Block diagram manipulation. (a) Combining blocks in cascade. (b) Moving a takeoff point ahead of a block. (c) Moving a takeoff point behind a block. (d) Moving a summing junction ahead of a block. (e) Moving a summing junction behind a block. (f) Reduction of feedback loops. (g) Reduction of feedforward paths

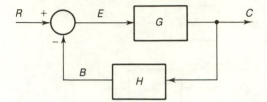

Figure 2.17   Negative feedback loop

accomplishing this. It is easy to memorize these rules if you realize that the output signal must be unchanged by the move. For example, looking at Fig. 2.16(b), simply moving the takeoff point ahead of the block would mean that the path from X to Z no longer contained the transfer function G, so in order to restore the situation the extra block shown must be incorporated. Try to use similar reasoning to verify the other manipulations shown.

Summing junctions can be moved past blocks and past each other and takeoff points can be moved past blocks and past each other. But **a takeoff point cannot be moved past a summing junction or vice versa**.

Feedback Loops

Figure 2.17 shows the block diagram of a *feedback loop* which commonly occurs in practical systems. Feedback control systems get their name from such a configuration. Feedback loops are special in that a signal can travel around the loop to return to its starting point. The *forward path* contains the transfer function, G, and the *feedback path H*. This particular configuration is called a *negative feedback loop* in that the feedback signal is subtracted from the input signal. Being such a common configuration, it is worth being familiar with the equivalent transfer function relating input and output signals.

The signal, E, is obtained by subtracting the feedback signal, B, from the input, R:

$$E = R - B$$

but the block diagram shows that

$$B = HC$$

and so,

$$C = GE = G(R - B) = G(R - HC)$$

rearranging we have

$$C(1 + GH) = GR$$

or

$$\frac{C}{R} = \frac{G}{1 + GH} \tag{2.38}$$

Figure 2.16(f) summarizes the reduction of both negative and positive feedback loops.

Feedforward Branches

Figure 2.16(g) shows a construction which at first sight looks similar to the feedback

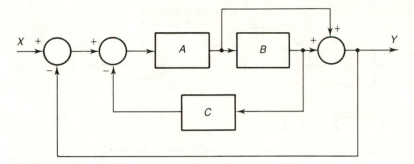

Figure 2.18 Block diagram reduction example

loop. However, there is a major difference: try to follow the signal around the loop and you can't because there is no loop! Both paths **feed forward** towards the output.

The equivalent reduced block is obtained by adding the individual path transfer functions (with appropriate sign).

Example 2.8 Block diagram reduction Reduce the block diagram of the system shown in Fig. 2.18 and obtain the overall transfer function relating Y and X.

SOLUTION We must look for structures that can be reduced using the rules of Fig. 2.16. Blocks A and B are in cascade but there is a takeoff point in between, so let's move it. Figure 2.19(a) shows the takeoff point being moved to the right; notice that to keep the overall diagram correct, the block '$1/B$' has been introduced in order to nullify the extra multiplication by 'B' which has been introduced into that path.

Figure 2.19(b) shows that the blocks A and B can then be multiplied together. At the same time there is a feedforward branch consisting of the block $1/B$ and a simple path with no block. This path effectively has a gain of unity, so the equivalant block for the overall feedforward branch should contain $(1/B + 1)$. You should recognize the negative feedback loop in Fig. 2.19(c) formed with AB in the forward path and C in the feedback path.

We are left with Fig. 2.19(d) which contains two blocks in cascade, within a second feedback loop which has unity negative feedback. Figure 2.19(e) shows the final reduction, and the resulting overall transfer function.

Example 2.9 Aircraft hydraulic flight control The flight controls in large or fast flying aircraft are often actuated by a hydraulic servomechanism in order to reduce the control column force required to manoeuvre the aircraft. Figure 2.20 shows a diagram of such a servomechanism. This example will show how a model for this system can be developed using a block diagram approach.

Before attempting to write down equations, you should first be familiar with the working principle of the servo. Consider initially the operation of the spool valve and cylinder in isolation, removed from the aircraft but still attached to the hydraulic supply.

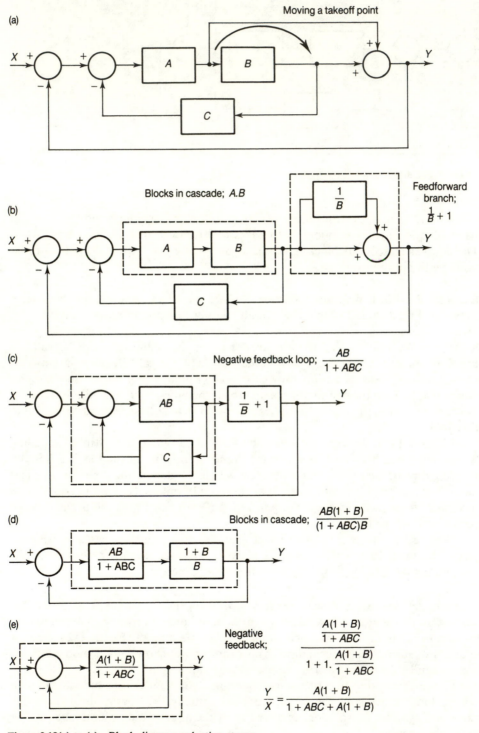

Figure 2.19(a) to (e)   Block diagram reduction stages

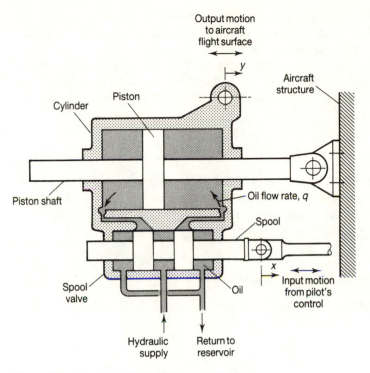

Figure 2.20 Aircraft hydraulic flight control servo

The spool valve controls the flow of hydraulic oil from the supply to the cylinder and eventually back to the reservoir. When the spool is centred, the ports which feed oil to the cylinder are shut off and no oil flows. If the spool is moved by a small amount to the right then oil is able to flow from the supply into the right hand side of the cylinder. The pressure of oil will push the piston towards the left and the more oil that flows, the faster the piston will move. Of course, as the piston moves to the left the oil on the left hand side will escape through the other spool port to drain back to the hydraulic reservoir. You will probably realize that a small movement of the spool to the left will have an opposite effect, causing oil to flow in the reverse direction, so moving the piston to the right.

Imagine that we replace the cylinder assembly into the aircraft so that the piston rod is attached to the aircraft structure and so cannot move to the left. Now, when the spool is moved to the right, oil which flows into the right side of the cylinder will force the cylinder body to move to the right, rather than the piston to the left. The spool valve body is, however, attached to the cylinder body and so must move with it. Thus as the cylinder moves to the right, the relative displacement between the spool valve and its body will be reduced so restricting the oil flow rate. The oil will none the less continue to flow (albeit at reduced rate) and the cylinder continue to move until eventually, the spool is again centred in the body stopping the oil flow. The overall system operates in such a way as to make the cylinder body position, y, attempt to follow the spool position, x.

A simplified but realistic linear model of the system will now be derived. A straightforward choice is to consider the spool displacement, x, as the input and the cylinder position, y, as the output.

The dynamic behaviour of the cylinder is fairly straightforward. Since hydraulic oil, like most liquids, is virtually incompressible the principle of continuity of volume can be applied. The flow rate of oil into the cylinder will be equal to the rate of change of cylinder volume

$$q = \frac{dv}{dt}$$

The enclosed volume in one side of the cylinder is the effective piston area, A, (i.e. the piston area minus the shaft area) multiplied by the distance between the piston and the cylinder end. As the piston moves relative to the cylinder by an amount, δy, the enclosed volume will increase or decrease by

$$\delta v = A \, \delta y$$

and so, the rate of change of volume is given by

$$\frac{dv}{dt} = A \frac{dy}{dt}$$

and we can write

$$q = A \frac{dy}{dt}$$

Which upon Laplace transformation gives

$$Y(s) = \frac{1}{As} Q(s) \tag{2.39}$$

The spool valve behaviour is rather more complex since the oil flow to the cylinder will not only depend upon the spool opening, e, but also the oil pressure across the valve. We will assume that the pressure drop across the valve is constant, this being justified if the load being moved is light and requires only a low pressure difference across the piston relative to the supply pressure.

Assuming constant spool valve pressure drop and a linear characteristic relating the oil flow to spool displacement we can write

$$q = K_v e$$

where K_v is the linearized slope of the valve characteristic at normal operating conditions. This is a static equation, and so is unchanged by Laplace transformation

$$Q(s) = K_v E(s) \tag{2.40}$$

Finally, the displacement of the spool relative to the valve body is simply the difference between x and y, i.e.

$$e = x - y$$

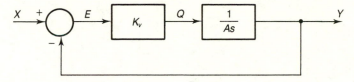

Figure 2.21   Block diagram of hydraulic servo

or
$$E(s) = X(s) - Y(s) \qquad\qquad (2.41)$$

Figure 2.21 shows these equations expressed as a block diagram. The result is a classic unity negative feedback loop. The system is driven, not by the input alone, but by the difference between the input, X, and the output, Y.

Block diagram reduction gives us the overall transfer function:

$$\frac{Y}{X}(s) = \frac{1}{1 + (A/K_v)s}$$

yet another first-order transfer function with time constant

$$\tau = \frac{A}{K_v}$$

2.6 STATE-VARIABLE MODELS

So far differential equations and transfer functions or operators have been used to describe system dynamics. Another type of representation is by *state-variable* or *state-space* models. State-variable models are based on describing the system dynamics as a **set** of simultaneous first-order differential equations called the *state equations*. The state equations can be expressed in standard matrix or vector form and much of the analysis and design can be done using matrix manipulation techniques.

The fundamental concept behind state-variable models is that the dynamic condition of a system at any instant is completely described by its *state*. The state is defined in terms of a set of state variables, $x_1(t), x_2(t) \cdots x_n(t)$. Knowledge of these state variables together with any inputs allows the future state of the system to be found from the state equations. The system output can be expressed in terms of the state variables. An nth order system requires 'n' state variables and 'n' state equations to model its dynamics.

Example 2.10 State-variable model of a mechanical system Figure 2.22(a) shows a schematic diagram for a spring/mass/damper system. This second-order system requires two state variables. A simple choice is:

$$x_1 = y \qquad \text{and} \qquad x_2 = \frac{dy}{dt}$$

The state equations can be obtained from the system free body diagram

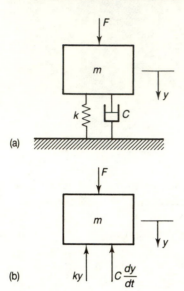

(a)

(b)

Figure 2.22   (a) Spring/mass/damper system. (b) Free body diagram of mass

(Fig. 2.22(b)). Applying Newton's second law to the mass leads to the equation of motion

$$F - ky - C\frac{dy}{dt} = m\frac{d^2y}{dt^2} \tag{2.42}$$

Substituting the chosen state variables into the equation of motion produces

$$m\frac{dx_2}{dt} = -Cx_2 - kx_1 + F$$

or

$$\frac{dx_2}{dt} = -\frac{Cx_2}{m} - \frac{kx_1}{m} + \frac{F}{m} \tag{2.43(a)}$$

and also

$$\frac{dx_1}{dt} = x_2 \tag{2.43(b)}$$

The two state equations of Example 2.10 can be expressed as a matrix equation

$$\begin{bmatrix} \dfrac{dx_1}{dt} \\[2mm] \dfrac{dx_2}{dt} \end{bmatrix} = \begin{bmatrix} 0 & 1 \\[2mm] -\dfrac{k}{m} & -\dfrac{C}{m} \end{bmatrix} \begin{bmatrix} x_1 \\[2mm] x_2 \end{bmatrix} + \begin{bmatrix} 0 \\[2mm] \dfrac{1}{m} \end{bmatrix} F \tag{2.44}$$

Generally, this *matrix state equation* is written in standard form as

$$\frac{d}{dt}\mathbf{X}(t) = \mathbf{A}\mathbf{X}(t) + \mathbf{B}u(t) \tag{2.45}$$

The vector **X** is called the *state vector* and '*u*' is the system input (in this case the force, *F*). The system output in this example is actually one of the state variables but generally the system output is a function of the state vector and possibly the input, i.e.

$$y(t) = \mathbf{C}\mathbf{X}(t) + \mathbf{D}u(t) \tag{2.46}$$

Equation (2.46) is called the *matrix output equation*. For the example system the output equation is simply

$$y = \begin{bmatrix} 1 & 0 \end{bmatrix} \begin{bmatrix} x_1 \\ x_2 \end{bmatrix} \tag{2.47}$$

or just
$$y = x_1$$

The state variables are by no means unique. A different choice leads to completely different matrices for **A**, **B**, **C** and **D**. Returning to Example 2.10, suppose that the two state variables were chosen as the forces in the spring and damper, i.e.

$$x_1 = ky \qquad \text{and} \qquad x_2 = C\frac{dy}{dt}$$

The matrix state equation now becomes

$$\begin{bmatrix} \dfrac{dx_1}{dt} \\ \dfrac{dx_2}{dt} \end{bmatrix} = \begin{bmatrix} 0 & \dfrac{k}{C} \\ -\dfrac{C}{m} & -\dfrac{C}{m} \end{bmatrix} \begin{bmatrix} x_1 \\ x_2 \end{bmatrix} + \begin{bmatrix} 0 \\ \dfrac{C}{m} \end{bmatrix} F \tag{2.48(a)}$$

and the output equation is

$$y = \begin{bmatrix} \dfrac{1}{k} & 0 \end{bmatrix} \begin{bmatrix} x_1 \\ x_2 \end{bmatrix} \tag{2.48(b)}$$

The main reasons for the development of state-variable methods include:

(a) State equations can be solved on a digital computer using standard matrix methods and/or packages.
(b) State-variable models can readily be applied to systems with more than one input and/or output (multi-variable systems).
(c) Optimum controller design methods are normally based on state-variable models. Optimum control strives to achieve the best possible control in terms of minimizing or maximizing some cost function.

State-variable techniques are often referred to as *modern control* methods whereas transfer function methods are called *classical*. This distinction is somewhat misleading since state-variable methods date back over 100 years. However, state-variable models are widely used in modern control engineering literature and so an appreciation of the terminology is important.

State-variable models are not ideal for simple single-input/single-output systems mainly because it is difficult to gain insight into system dynamics from the matrix equations. In addition transfer functions have the advantage of being more concise than state-variable models. This book, together with CODAS, deals exclusively with transfer function models. There are parallels between the design and analysis methods which apply to transfer functions and state-variable models and most of the techniques covered in this book have state-variable equivalents.

Obtaining Transfer Functions from State-variable Models

Given the state-variable model

$$\frac{d}{dt}\mathbf{X}(t) = \mathbf{A}\mathbf{X}(t) + \mathbf{B}u(t) \qquad (2.49(a))$$

$$y(t) = \mathbf{C}\mathbf{X}(t) + \mathbf{D}u(t) \qquad (2.49(b))$$

where $u(t)$ and $y(t)$ are the system input and output, how can the transfer function relating $Y(s)$ to $U(s)$ be obtained?

The first step is to Laplace transform the state equation. Zero initial conditions will be assumed as usual when obtaining transfer functions. Equation (2.49(a)) thus becomes:

$$s\mathbf{X}(s) = \mathbf{A}\mathbf{X}(s) + \mathbf{B}U(s)$$

or

$$s\mathbf{X}(s) - \mathbf{A}\mathbf{X}(s) = \mathbf{B}U(s)$$

It should not be forgotten that this is a matrix equation and therefore to factorize the state vector we must write

$$[s\mathbf{I} - \mathbf{A}]\mathbf{X}(s) = \mathbf{B}U(s) \qquad (2.50)$$

where $\mathbf{I}$ is the identity matrix:

$$\mathbf{I} = \begin{bmatrix} 1 & 0 & \cdots & 0 & 0 \\ 0 & 1 & & & 0 \\ \vdots & & \ddots & & \vdots \\ \vdots & & & 1 & 0 \\ 0 & \cdots & \cdots & 0 & 1 \end{bmatrix}$$

Now by pre-multiplying both sides of Eq. (2.50) by the inverse of $[s\mathbf{I} - \mathbf{A}]$ we obtain an expression for the Laplace transform of the state vector.

$$\mathbf{X}(s) = [s\mathbf{I} - \mathbf{A}]^{-1}\mathbf{B}U(s) \qquad (2.51)$$

Finally the output, Eq. (2.49(b)), can be used to find the Laplace transform of the output

$$Y(s) = \mathbf{C}\mathbf{X}(s) + \mathbf{D}U(s)$$

$$Y(s) \qquad\qquad = \mathbf{C}[s\mathbf{I} - \mathbf{A}]^{-1}\mathbf{B}U(s) + \mathbf{D}U(s) \qquad (2.52)$$

and so the transfer function can be obtained as

$$\frac{Y}{U}(s) = \mathbf{C}[s\mathbf{I} - \mathbf{A}]^{-1}\mathbf{B} + \mathbf{D} \qquad (2.53)$$

Example 2.11 Transfer function from a state-variable model The spring/mass/damper system of Example 2.10 can be described by the state-variable Eqs (2.44) and (2.47). Obtain the transfer function relating mass displacement, y, to the applied force, F.

SOLUTION The system equations are given by

$$\frac{d}{dt}\mathbf{X}(t) = \mathbf{A}\mathbf{X}(t) + \mathbf{B}F(t)$$

$y(t)$ and

$$= \mathbf{C}\mathbf{X}(t) + \mathbf{D}\dot{F}(t)$$

with
$$\mathbf{A} = \begin{bmatrix} 0 & 1 \\ -\dfrac{k}{m} & -\dfrac{C}{m} \end{bmatrix} \quad \mathbf{B} = \begin{bmatrix} 0 \\ \dfrac{1}{m} \end{bmatrix} \quad \mathbf{C} = \begin{bmatrix} 1 & 0 \end{bmatrix} \quad \mathbf{D} = 0$$

so that
$$[s\mathbf{I} - \mathbf{A}] = \begin{bmatrix} s & 0 \\ 0 & s \end{bmatrix} - \begin{bmatrix} 0 & 1 \\ -\dfrac{k}{m} & -\dfrac{C}{m} \end{bmatrix} = \begin{bmatrix} s & -1 \\ \dfrac{k}{m} & s + \dfrac{C}{m} \end{bmatrix}$$

Inverting this two by two matrix gives

$$[s\mathbf{I} - \mathbf{A}]^{-1} = \frac{1}{s(s + C/m) + k/m} \begin{bmatrix} s + \dfrac{C}{m} & 1 \\ -\dfrac{k}{m} & s \end{bmatrix}$$

and

$$\mathbf{C}[s\mathbf{I} - \mathbf{A}]^{-1}\mathbf{B} = \frac{1}{s(s + C/m) + k/m} \begin{bmatrix} 1 & 0 \end{bmatrix} \begin{bmatrix} s + \dfrac{C}{m} & 1 \\ -\dfrac{k}{m} & s \end{bmatrix} \begin{bmatrix} 0 \\ \dfrac{1}{m} \end{bmatrix}$$

$$\frac{Y}{F}(s) = \frac{1/m}{s(s + C/m) + k/m}$$

Exercise Verify that the alternative state-variable model for the spring/mass/damper system given by Eqs (2.48(a) and (b)) gives the same transfer function.

PROBLEMS

2.1 The liquid level system of Example 2.1 can be represented by the transfer function

$$\frac{H}{Q_i}(s) = \frac{1}{As}$$

Assuming an area, A, of 10 m^2, simulate the open-loop response of the system with the input defined by Eq. (2.2). Verify that the response obtained corresponds to that obtained analytically.

   Hint: set the x-axis to cover 0 to 60 seconds, the y-axis to cover 0 to 0.03 m and use a user defined input set up as '0.01*(1 − t/60)'.

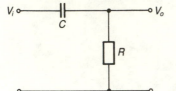

Figure P2.2

2.2 Show that the R/C network in Fig. P2.2 has the transfer function

$$\frac{V_o}{V_i}(s) = \frac{RCs}{1 + RCs}$$

Using the Laplace method show that the unit step response of the network is

$$v_o(t) = e^{-t/RC}$$

Simulate the system on CODAS with RC set to 1 second and confirm the result.

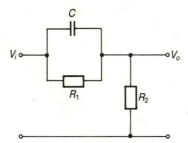

Figure P2.3 Phase lead network

2.3 Show that the electrical phase lead network of Fig. P2.3 can be modelled by the transfer function

$$\frac{V_o}{V_i}(s) = \frac{1}{a}\frac{(1 + a\tau s)}{(1 + \tau s)}$$

and hence determine the values of a and τ in terms of R_1, R_2 and C.

2.4 The manufacturers of a dc servomotor quote an armature moment of inertia equal to 7.7×10^{-3} kgm^2 and a mechanical time constant of 7.2 ms. Determine the rotational damping coefficient of the motor. Given that the motor torque constant is 0.334 Nm/A determine the transfer function relating the motor speed to the armature current. Obtain the open-loop step response of the motor to a suddenly applied current of 1 amp. The load torque may be taken to be zero.

If the motor is used to drive a load consisting of a pure inertia of 20×10^{-3} kgm^2 determine the effect on the motor time constant.

Hint: use the equations derived in Example 2.6.

2.5 Figure P2.5 shows a liquid hold-up tank which has been extended by adding a second tank connected by the valve R_2. Assuming the valve flow rates are proportional to the head across the valve, i.e. $q = \delta h/R$,

(a) Show that the transfer function relating outflow to inflow is

$$\frac{q_3}{q_1}(s) = \frac{1 + A_2 R_2 s}{A_1 R_1 A_2 R_2 s^2 + (A_1 R_1 + A_2 R_2 + A_2 R_1)s + 1}$$

(b) Show that when the valve R_2 is fully opened so that $R_2 \to 0$ the system behaves as first order.

(c) Show that when the valve R_2 is fully closed so that $R_2 \to \infty$ the system again behaves as first order.

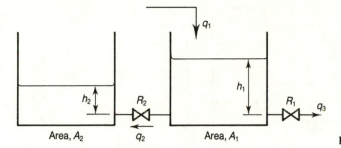

Figure P2.5 Extended tank problem

(d) Show that the ratio of the time constants with the valve R_2 fully opened and fully closed is

$$\frac{\tau_{\text{open}}}{\tau_{\text{closed}}} = \frac{A_1 + A_2}{A_1}$$

2.6 Figure P2.6 shows a spring return hydraulic actuator. The flow through the restriction, R, can be assumed to be proportional to the pressure drop. Assuming the piston is frictionless and leak free and that the hydraulic fluid is incompressible, show that the system can be modelled by the second-order transfer function

$$\frac{x}{P}(s) = \frac{A}{k + A^2 Rs + ms^2}$$

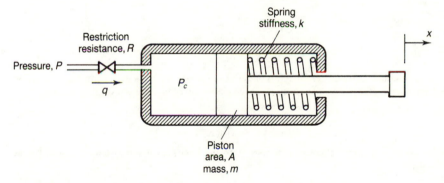

Figure P2.6 Spring return hydraulic actuator

Hint: write down the volumetric continuity equation for the hydraulic oil flowing into the cylinder. Express the oil flow rate into the cylinder in terms of the inlet and cylinder pressures. Finally, draw the free body diagram of the piston showing the forces resulting from the spring and the hydraulic cylinder pressure.

2.7 Using block diagram reduction techniques show that the system in Fig. P2.7 can be represented by

$$\frac{Y}{X}(s) = \frac{2}{s^2 + 8s + 2}$$

2.8 Figure P2.8 shows the essential characteristics of an electrodynamic shaker. The voltage applied to the shaker coil must overcome the back emf and the coil resistance in order to produce the current, i. The current produces a proportional force, F, which tends to move the armature. The back emf is proportional to the armature velocity.

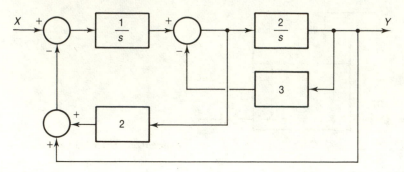

Figure P2.7

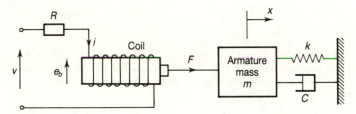

Figure P2.8   Electrodynamic shaker

Back emf,
$$e_b = K_b \frac{dx}{dt}$$

and force,
$$F = K_f i$$

Write down an equation for the current in terms of the input voltage and back emf. Obtain an expression for the displacement in terms of the force. Hence show that the shaker can be modelled by the transfer function

$$\frac{X}{V}(s) = \frac{K_f}{R(ms^2 + Cs + k) + K_b K_f s}$$

2.9 Determine the transfer function relating the output, y, to the input, u, for the system described by the state-variable model

$$\begin{bmatrix} \dfrac{dx_1}{dt} \\ \dfrac{dx_2}{dt} \end{bmatrix} = \begin{bmatrix} 2 & -1 \\ 3 & -4 \end{bmatrix} \begin{bmatrix} x_1 \\ x_2 \end{bmatrix} + \begin{bmatrix} 1 \\ -1 \end{bmatrix} u$$

$$y = \begin{bmatrix} 1 & 0 \end{bmatrix} \begin{bmatrix} x_1 \\ x_2 \end{bmatrix}$$

THREE

SYSTEM TIME RESPONSE

3.1 INTRODUCTION

Chapter 2 developed the ideas of linear models for dynamic systems. Here the response of such models to various inputs will be examined. The time domain is the natural way to look at the behaviour of such systems since this is how we see the response of real systems. We may not be particularly interested in the mathematical expression for the system response but rather in answers to questions such as 'How long will it take for a system to respond to a change in the input?'; 'How accurate will the response be?'; 'Will the response be oscillatory, and if so how much overshoot will be present?'.

The time response can be obtained by analytical means such as the Laplace transform method (explored in Chapter 2), alternatively computer simulation can be used. The approach adopted here is to develop an understanding of the association between a transfer function and the time response using a simple set of analytical relationships. You may ask why we need to bother with analytical methods if accurate computer simulation is available. The main reason is that simulation can only give us the response of a given system, it cannot tell us how to modify the system in order to obtain a specified response. In other words simulation by itself is not a design tool. Of course, 'trial and error' methods can be used in simple instances but even so, a knowledge of the connection between the parameters in a transfer function and the characteristics of its response is essential.

Until recently, analogue computers were widely used to solve differential equations and so simulate dynamic systems. An analogue computer consists of several operational amplifiers which can be configured with resistors and capacitors to behave in the same way as the physical problem being investigated. The voltages in the analogue computer mimic the variables in the problem and an oscilloscope or pen recorder is used to display the results. Digital computers have now overtaken their analogue predecessors to provide fast, flexible and accurate simulation tools.

3.2 SYSTEM TEST SIGNALS

The response of a dynamic system depends not only upon its transfer function but also upon the input signal and the initial conditions. The actual input to the system will not generally be known in advance. However, if the performance is favourable with certain well selected test signals then acceptable performance can be expected when subjected to actual operating inputs.

A Note on Units

A transfer function is a representation of the dynamic sensitivity of a system and as such it will have units of system output divided by system input. The units associated with input and output signals may represent displacement, voltage, pressure, temperature or other diverse quantities. In order to simplify the situation, the signals are often *normalized* by expressing the amplitude as a fraction of some convenient reference such as the normal range of the signal. In this way we can deal with diverse signals as though they were all dimensionless.

The Laplace operator, 's', has units of time^{-1} and when used within a transfer function, the associated coefficients must have appropriate units to maintain dimensional homogeneity. Strictly speaking, the units of 's' should always be specified when writing down a transfer function. However, conventionally the units of time in a transfer function are taken as seconds. For slow signals or systems, units of minutes or hours may be appropriate and in such cases there is no need to convert time constants, etc. into seconds, as long as the same time units are used when expressing the input and output functions. The units of time can be normalized relative to some time constant if desired. In this book, unless otherwise stated, units of time are taken as seconds.

Time Scaling

Practical systems have a wide range of time scales associated with them. For example, many electronic circuits have response times measured in fractions of a microsecond. On the other hand, a large process plant may have a time constant of over an hour. When systems have very fast or slow dynamics the coefficients in the transfer functions can be excessively small or large. Such extreme values are a nuisance to write down and manipulate, also they can give rise to inaccuracies in simulation. One way of overcoming these effects is to time scale the problem.

The simulation can be slowed down by a factor 'a' if each coefficient of 's' in the transfer function is multiplied by 'a'. The response will be speeded up if 'a' is less than unity.

For example, suppose the open-loop transfer function of a system is:

$$G(s) = \frac{5 \times 10^{12}}{s(3000 + s)(10^6 + 2000s + 3s^2)}$$

This system has a very fast response and it can be slowed down by a factor of 1000 by multiplying every 's' in the transfer function by 1000,

i.e.
$$G'(s) = \frac{5 \times 10^{12}}{1000s(3000 + 1000s)(10^6 + 2 \times 10^6 s + 3 \times 10^6 s^2)}$$

Therefore the response of the system

$$G'(s) = \frac{5}{s(3 + s)(1 + 2s + 3s^2)}$$

will be identical but on a time scale which is 1000 times slower.

Standard Input Signals

Owing to its simplicity, the step input is the most widely used signal for testing dynamic systems. A step input is achieved by suddenly changing the input from one steady value to another (Fig. 3.1(a)). In practice the transition will take some time. However as long as the transition time is fast compared to the response time of the system being tested the step change can be considered to be instantaneous. A step input is often idealized by assuming that it starts from zero and changes instantaneously at time $t = 0$. If the amplitude of the step is normalized to unity then we have the *unit step* (Fig. 3.1(b)).

The response to progressively changing inputs may be of interest, in which case a ramp input is a useful test signal (Fig. 3.2). Again, the ramp can be standardized to start from a value of zero at time zero and for a *unit ramp* the slope is normalized to one unit per second.

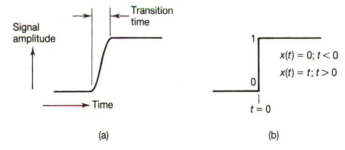

Figure 3.1 Step input signal. (a) Practical step. (b) Idealized unit step

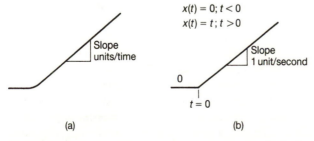

Figure 3.2 Ramp input signal. (a) Practical ramp. (b) Idealized unit ramp

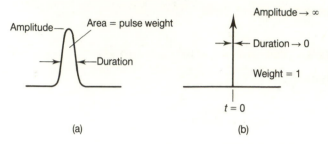

Figure 3.3   Pulse input signal. (a) Practical pulse. (b) Idealized unit impulse

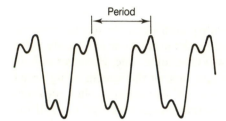

Figure 3.4   General periodic input signal

The response to impacts or shocks is of both theoretical and practical interest in investigating system dynamics. Such sharp input signals are generally called pulses (Fig. 3.3(a)). To qualify as an impact or shock the pulse duration must be short compared to the response time of the system. It is interesting that when the duration is sufficiently short, the shape of the pulse is unimportant. The area under the pulse is however most important and is called the pulse weight. The pulse weight has units of amplitude multiplied by time. The idealized pulse has a duration approaching zero, but in order to have a finite pulse weight, its amplitude must grow as the duration contracts. In the limit, as the duration tends toward zero, the amplitude will tend toward infinity. Such an idealized pulse is called an *impulse* (Fig. 3.3(b)). When normalized, a *unit impulse* results which has a weight of one unit-second. Reference to Appendix B shows that the Laplace transform of a unit impulse is 1.

The unit impulse, step and ramp are related by integration and differentiation. A unit step is obtained by integrating a unit impulse. Similarly, the unit ramp is the integral of the unit step. This means that the different responses obtained when these signals are applied to a linear system will also be related by integration and differentiation. In other words the impulse response of a system is the derivative of its step response. Similarly integrating the step response of a system gives its ramp response.

In many applications the system will be subjected to a periodic input as shown in Fig. 3.4. Vibrating equipment and rotating machines are good examples. The sine wave (or cosine wave) is a simple test signal for such cyclic inputs. The response of systems to this type of input is called *frequency response* and is of fundamental importance in dynamic system design and testing. Chapter 4 is dedicated to frequency response methods.

3.3 STEADY-STATE AND TRANSIENT RESPONSES

The overall response of any dynamic system will consist of two distinct parts. The *transient* part which, as its name suggests, normally dies away as time proceeds and a *steady-state* part that is left when the transient has indeed died away to zero.

$$\begin{matrix} \text{Overall} \\ \text{response} \end{matrix} = \begin{matrix} \text{Steady-state} \\ \text{response} \end{matrix} + \begin{matrix} \text{Transient} \\ \text{response} \end{matrix}$$

$$y(t) \quad = \quad \bar{y}(t) \quad + \quad \tilde{y}(t)$$

Figure 3.5(a) shows the response of a typical dynamic system subjected to a step input. The initial transient is seen to die away leaving the steady-state response which here is a constant. Figure 3.5(b) shows the same system with a ramp input. The transient again dies away leaving this time a steady-state response which is varying with time (a ramp).

Mathematicians use the terms complementary function and particular integral to describe the transient and steady-state parts of the response. Both the transient and steady-state responses depend upon the input signal which is applied to the system. However, the **form** of the transient depends only upon the system equations and is

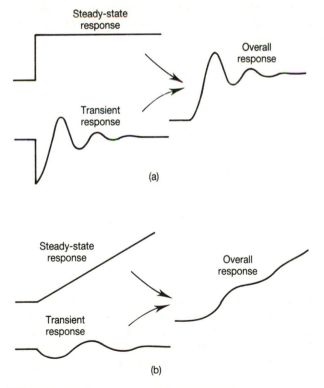

Figure 3.5 Steady-state and transient components of system response. (a) Typical step response. (b) Typical ramp response

independent of the input signal. By 'form' we mean the underlying function which determines the shape or structure of the transient.

3.4 SYSTEM CLASSIFICATION

Transfer functions can be classified in several ways. We have already seen that the *order* of a system is the highest power of '*s*' in the transfer function. The system order is very important in determining the type of response obtained. Examining the steady-state and instantaneous response of systems will enable further useful classifications to be applied.

Steady-state Response to a Constant Input

When the input to a system is held at some steady value (i.e. after a step input has been applied) the output will generally settle down to a constant steady-state value. The relationship between this steady-state output and the steady-state input is termed the *steady-state gain* or *static sensitivity* of the system. In such a situation when the input and output are both steady at constant values, then the derivatives of input and output signals will be zero. Within a transfer function, the Laplace operator can be considered to represent differentiation, therefore the steady-state gain of the system can be obtained by letting the Laplace operator tend towards zero in the transfer function, i.e., in steady-state, with input and output constant

$$\frac{d}{dt} = 0 \quad \text{is equivalent to} \quad s = 0$$

thus

$$\text{Steady-state gain} = G(s)|_{s \to 0} = G(0) \tag{3.1}$$

The steady-state response of a system to a **constant** input can be obtained by multiplying the input by the steady-state gain, i.e.

$$\bar{y} = G(0)\bar{x}$$

It must be remembered that the correct answer will only be obtained when the input is steady or unchanging.

System Instantaneous Response

The *instantaneous response* of a system describes how its output changes at the instant that the input is changed by a sudden amount. The instantaneous response shows how a system will respond when the input is altered very rapidly. During the rising edge of a step input, for example, the rate of change is very large and for an ideal step approaches infinity. Some transfer functions will show an instantaneous response when a step is applied.

The size of this instantaneous response can be obtained from the transfer function in a similar way to the steady-state response. Because the rate of change of variables during such rapid transitions approaches infinity, the value of the transfer function with 's' tending toward infinity will give the relationship between input and output.

$$\text{Instantaneous gain} = G(s)|_{s \to \infty} = G(\infty) \tag{3.2}$$

The instantaneous response to the rising edge of a step input is obtained by multiplying the change at the input by the instantaneous gain. Remember that the instantaneous gain of a system respresents the relationship between the **change** at the input and the **change** at the output.

Example 3.1 Phase lead network response A phase lead network has the general transfer function given by

$$\frac{Y}{X}(s) = \frac{1 + a\tau s}{1 + \tau s}$$

Find the steady-state and instantaneous responses when the system is subjected to a unit step input.

SOLUTION The steady-state gain is given by

$$G(0) = \frac{1 + a\tau s}{1 + \tau s}\Big|_{s \to 0} = 1$$

and so the steady-state change of output is equal to the change in the input, i.e. $y = x$ in steady-state.

The instantaneous gain is given by

$$G(\infty) = \frac{1 + a\tau s}{1 + \tau s}\Big|_{s \to \infty} = \frac{a\tau s}{\tau s} = a$$

Note, for the case when $a > 1$ the instantaneous response is larger than the steady-state response. Simulate the open-loop step response of this system using CODAS and verify the answers. Try an initial condition at the plant output to see its effect on the instantaneous and steady-state output.

Exercise Find the instantaneous and steady-state gain for each of the following first-order transfer functions. Verify your answers using CODAS.

(a) $\dfrac{1}{2 + s}$ (b) $\dfrac{1 + s}{2 + s}$ (c) $\dfrac{s}{2 + s}$ (d) $\dfrac{1 + s}{s}$

Proper and Strictly Proper Systems

There are certain practical limitations to the dynamic behaviour of real systems. Strictly, no real system can respond instantaneously since this would require an infinite acceleration. So-called *strictly proper* systems have an instantaneous gain of zero.

$$G(\infty) = 0 \qquad \text{for a strictly proper system}$$

Consider a general transfer function with polynomial numerator of order m and polynomial denominator of order n

$$G(s) = \frac{b_0 + b_1 s + \cdots b_m s^m}{a_0 + a_1 s + \cdots a_n s^n} \tag{3.3}$$

Now $G(\infty)$ is dominated by the terms involving the highest power of 's', i.e. s^n or s^m. Clearly, for $G(\infty) = 0$ 'm' must be less than 'n'. This means that the numerator of a strictly proper system must be of lower order than the denominator.

This rather restrictive rule can be relaxed slightly by allowing the transfer function to have an instantaneous gain which is finite. Such a system is called *proper*.

$$G(\infty) = c \qquad \text{for a proper system}$$

Thus proper systems can have a numerator of the same order as the denominator but cannot have a numerator of higher order.

For example,

$$\frac{1+s}{s(1+2s)} \qquad \text{is strictly proper}$$

$$\frac{s}{1+s} \qquad \text{is proper}$$

and $\qquad \dfrac{1+s+s^2}{2+s} \qquad$ is not a proper transfer function

All real systems are strictly proper. However, models may be used which are merely proper that approximate the system behaviour quite closely. Note that CODAS cannot be used to obtain the time response of transfer functions which are not proper.

Exercise Examine the open-loop step response of the following (strictly proper) phase lead transfer function. Compare the response with that from the normal proper transfer function obtained by removing the lag with the small time constant.

$$\frac{Y}{X}(s) = \frac{1 + 10s}{(1+s)(1+0.1s)}$$

Example 3.2 Implementation of a differentiator A differentiator is a device for obtaining the rate of change of signals; for example it can be used to derive velocity from a position signal. The relationship between output, y, and input, x, for a pure differentiator is

$$y(t) = \frac{dx}{dt}(t)$$

giving the transfer function

$$\frac{Y}{X}(s) = s$$

This is not a proper transfer function and thus cannot be implemented as it stands (imagine the response of this pure differentiator to a step input). Any practical differentiator must be strictly proper and the transfer function will be accompanied by other dynamic terms in the denominator.

Equations (3.4(a) and (b)) represent successively more realistic differentiators each giving a worse approximation of pure differentiation. For signals which change slowly compared to the lag time constant(s), the approximation will be good. However with rapidly changing signals the presence of the first order lags will provide smoothing of the differentiated signal.

$$\frac{Y}{X}(s) = \frac{s}{1 + \tau_1 s} \tag{3.4(a)}$$

$$\frac{Y}{X}(s) = \frac{s}{(1 + \tau_1 s)(1 + \tau_2 s)} \tag{3.4(b)}$$

Note that no such restrictions exist for the practical implementation of pure integrators (free integrators) since the transfer function of a pure integrator

$$\frac{Y}{X}(s) = \frac{1}{s}$$

is strictly proper.

The Bode Form of Transfer Functions

It is often convenient to express transfer functions in a standardized form so that the essential characteristics are clear.

Transfer functions can be expressed in a standardized way called the *Bode form* named after H. W. Bode who made major contributions to frequency response theory. The Bode form of a transfer function is written as

$$G(s) = \frac{K_B G'(s)}{s^n} \tag{3.5}$$

where the function $G'(s)$ has a steady-state gain of unity, i.e.

$$G'(0) = 1$$

The term s^n in the denominator corresponds to the number of free integrators in the system. The number, n, is called the system *Type number*.

The constant, K_B, is called the *Bode gain* and for a Type-0 system (i.e. no free integrators) it is the same as the steady-state gain. Type-1 and higher systems will have a steady-state gain of infinity since

$$\lim_{s \to 0} \frac{1}{s} = \infty$$

The infinite steady-state gain can be visualized by considering a free integrator with a constant input. The output will ramp up indefinitely and in fact never reach steady-state.

Example 3.3 Bode form of a system Determine the order, type number and Bode gain of the following transfer function.

$$G(s) = \left(2 + \frac{3}{s}\right)\left[\frac{5}{(s+4)(s+10)}\right]$$

SOLUTION The transfer function can be expressed as the ratio of two polynomials

$$G(s) = \frac{(2s+3)5}{s(s+4)(s+10)}$$

We can see that the system is third order and, because of the existence of a free integrator, is of Type-1. The Bode gain is obtained by finding the steady-state gain with the free integrator removed.

$$K_B = \lim_{s \to 0}\left[\frac{(2s+3)5}{(s+4)(s+10)}\right] = 0.375$$

Exercise Put the following transfer functions in their Bode form; determine their order, Type number and Bode gain.

(a) $\dfrac{2s+3}{(s^2+6s+20)(s^2+0.1s)}$ (b) $\dfrac{2+(3/s)}{s(4+7s)}$

3.5 THE CHARACTERISTIC EQUATION AND THE TRANSIENT RESPONSE

Consider a general transfer function

$$\frac{Y}{X}(s) = G(s) = \frac{N(s)}{D(s)} \tag{3.6}$$

where $D(s)$ is the denominator of the transfer function and $N(s)$ the numerator. Cross multiplying we have

$$Y(s)D(s) = X(s)N(s) \tag{3.7}$$

If the input to the system is zero the resulting response must contain only a transient component (since the steady-state response would be zero). The form of the transient response of any system can therefore be found by setting $X(s)$ to zero in Eq. (3.7) which gives

$$Y(s)D(s) = 0 \tag{3.8}$$

Ignoring the trivial case of $Y(s) = 0$ we find that the transient response for a general system is determined by the equation

$$D(s) = 0 \tag{3.9}$$

This important equation is called the *characteristic equation* and its solution determines the **form** of the transient response for **any input**. Notice that only the denominator of the transfer function appears in the characteristic equation. Solving the characteristic equation involves finding the values of 's' which make $D(s)$ zero. These *roots* of the characteristic equation are intimately related to the resulting transient response.

To see how the roots of the characteristic equation relate to the transient response consider Eq. (3.8) in a general form where $D(s)$ is an nth order polynomial and $Y(s)$ is the Laplace transform of the transient ($\mathscr{L}\{\tilde{y}(t)\} = \tilde{Y}(s)$):

$$\tilde{Y}(s)(a_0 + a_1 s + \cdots a_n s^n) = 0$$

or

$$a_0 \tilde{Y}(s) + a_1 s \tilde{Y}(s) + \cdots a_n s^n \tilde{Y}(s) = 0$$

Using the idea that multiplication by 's' in the s-domain is equivalent to differentiation in the time domain we can see that the transient response must be of a form such that

$$a_0 \tilde{y}(t) + a_1 \frac{d\tilde{y}}{dt}(t) + \cdots a_n \frac{d^n \tilde{y}}{dt^n}(t) = 0 \qquad (3.10)$$

The transient, $\tilde{y}(t)$, must be such that when it is combined with its derivatives in Eq. (3.10) the terms add up to zero. The exponential function, $e^{\alpha t}$, is a candidate for the solution of Eq. (3.10) since differentiation of an exponential function results in yet another exponential function.

Hence a possible transient response function is

$$\tilde{y}(t) = C e^{\alpha t} \qquad (3.11)$$

where C is an arbitrary constant. Successive derivatives are given by

$$\frac{d\tilde{y}}{dt}(t) = \alpha C e^{\alpha t} = \alpha \tilde{y}(t)$$

and

$$\frac{d^2 \tilde{y}}{dt^2}(t) = \alpha^2 C e^{\alpha t} = \alpha^2 \tilde{y}(t) \qquad \text{etc.}$$

Substituting for $\tilde{y}(t)$ and its derivatives in Eq. (3.10)

$$a_0 C e^{\alpha t} + a_1 \alpha C e^{\alpha t} + \; .. \; a_n \alpha^n C e^{\alpha t} = 0$$

and finally dividing out $C e^{\alpha t}$ we obtain

$$a_0 + a_1 \alpha + \cdots a_n \alpha^n = 0$$

or

$$D(\alpha) = 0 \qquad (3.12)$$

Equation (3.12) is the characteristic equation with 'α' replacing 's'. Thus the value of α in the proposed transient response (Eq. (3.11)) will be a root of the characteristic equation.

The above justification is independent of the value of 'C' in Eq. (3.11), and thus it seems that C can have any value. In practice, the value of C is determined by the system input, initial conditions and, as we shall see, the transfer function numerator.

Example 3.4 First-order system transient response For the well-worn example of the first-order thermometer, the characteristic equation is obtained by equating the denominator of its transfer function (Eq. (2.16)) to zero

$$1 + \tau s = 0 \tag{3.13}$$

The characteristic equation has a single solution, i.e. one root at

$$s = -1/\tau$$

The transient response of the thermometer is thus a single exponential term with α equal to $-1/\tau$.

$$\tilde{y}(t) = C \, e^{-t/\tau}$$

Relationship of Characteristic Equation to State-variable Models

The state-variable notation for system equations was introduced in Chapter 2. It is worth returning briefly to state-variable models to examine their relationship to the ideas developed on the characteristic equation.

Equations (2.45) and (2.46) gave the general state-variable formulation for a dynamic system

$$\frac{d}{dt} \mathbf{X}(t) = \mathbf{A}\mathbf{X}(t) + \mathbf{B}u(t)$$

$$y(t) = \mathbf{C}\mathbf{X}(t) + \mathbf{D}u(t)$$

The transfer function for this system was shown to be

$$\frac{Y}{U}(s) = \mathbf{C}[s\mathbf{I} - \mathbf{A}]^{-1}\mathbf{B} + \mathbf{D}$$

The inverse of a matrix is obtained by dividing the adjoint by the determinant:

$$[s\mathbf{I} - \mathbf{A}]^{-1} = \frac{\mathrm{adj}(s\mathbf{I} - \mathbf{A})}{|s\mathbf{I} - \mathbf{A}|}$$

Therefore the denominator of the transfer function is determined by the determinant of the matrix $[s\mathbf{I} - \mathbf{A}]$. The system characteristic equation is obtained by equating this determinant to zero

$$|s\mathbf{I} - \mathbf{A}| = 0 \tag{3.14}$$

The transient characteristics of a system are thus determined solely by the matrix $\mathbf{A}$ in the state variable model. The roots of the characteristic equation in this form are called the eigenvalues of the matrix $\mathbf{A}$.

System Impulse Response

The impulse response of a system is of theoretical interest since the Laplace transform of a unit impulse is 1, i.e.

$$\mathscr{L}\{\delta(t)\} = 1$$

The Laplace transform of the impulse response is therefore

$$Y(s) = G(s) \cdot 1$$

i.e., the impulse response of a system is simply the inverse transform of its transfer function

$$\text{Impulse response} = \mathscr{L}^{-1}\{G(s)\} = g(t) \qquad (3.15)$$

The impulse response and transfer function are thus one and the same viewed from the time domain and s-domain respectively.

Since the impulse function is zero for all times other than at $t = 0$, the impulse response has a zero steady-state component and so consists of only the transient.

Let's see how this ties in with Example 3.4. In Chapter 2 the Laplace transform of an exponential function was derived as

$$\mathscr{L}\{e^{-at}\} = \frac{1}{s + a}$$

Rearranging the transfer function of the first-order system we can see that

$$\frac{1/\tau}{1/\tau + s} = \mathscr{L}\left\{\frac{1}{\tau} e^{-t/\tau}\right\}$$

Thus the impulse response is

$$g(t) = \frac{1}{\tau} e^{-t/\tau}$$

Which agrees with the transient term derived from the characteristic equation.

3.6 THE s-PLANE

An nth order characteristic equation has n roots, each giving rise to a term in the transient response. The total transient is thus the sum of these terms.

In order to develop the ideas relating the roots of the characteristic equation to the transient, a graphical representation of the roots is appropriate. The roots of the characteristic equation will either be real or complex and as such can be plotted on the complex plane (Argand diagram). Such a representation is called an s-plane diagram. The axes of the s-plane are the real and imaginary parts of s, i.e.

$$s = \sigma + j\omega \qquad (3.16)$$

Poles and Zeros

The roots of the characteristic equation are the values of s which make the transfer function denominator zero (Eq. (3.9)). The characteristic equation roots are called the *poles* of the system, i.e.

$$D(s) = 0 \qquad \text{gives the system poles}$$

The poles of a transfer function are plotted as small crosses on the s-plane.

As well as plotting poles on an s-plane diagram it is usual to also plot the system *zeros*. The system zeros are the values of s which make the transfer function numerator zero, i.e.

$$N(s) = 0 \qquad \text{gives the system zeros}$$

Zeros are shown on an s-plane diagram as small circles.

The s-plane pole/zero plot will always be symmetrical about the real axis since the complex roots of real equations always occur in conjugate pairs. Real roots of course, lie on the real axis and so do not influence the symmetry. For this reason the bottom half of the s-plane is often omitted from pole/zero diagrams.

It must be remembered that it is the **poles** which are the roots of the characteristic equation and thus determine the form of the transient response. The zeros, together with the system input and initial conditions, merely modify the relative size or weighting of the terms in the transient.

Transport delay gives rise to transfer functions which include the term e^{-sT_D}. For open-loop systems with delay, the term appears only in the numerator of the transfer function and thus does not affect the characteristic equation. In such cases the transient terms are unaltered by the delay and the system response is simply delayed without being changed. This is not the case for feedback (closed-loop) systems where the delay term appears in the characteristic equation. The resulting irrational function then gives rise to an infinite set of s-plane poles. The s-plane diagrams of closed-loop systems with transport delay are dealt with in Chapter 6.

Example 3.5 s-plane representation of a transfer function The transfer function given by

$$G(s) = \frac{0.5(2 - s)}{(s + 2)(1 + s + s^2)(1 + s)^2}$$

is a fifth-order system with a single zero from

$$2 - s = 0 \qquad \text{at} \qquad s = +2$$

and five poles. One simple pole at

$$s = -2$$

a complex pair of poles from

$$1 + s + s^2 = 0$$

at
$$s = \frac{-1 \pm \sqrt{1 - 4}}{2} \qquad = -0.5 + j0.866$$

and finally a double pole at

$$s = -1$$

Figure 3.6 shows the s-plane diagram for the system. Notice that the repeated pole is shown as two slightly displaced crosses.

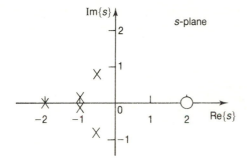

Figure 3.6 Pole/zero pattern for $G(s) = \dfrac{0.5(2 - s)}{(s + 2)(1 + s + s^2)(1 + s)^2}$

Note: CODAS can be used to display the s-plane diagram of a system by switching to the root locus domain. Try entering the above transfer function and confirming the results. The double pole must be entered into CODAS as $(1 + s)(1 + s)$ rather than $(1 + s)^2$.

3.7 RELATIONSHIPS BETWEEN POLE LOCATION AND TRANSIENT

So far we have seen that each system pole will contribute a term to the overall transient response. Poles can be *distinct* or *simple*, alternatively they can be *repeated* or *multiple* poles. A further complication is that poles can also be *complex*. Each type of pole will now be considered together with its related transient term.

Distinct Real Poles

Every distinct pole will give rise to a separate exponential term in the transient response, each with an associated arbitrary constant, i.e.

$$\begin{array}{ccc}
\text{Pole location} & & \text{Transient term} \\
s = \sigma & \rightarrow & \tilde{y}(t) = C\, e^{\sigma t}.
\end{array}$$

For the transient to decay away with time, σ must be negative, i.e. the pole must be located on the negative real axis. The further to the left the pole is located the faster the transient dies away.

Poles on the positive real axis give rise to exponential terms which increase with time. The more positive the pole the faster the rate of growth. However, any pole need be only slightly positive to produce a transient which eventually grows infinitely large. Such systems in which a transient term increases indefinitely with time are called *unstable systems*. Conversely systems in which all transient terms die away with time are called *stable systems*.

A single real pole corresponds to a first-order system, the relationship between pole

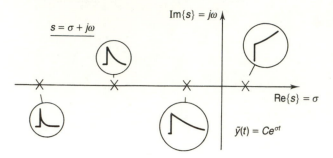

Figure 3.7 Transient corresponding to a distinct real pole

location and time constant, τ, being

$$\sigma = -1/\tau$$

The smaller the time constant the further to the left the pole and the faster the transient dies away. Figure 3.7 illustrates this idea.

A special case of a real pole occurs when we have a free integrator in the transfer function. Consider the transfer function

$$G(s) = \frac{1}{s}$$

which has a simple pole at the origin of the s-plane (at $s = 0$). The transient term corresponding to this pole location is

$$\tilde{y}(t) = C\,e^{0} = C$$

In other words the transient term is just a constant, neither decaying (stable) or increasing (unstable).

Distinct Complex Poles

The roots of second- or higher-order characteristic equations can be complex. For real systems these complex poles must occur in conjugate pairs

$$s_1 = \sigma + j\omega; \qquad s_2 = \sigma - j\omega$$

The transient terms generated by this pair of poles will be

$$\tilde{y}(t) = C_1\,e^{(\sigma + j\omega)t} + C_2\,e^{(\sigma - j\omega)t}$$

however, in order to keep $\tilde{y}$ real it turns out that the constants, C_1 and C_2, must also be a complex conjugate pair, i.e.

$$C_1,\,C_2 = a \pm jb = R\,e^{\pm j\theta}$$

therefore

$$\tilde{y}(t) = R\,e^{j\theta} \cdot e^{(\sigma + j\omega)t} + R\,e^{-j\theta} \cdot e^{(\sigma - j\omega)t}$$

$$= R\,e^{\sigma t}\{e^{j(\omega t + \theta)} + e^{-j(\omega t + \theta)}\}$$

or

$$\tilde{y}(t) = R\,e^{\sigma t}\,2\cos(\omega t + \theta) \tag{3.17}$$

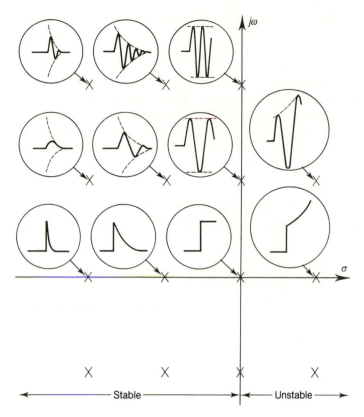

Figure 3.8 Relationship between pole location and transient shape

The transient term is therefore a sinusoid of frequency ω radians per second which is multiplied by an exponential function decaying (or increasing) at a rate governed by σ. The arbitrary constants, C_1 and C_2, have been replaced by two other arbitrary constants; R, which is related to the amplitude of the sinusoid and θ, which is related to its phase. Since the value of R is arbitrary, the factor of 2 in Eq. (3.17) can be ignored.

$$\text{Pole location} \qquad\qquad \text{Transient term}$$
$$s = \sigma \pm j\omega \qquad \rightarrow \qquad \tilde{y}(t) = R\, e^{\sigma t} \cos(\omega t + \theta)$$

The relationship between the location of a complex pole and the form of the transient is shown in Fig. 3.8. The larger the complex part of the pole the higher the frequency of the oscillation. Additionally, just like the first-order system, the more negative the real part the faster the transient dies away. A pole pair with a positive real part spells trouble in that the oscillation will increase exponentially with time. For a system to be stable any complex poles must be in the left hand half of the s-plane.

Repeated Poles

In the case where the characteristic equation has repeated roots, the simple relationships developed above break down. Repeated roots of the characteristic equation give rise to transient terms of the form

$$\tilde{y}(t) = (C_1 + C_2 t + C_3 t^2 + \cdots C_m t^m) e^{\alpha t} \qquad (3.18)$$

where m is the order of the repeated root.

The values of the arbitrary constants, C_1, C_2 etc., again depend upon the zeros, input and initial conditions, but the value of α depends only upon the location of the repeated pole. For large values of time, Eq. (3.18) is always dominated by the exponential function no matter how high the order, m. Thus as long as the pole has a negative real part the term will eventually decay to zero.

Example 3.6 Unit step response of two cascaded first-order lags Cascading two identical non-interacting standard first-order lags results in a system with the transfer function

$$\frac{Y}{X}(s) = \frac{1}{(1 + \tau s)(1 + \tau s)}$$

This system has a steady-state gain of unity and so the steady-state response to a unit step is unity.

The characteristic equation has a double root at

$$s = -1/\tau$$

which gives a transient term of

$$\tilde{y}(t) = (C_1 + C_2 t) e^{-t/\tau}$$

The total response is thus

$$y(t) = \bar{y}(t) + \tilde{y}(t)$$
$$= 1 + (C_1 + C_2 t) e^{-t/\tau} \qquad (3.19)$$

The initial conditions can be used to find values of the arbitrary constants. Two initial conditions are necessary; namely position and velocity. For this example they will both be taken as zero.

At $t = 0$ $\qquad\qquad\qquad\qquad y = 0 = 1 + C_1$

giving $\qquad\qquad\qquad\qquad C_1 = -1.0$

The velocity can be obtained by differentiating Eq. (3.19)

$$\frac{dy}{dt}(t) = -\frac{C_1}{\tau} e^{-t/\tau} + C_2(1 - t/\tau) e^{-t/\tau}$$

and at $t = 0$ $\qquad\qquad\qquad \frac{dy}{dt} = 0 = -\frac{C_1}{\tau} + C_2$

giving $\qquad\qquad\qquad\qquad C_2 = \frac{C_1}{\tau} = -\frac{1}{\tau}$

The final step response is therefore

$$y(t) = 1 - (1 + t/\tau)\, e^{-t/\tau}$$

General Stability on the *s*-plane

Any pole, whether distinct or repeated, real or complex, which has a positive real part leads to a transient term which increases with time. In order that a system settle into steady state after any input or disturbance all the terms in the transient response must decay to zero. Thus for a system to be stable **all** the system poles must lie in the negative (left) half of the *s*-plane. The location of zeros will not affect the stability of a system.

Poles with a real part of zero (i.e. lying on the imaginary axis) give rise to transient terms which neither die away nor increase indefinitely. Systems with no poles in the right half plane but with one or more on the imaginary axis are called *marginally stable*. Purely imaginary poles produce a non-decaying sinusoidal response of frequency equal to the imaginary part of the pole. A free integrator (pole at the origin) is an example of a marginally stable system with zero frequency, i.e. a non-decaying constant term.

The relationship between the characteristic equation roots and the transient response is now clear and the importance of the characteristic equation in system dynamics can be appreciated.

Exercise Comment on the stability of the system in Example 3.5. Examine the open-loop step response using CODAS to confirm your prediction.

3.8 THE STANDARD SECOND-ORDER SYSTEM

Just as we found it useful to have a standard form for a first-order system with the performance-related parameter of the time constant, we will now introduce a standardized second-order transfer function with its own performance-related parameters.

Consider the spring/mass/damper system modelled in Example 2.5 which has a transfer function given by

$$\frac{Y}{X}(s) = \frac{k}{k + Cs + ms^2}$$

where k is the spring stiffness, C the damping coefficient and m the mass. Dividing throughout by k, we obtain the Bode form of the transfer function

$$\frac{Y}{X}(s) = \frac{1}{1 + (C/k)s + (m/k)s^2} \tag{3.20}$$

The performance-related parameters which are generally used are:

(a) The natural frequency, ω_n, which relates to the speed of oscillation of the transient.
(b) The damping ratio, ζ, (pronounced zeta) which is a dimensionless representation of the amount of damping in the system.

Using these parameters, the standard form for a second-order transfer function is

$$\frac{Y}{X}(s) = \frac{1}{1 + 2\zeta s/\omega_n + (s/\omega_n)^2} \tag{3.21}$$

The factor 2 associated with ζ is included for convenience.

By comparing Eqs (3.20) and (3.21) which are both in standard (i.e. Bode) form, we can see how the parameters of spring stiffness, mass and damping coefficient relate to natural frequency and damping ratio. Comparing coefficients we have

$$m/k = 1/\omega_n^2 \quad \text{giving} \quad \omega_n = \sqrt{k/m} \tag{3.22}$$

also
$$C/k = 2\zeta/\omega_n \quad \text{giving} \quad \zeta = \frac{C\omega_n}{2k} = \frac{C}{2\sqrt{km}} \tag{3.23}$$

Relationship between ζ and ω_n and Pole Location

The poles of the standard second-order system from Eq. (3.21) are at

$$s = \frac{-2\zeta/\omega_n \pm \sqrt{(2\zeta/\omega_n)^2 - 4/\omega_n^2}}{2/\omega_n^2}$$

which simplifies to

$$s = -\zeta\omega_n \pm \omega_n\sqrt{\zeta^2 - 1} \tag{3.24}$$

There are four distinct cases to consider:

Systems with no damping, $\zeta = 0$ If the system has no damping whatsoever (i.e. $C = 0$ in Eq. (3.20)) then the poles are located on the imaginary axis with values of

$$s = \pm j\omega_n$$

In such a situation the transient is a continuous sinusoid of frequency ω_n which neither increases or decays with time. Zero damping thus produces *simple harmonic motion*. The natural frequency, ω_n, is thus the frequency at which the system oscillates when there is no damping present.

Underdamped systems, $0 < \zeta < 1.0$ Examining Eq. (3.24) you will see that the poles are complex for all values of damping ratio less than unity. The real and imaginary parts of the poles are given by

real part, $\qquad\qquad\qquad \sigma = -\zeta\omega_n \tag{3.25}$

and imaginary part, $\qquad\qquad jω = j\omega_n\sqrt{1 - \zeta^2} \tag{3.26}$

Corresponding to a transient which oscillates at a frequency of

$$\omega_d = \omega_n\sqrt{1 - \zeta^2} \tag{3.27}$$

This frequency, ω_d, is called the *damped natural frequency* and is always lower than the natural frequency, ω_n, the difference becoming more pronounced at larger values of ζ.

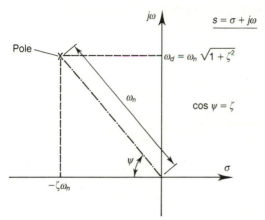

Figure 3.9 *s*-plane geometry for an underdamped second-order system

Figure 3.9 shows the geometry of the *s*-plane diagram for an underdamped second-order system. The radius of the pole from the *s*-plane origin is given by

$$r = \sqrt{\sigma^2 + \omega^2} = \sqrt{(\zeta\omega_n)^2 + \omega_n^2(1 - \zeta^2)} = \omega_n \qquad (3.28)$$

i.e., the natural frequency is the distance from the pole to the *s*-plane origin.

Furthermore the cosine of the angle between the negative real axis and the pole, ψ, is

$$\cos(\psi) = \frac{\zeta\omega_n}{\omega_n} = \zeta \qquad (3.29)$$

Critically damped systems, $\zeta = 1.0$ When the damping ratio, ζ, is unity the imaginary part becomes zero and the system has a double pole at

$$s = -\omega_n$$

There is just enough damping to prevent the poles from being complex and so there is no oscillatory component to the transient response. Such a value of damping is called *critical damping* and the system is said to be *critically damped*.

Overdamped systems, $\zeta > 1.0$ The poles given by Eq. (3.24) become real and distinct if the damping exceeds the critical value. The transient thus consists of two real exponentials

$$y(t) = C_1 e^{\alpha_1 t} + C_2 e^{\alpha_2 t}$$

where $\qquad \alpha_1 = -\zeta\omega_n + \omega_n\sqrt{\zeta^2 - 1} \quad$ and $\quad \alpha_2 = -\zeta\omega_n - \omega_n\sqrt{\zeta^2 - 1}$

Figure 3.10 shows the unit step response of a standard second-order system with a range of damping ratios.

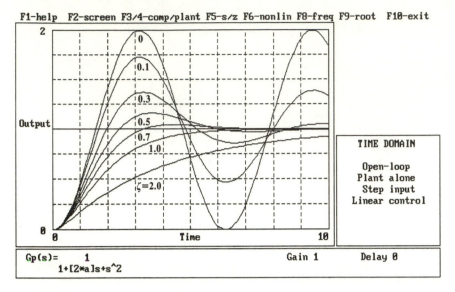

F1-help F2-screen F3/4-comp/plant F5-s/z F6-nonlin F8-freq F9-root F10-exit

Figure 3.10   Second-order system step response

3.9 TIME DOMAIN PERFORMANCE SPECIFICATIONS

Step Response Criteria

The step response is arguably the most relevant test of general system performance and as such it is worth discussing some of the related criteria which are used to specify system performance.

Figure 3.11(a) and (b) show the general step response of two systems, one oscillatory and the other with a monotonic response showing no oscillation. Accepted dynamic performance measures relate to the degree of oscillation, the time for the transient to die away and the response time or time for the response to reach some fraction of the final value.

For oscillatory systems the amount of oscillation can be specified by the *percentage peak overshoot*, or just *percentage overshoot* which, referring to Fig. 3.11(a), is defined as

$$\% \text{ overshoot} = 100\,\frac{m_1}{m_0} \tag{3.30}$$

The *decay ratio* is defined as the ratio of two successive overshoots, normally the first two are used, i.e.

$$\text{Decay ratio} = \frac{m_3}{m_1} \tag{3.31}$$

The *settling time*, t_s, tells us the time for the transient to die away to within some specified tolerance band. The response must not only reach but also stay within the band. Typical tolerance bands are $\pm 2\%$ and $\pm 5\%$.

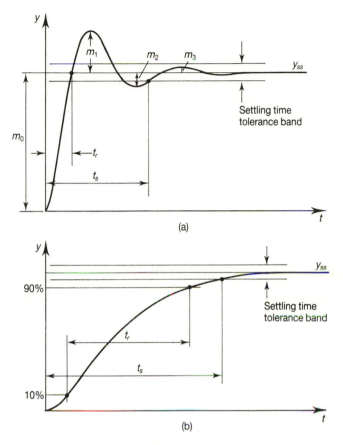

Figure 3.11 Step response performance criteria. (a) Oscillatory response. (b) Monotonic response

For oscillatory systems the speed of oscillation is expressed in terms of the *time to first peak* or the *period of oscillation*.

The response time is normally defined in terms of the rise time, t_r, which for an oscillatory response is the time to rise from the initial value to the final value. This definition however is of no use for monotonic responses which approach the final value asymptotically. The rise time of non-oscillatory systems is usually defined as the time to rise from 10% to 90% of the final value.

The 63% rise time is also sometimes used to specify the response time. The odd value of 63% originates from the step response of a first-order system.

First-order Systems

A first-order system has a monotonic step response, the transient being a simple exponential.

$$\tilde{y}(t) = C \, e^{-t/\tau}$$

After one time constant (i.e. $t = \tau$) the transient will die away to

$$\frac{\tilde{y}}{C}(\tau) = e^{-1} = 0.37$$

and thus the response will be within $(1-0.37)$ or 63% of the steady-state value. Hence the 63% rise time corresponds to one time constant for a first-order system.

The 5% settling time, t_s, can be similarly found from

$$0.05 = e^{-t_s/\tau}$$

taking natural logs of this expression gives

$$t_s = -\tau \ln(0.05) \approx 3\tau$$

Thus the 5% settling time of a first-order system is approximately three time constants.

Second-order Systems, Logarithmic Decrement and Decay Ratio

The step response of an underdamped second-order system reaches the first peak in half the oscillation period. The transient oscillates at the damped natural frequency which is equal to the imaginary part of the poles. Thus the time to first peak

$$t_p = \frac{1}{2}\frac{2\pi}{\omega_d} = \frac{\pi}{\omega_d} \tag{3.32}$$

The rise time is approximately half this value and is therefore shortened if the damped natural frequency is increased.

For the case of a second-order underdamped system a simple but useful relationship can be developed between the rate of decay of the oscillatory transient and the damping ratio.

The transient response is expressed by Eq. (3.17) with the real and imaginary parts of the poles given by Eqs (3.25) and (3.26) respectively.

$$\tilde{y}(t) = R\,e^{-\zeta\omega_n t}\cos(\omega_d t + \theta) \tag{3.33}$$

Figure 3.12 shows the general shape of this function.

The positive and negative peaks of the oscillations occur at times separated by half the period of the sinusoid, π/ω_d. The amplitude of these peaks have been sequentially labelled as m_1, m_2, etc. Each peak occurs at approximately the time when the cosine function in Eq. (3.33) is at its maximum of ± 1.0.

The absolute magnitude of any peak, m_k, will therefore be given by

$$m_k = R\,e^{-\zeta\omega_n t_k}$$

and the ratio of any two peaks separated by 'n' half periods of oscillation will thus be

$$\frac{m_{k+n}}{m_k} = \frac{R\,e^{-\zeta\omega_n(t_k + n\pi/\omega_d)}}{R\,e^{-\zeta\omega_n t_k}} = \frac{e^{-\zeta\omega_n t_k} \cdot e^{-\zeta\omega_n n\pi/\omega_d}}{e^{-\zeta\omega_n t_k}}$$

$$= e^{-\zeta\omega_n n\pi/\omega_d}$$

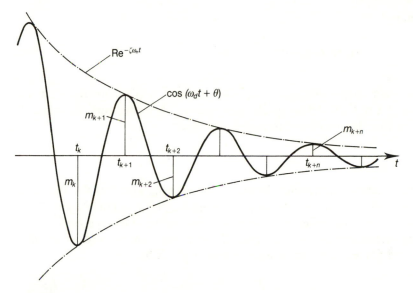

Figure 3.12 Second-order underdamped transient

which using Eq. (3.27) gives

$$\frac{m_{k+n}}{m_k} = e^{-n\pi\zeta/\sqrt{1-\zeta^2}} \tag{3.34}$$

This expression gives the ratio of any two peaks of the transient which are separated by 'n' half cycles of oscillation. The result is a function of damping ratio only. Thus it is the damping ratio alone that determines the shape of the transient.

For a second-order system the decay ratio is given by the ratio of two successive overshoots, i.e. separated by one whole cycle

$$\text{Decay ratio} = \frac{m_{k+2}}{m_k} = e^{-2\pi\zeta/\sqrt{1-\zeta^2}} \tag{3.35}$$

The exponent in this function (without the negative sign) is called the *logarithmic decrement* or *log-dec* for short

$$\begin{matrix}\text{Logarithmic} \\ \text{decrement}\end{matrix} = \ln\frac{m_k}{m_{k+2}} = \frac{2\pi\zeta}{\sqrt{1-\zeta^2}} \tag{3.36}$$

Percentage Overshoot

Equation (3.34) can be used to give the percentage overshoot in terms of the damping ratio for a second-order system. The transient oscillates about the steady-state value and since the first peak starts at an absolute value of zero, m_0 is the steady-state response. The first overshoot will peak at m_1 units above the steady-state value. The percentage overshoot is therefore

$$\% \text{ overshoot} = 100 \, \frac{m_1}{m_0} = 100 \, e^{-\pi\zeta/\sqrt{1-\zeta^2}} \tag{3.37}$$

Exercise A second-order system exhibits a step response with a 10% overshoot, what is the damping ratio? Use CODAS to verify your answer. Try changing the natural frequency and check that the overshoot remains unchanged.

Settling Time

There is no simple, accurate relationship between settling time and transfer function parameters for a second-order system. However, the settling time is mainly governed by the time for the exponential envelope to decay to within the tolerance band. This as we have seen is determined by the real part of the poles, $-\sigma$. So just as for the first-order system, the 5% settling time is approximately given by

$$t_s \approx 3\tau = \frac{3}{-\sigma} = \frac{3}{\zeta\omega_n} \tag{3.38}$$

Intuitively, we know that a 'best damping ratio' will exist since lightly damped systems will overshoot many times before settling down and heavily damped systems will be sluggish. The optimum damping ratio for a given settling time is the minimum which can be used without the overshoot exceeding the tolerance limit. For example, the 5% settling time is optimized when the percentage overshoot is **just** 5%, resulting in a damping ratio of 0.69.

Design Regions on s-plane

The above step response performance criteria are related to the position of the poles on the s-plane. Using the results derived for first- and second-order systems enables a bounded design region to be drawn on the s-plane within which the poles produce an acceptable performance.

The percentage overshoot or decay ratio is generally required to be within some specified limits. This corresponds to a second-order system with poles located between two lines of constant damping ratio as shown in Fig. 3.13.

A maximum settling time is often part of a performance specification and as we have seen settling time is related to the real part of the pole or poles. Thus for the settling time to be less than some specified figure the poles must lie to the left of a vertical line on the s-plane.

Finally the rise time depends on the damped natural frequency which is the imaginary part of the poles. Thus for the rise time to be no more than some specified value the upper half plane pole must lie above a horizontal line. Figure 3.13 shows these design constraints on an s-plane diagram.

Exercise Draw the s-plane region in which the poles of an underdamped second-order system must lie to meet the following specifications:

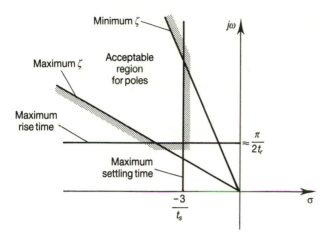

Figure 3.13 s-plane design region

(a) percentage overshoot not more than 20%;
(b) time to first peak less than 1 second; and
(c) 5% settling time less than 1 second.

Determine the location of a pair of poles which lie within the required region and have the minimum possible natural frequency. Which criteria design the system? Express the result as a second-order transfer function and use CODAS to simulate its open-loop step response and thus check the three criteria.

Steady-state Ramp Response of Systems

The steady-state response of a system to a constant input is obtained by evaluating its transfer function at $s = 0$. However, when the input is changing, $G(0)$ gives the incorrect answer because the derivatives of the input and output are non-zero. A simple and useful result can be obtained by considering the steady-state ramp response of a general system.

Consider a general transfer function expressed as the ratio of two polynomials

$$\frac{Y}{X}(s) = \frac{b_0 + b_1 s + \ldots b_m s^m}{a_0 + a_1 s + \ldots a_n s^n} \tag{3.39}$$

corresponding to the differential equation

$$a_0 y + a_1 \frac{dy}{dt} + a_2 \frac{d^2 y}{dt^2} \ldots = b_0 x + b_1 \frac{dx}{dt} + b_2 \frac{d^2 x}{dt^2} \ldots \tag{3.40}$$

The input is a ramp of slope A

$$x(t) = At$$

After the initial transient has died away, the steady-state output will be another ramp, i.e.

$$y(t) = \alpha t + \beta$$

Substituting for x, y and their derivatives in Eq. (3.40) results in

$$a_0(\alpha t + \beta) + a_1\alpha = b_0 At + b_1 A \tag{3.41}$$

Higher-order derivatives of the input and output are zero.

Now the steady-state system gain (for constant inputs) is given by

$$G(0) = \frac{b_0}{a_0}$$

so dividing Eq. (3.41) by a_0 gives

$$\alpha t + \beta + \frac{a_1}{a_0}\alpha = G(0)At + \frac{b_1}{a_0}A \tag{3.42}$$

By comparing coefficients in Eq. (3.42) we can see that

$$\alpha = G(0)A$$

The slope of the steady-state response is, therefore, not surprisingly the steady-state gain times the input slope. When the steady-state gain is unity the input and output slopes are the same and the difference between the input and the steady-state response will be a constant.

$$\beta = A\left[\frac{b_1}{b_0} - \frac{a_1}{a_0}\right] \tag{3.43}$$

Figure 3.14 illustrates the ramp response of a unity steady-state gain system. The response has the same slope or velocity as the input. The steady-state output typically lags behind

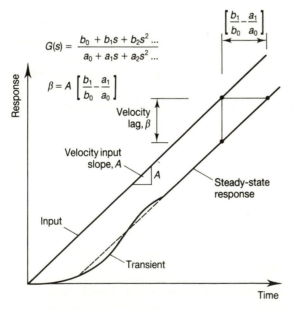

Figure 3.14 Ramp response of a unity steady-state gain system

the input and the difference or error, β, is usually called the *velocity lag*. The velocity lag has identical units to the system input (and output) and is proportional to the input slope, A. However the quantity

$$\frac{\beta}{A} = \left[\frac{b_1}{b_0} - \frac{a_1}{a_0} \right]$$
(3.44)

has units of time, and represents the steady-state time delay due to the velocity lag. This time delay depends only on the system and is independent of the input slope.

Velocity Lag of Standard First- and Second-order Systems

The standard first-order system

$$\frac{Y}{X}(s) = \frac{1}{1 + \tau s}$$

when compared with the general transfer function (Eq. (3.39)) has coefficients given by

$$a_0 = 1 \qquad a_1 = \tau$$
$$b_2 = 1 \qquad b_1 = 0$$

resulting in the velocity lag for a standard first-order system being

$$\beta = -A\tau$$
(3.45)

Similarly the standard second-order system

$$\frac{Y}{X}(s) = \frac{1}{1 + 2\zeta s/\omega_n + (s/\omega_n)^2}$$

has a velocity lag

$$\beta = -\frac{2\zeta A}{\omega_n}$$
(3.46)

Example 3.7 Thermometer velocity lag A first-order thermometer with time constant 30 seconds is used to measure the temperature of a bath of liquid which is being heated at the rate of 2.5°C/minute. Assuming the thermometer has reached steady-state determine the actual bath temperature when the thermometer reads 75°C.

SOLUTION The velocity lag is given as

$$\beta = -A\tau = \frac{2.5 \times 30}{60}$$

$$\beta = -1.25°C$$

As the bath temperature is rising the reading lags below the actual bath temperature so when the reading is 75°C the actual bath temperature must be 1.25°C higher.

$$\text{Bath temperature} = 76.25°C$$

3.10 MODEL ESTIMATION FROM TIME RESPONSE

Given a transfer function we can predict the system response. The reverse process of formulating a transfer function from a measured response is called *system identification*. System identification is a complex subject and many text books are exclusively devoted to it. We will however look at some simple techniques of estimating system models from their step response.

First-order Systems

A quick estimate of the time constant of a first-order system can be obtained from the time to reach the 63% point. Other schemes based upon the initial slope of the response are notoriously inaccurate and are not recommended. The 63% point method uses only one point from the response and ignores the remainder. A more accurate method, which utilizes more of the response, follows.

Logarithmic Response Plots

The measured transient response can be plotted logarithmically in order to achieve a more accurate estimate of the time constant. The transient of a first-order system is given by

$$\tilde{y}(t) = C\,e^{-t/\tau}$$

Dividing by C and taking natural logs we arrive at

$$\ln\left(\frac{\tilde{y}}{C}\right) = -\frac{t}{\tau}$$

which indicates that if the log function is plotted against time, a first-order system will produce a linear graph of slope $-1/\tau$. Not only does the graph give a much better estimate of τ than that obtained from the 63% point but the linearity of the graph can be used to confirm that the system is indeed first order.

> **Example 3.8 Experimental determination of time constant** The liquid outlet temperature from an air-cooled heat exchanger is controlled by varying the speed of the cooling fan. Figure 3.15 shows how the outlet temperature varies when the fan speed is suddenly changed from 50% to 70% of the maximum speed. The response clearly shows some transport delay and an exponential response which could be first order. Subtracting the steady-state value from the recorded response gives the transient. The transient starts to decrease after about 0.75 minutes from an initial value of 17.5°C. This initial value of the transient is equal to the constant C.
>
> Figure 3.16 shows the resulting log plot of the transient. This shows that for times between about 0.75 and 2 minutes the system behaves as first order. As the transient becomes small errors due to measurement inaccuracy and noise are introduced, thus the resulting deviation from linearity at larger times is not significant. At times less than the transport delay the transient is constant which results in the horizontal portion of the graph. The system can thus be reasonably

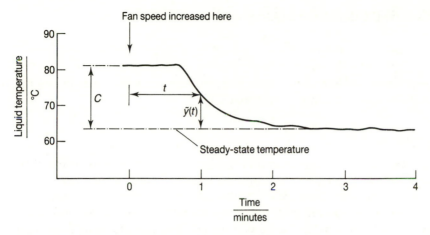

Figure 3.15 Measured step response of an air-cooled heat exchanger

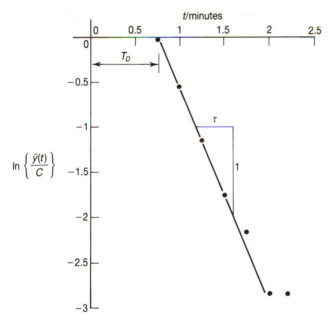

Figure 3.16 Logarithmic plot of heat exchanger transient

accurately modelled as first order with a time constant of 0.55 minutes and a transport delay of 0.77 minutes. The steady-stage gain is given by the steady-state temperature change divided by the change in fan speed, i.e.

$$\frac{-17.5}{20} = -0.875°C \text{ per } \%.$$

The estimated transfer function for this process is

$$G(s) = \frac{-0.875 \, e^{-0.77s}}{1 + 0.55s} \qquad °C/\%$$

where 's' has units of $(\text{minutes})^{-1}$. The negative gain is a consequence of the temperature falling with increasing fan speed.

Second-order Systems

The logarithmic decrement theory developed in Section 3.9 can be used to estimate the damping ratio of second-order systems from their decay ratio. Again accuracy is improved if a logarithmic plot of the peak amplitudes against time is employed. The frequency of the transient oscillation gives the damped natural frequency of the system, and knowledge of the damping ratio can be used to relate this to the system's natural frequency.

3.11 HIGHER-ORDER SYSTEMS AND SYSTEMS WITH ZEROS

Dominant Poles and System Simplification

The response of higher-order systems with many poles can be dealt with by using the idea of dominant poles. Consider the general transient response of a stable system which results from a mixture of real and complex poles. Each pole will produce a transient term which decays away at a rate dependent upon the real part of the pole. The further to the left the pole, the more quickly the term dies away. The distance from the imaginary axis to a pole is termed the *pole attenuation*. Highly attenuated poles have a transient which dies away rapidly. One might therefore expect the overall response to be dominated by the last term to die away, i.e. the pole with the least attenuation.

> **Example 3.9 Third-order system response** As an example consider the response of a third-order system with a pair of complex poles located at
>
> $$s = -1 \pm j\sqrt{3}$$
>
> and a single real pole located at
>
> $$s = -a$$
>
> This corresponds to a transfer function
>
> $$G(s) = \frac{1}{(1 + s/a)(1 + s/2 + s^2/4)}$$
>
> Figure 3.17 shows the step response for various real pole positions, a.
>
> When the real pole has four times the attenuation of the complex pole the response is very similar to that predicted by the complex pole pair alone. As the real pole becomes less attenuated its influence on the response becomes greater.

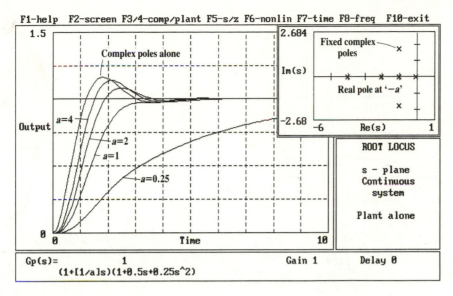

Figure 3.17 Effect of real pole location on the step response of a third-order system

Table 3.1 Performance of a third-order system

Real pole attenuation	∞	4	2	1	0.25
% overshoot	16.3	13.9	8.1	—	—
63% rise time/second	0.77	1.02	1.24	1.62	4.45

The response has no overshoot when the attenuation of the real pole is less than 1.2. When all poles are equally attenuated the 63% time is considerably longer than the real part of the poles would suggest (Table 3.1).

As the attenuation of the real pole reduces further, the response starts to look like that of the real pole alone since it is this which now dominates the overall response. With $a = 0.25$ the pole has a time constant of 4 seconds and the 63% response time ties in reasonably well with that predicted from the real pole alone.

The results from this example are typical of higher-order systems. When a pole or pole pair is clearly in a dominant position, the response can be approximated by the dominant pole(s) alone. The results which were derived for first- and second-order systems can then be used to give approximate response times, etc. When there are several poles in a dominant position the response will be a mix of the individual pole contributions and will always be slower than the attenuation suggests. Complex poles will always show less oscillation and hence overshoot when accompanied by a real pole of similar attenuation.

A pair of poles with little damping will be more affected by an approaching real

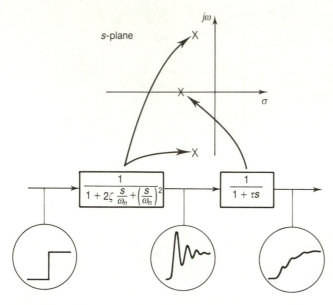

Figure 3.18 Real pole acts as a filter in oscillatory systems

pole than if the damping ratio were higher. The reason for this can be appreciated by considering the complex poles and the real pole as two separate cascaded transfer functions (Fig. 3.18). The real pole acts as a filter attenuating the rapidly changing (high frequency) transient from the complex poles. The higher the transient frequency the more it will be filtered by the approaching pole. The frequency of the transient arising from a lightly damped complex pair of poles is equal to the imaginary part of the poles. Now a first-order system (real pole) has a significant filtering effect on components at frequencies above $1/\tau$ rad/s (see Section 4.3). This cutoff frequency is equal to the distance from the pole to the origin of the s-plane. Therefore a real pole will still have a significant effect on the transient if its distance from the origin is similar to the imaginary part of the complex poles. This applies even if the real pole has an attenuation very much greater than the complex pole pair.

Rather than just using the attenuation as a criterion for dominance, the damping ratio of any complex poles should also be taken into account. Attenuation can be used to decide which poles will have the longest lasting effect. However, the radius of the pole is usually a better guide to its overall dominance. As a general rule of thumb the response of system will be dominated by the pole or pole pair which is closest to the s-plane origin. Poles which are at a radius of more than three times that of the dominant pole(s) will generally have a small contribution to the overall response. The factor of three is somewhat arbitrary and plainly less error will occur in approximations if larger factors are used.

Poles which have a negligible contribution to the transient can be disregarded; however, it must be remembered to retain any steady-state gain associated with the neglected factor.

Example 3.10 Simplification of a higher-order system Using dominant pole simplification methods, estimate the characteristics of the step response of the fifth-order system

$$G(s) = \frac{100}{(1 + s + s^2)(9 + s + s^2)(3 + s)}$$

Estimate the peak amplitude of the response and the time at which it occurs. Compare your estimate with the values obtained for the full system.

SOLUTION The system poles are located at

$$s = -0.5 \pm j0.866, \ -0.5 \pm j2.96 \text{ and } -3$$

or
$$s = -e^{\pm j1.05}, \ -3 e^{\pm j1.4} \text{ and } -3$$

The dominant poles are those with the smallest radius, i.e. the first complex pair. All the other poles are at a radius of three times that of the dominant poles and so can be approximated by their steady-state gain. The system can therefore be approximated by the second-order model

$$G(s) = \frac{100}{(1 + s + s^2)9 \times 3} = \frac{3.7}{1 + s + s^2}$$

The damping ratio of this equivalent system is 0.5 and so the logarithmic decrement (Eq. (3.37)) predicts a percentage overshoot as follows:

$$\% \text{ overshoot} = 100 \, e^{-\pi 0.5/\sqrt{1 - 0.5^2}} = 16\%$$

The peak amplitude of the response will be the steady-state gain multiplied by 116%, i.e.

$$\text{peak amplitude} = 1.16 \times 3.7 = 4.29$$

The time to reach the first peak is half the period of oscillation, i.e.

$$t_p = \frac{\pi}{\omega_d} = \frac{\pi}{\omega_n\sqrt{1 - \zeta^2}} = 3.63 \text{ second}$$

Figure 3.19 shows the response of the full fifth-order system together with the second-order approximation. The estimates of the percentage overshoot and time to first peak are reasonably accurate.

The Contribution of Zeros to System Response

The poles of a transfer function are solely responsible for the terms in a system transient response. However the zeros do have a marked effect on the magnitude or weighting of these terms. The significance of the numerator of a transfer function can be seen from Eq. (3.7),

$$Y(s)D(s) = X(s)N(s)$$

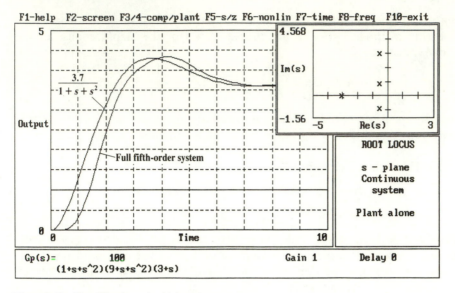

Figure 3.19 Simplification of a high-order system

This shows that the excitation which causes the system response comes from the system input together with the transfer function numerator.

Introducing a modified input

$$V(s) = X(s)N(s)$$

and expressing $N(s)$ as an mth order polynomial

$$V(s) = X(s)(b_0 + b_1 s + \cdots b_m s^m)$$

or

$$v(t) = b_0 x(t) + b_1 \frac{dx}{dt}(t) + \cdots b_m \frac{d^m x}{dt^m}(t)$$

The modified input is therefore made up from the input and its derivatives, each weighted by the appropriate coefficient of the numerator. The actual response can thus be considered as being made up from the sum of the responses to each of these successive derivatives. Remember that each of these individual responses is made up from the same basic transient terms.

Example 3.11 Effect of a zero on a second-order system Consider the transfer function

$$G(s) = \frac{1 + as}{1 + 5s + 25s^2}$$

The numerator produces a modified input

$$v(t) = x(t) + a \frac{dx}{dt}(t)$$

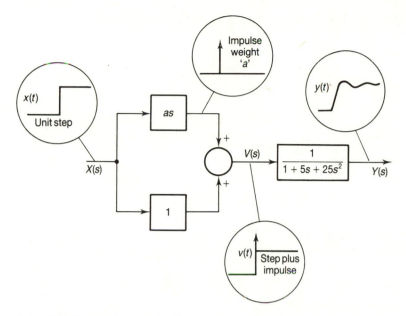

Figure 3.20 Showing how a zero gives an extra kick to the system step response

Now suppose the system is subjected to a unit step input. The modified input will consist of a unit step together with the weighted derivative of the step, i.e. an impulse of weight '*a*' (Fig. 3.20).

The overall response of the system is thus a combination of the step and impulse responses. If the value of '*a*' is positive then the impulse gives an initial kick to the system which makes the system get moving more quickly. The larger the value of '*a*' the more effective the kick. A negative '*a*' value produces a kick in the opposite direction to the step thus causing the overall response to initially go in the wrong direction! Figure 3.21 summarizes the results for this system.

It is important to realize that the effect of zeros can be duplicated by introducing initial conditions or indeed by modifying the system input. The kick introduced by the zero in Example 3.10 is exactly equivalent to the system having an initial velocity at $t = 0$. This ties in with the idea that zeros do not alter the system transient characteristics but simply alter the weighting of the transient terms. Figure 3.21 shows that the damped natural frequency and decay ratio are unaffected by the zero.

Examined from the *s*-plane perspective, when a zero or zero pair is far to the left of the dominant poles it has little effect on the response. As the zero approaches the dominant poles its effect becomes more pronounced and eventually as the zero moves to the right of the dominant poles its effect is overwhelming. Systems with zeros in the right half plane belong to a class of system called *non-minimum phase* (a name derived from their frequency response). Right half plane zeros do not of course give rise to instability but they do give rise to responses which initially tend to go in the opposite direction to the steady-state response.

The ideas developed for evaluating the dominance of poles can also be applied to

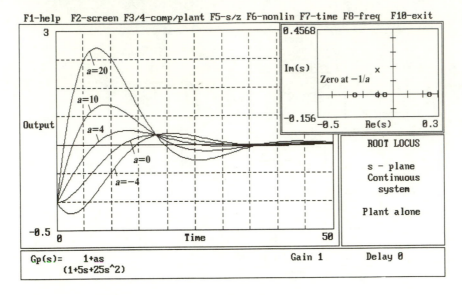

Figure 3.21 Effect of a zero on a second-order system

zeros. Again it is the radius of the zero relative to the dominant pole(s) which determines whether the zero will have a significant effect on the system response. A zero will have little effect if it is located at a radius of more than three times that of the dominant pole(s). Zeros which are located close to poles tend to reduce the weighting of the transient term associated with the poles. A zero can cancel a pole completely and remove the associated transient term from the response. The existence of a zero close to a dominant pole can give rise to confusing responses since the approximate cancellation of a dominant pole can remove its dominance (see Problem 3.9). The interpretation of pole/zero patterns borders on being an art rather than a science. Finally, since the effect of non-zero initial conditions can be duplicated by introducing system zeros, the estimation of step response characteristics from an *s*-plane diagram must apply to systems with zero initial conditions.

To summarize:

(a) For the system to be stable and the transient to decay to zero all poles must be located in the left half of the *s*-plane.
(b) The transient derived from the pole with the least attenuation will take the longest to decay away but the response of a stable system will generally be dominated by the pole(s) closest to the *s*-plane origin.
(c) The dynamic contribution of poles and zeros in the left half of the *s*-plane can generally be neglected if they are at greater than three times the radius of the dominant poles.
(d) The dominance of a pole will be reduced by the existence of a close zero. But even exact cancellation will not remove a pole from the system characteristic equation.

3.12 CLOSED-LOOP SYSTEMS

All the material covered so far applies to open-loop systems, i.e. systems without feedback. The main idea behind feedback systems was introduced in Chapter 1. Now sufficient analytical tools have been developed to examine the performance of closed-loop systems.

The general representation of a feedback control system is shown in Fig. 3.22. The forward path transfer function is $G(s)$ representing the process or plant being controlled together with any controller dynamics. The feedback path transfer function, $H(s)$, represents the measurement system or transducer which provides the feedback signal. The overall transfer function relating the controlled variable C to the desired value or reference, R is

$$\frac{C}{R}(s) = \frac{G(s)}{1 + G(s)H(s)}$$

A useful tip when modelling feedback systems is that the transfer function around the loop, $G(s)H(s)$, must be dimensionless, otherwise the signals around the loop would not be compatible. In addition the forward path transfer function, $G(s)$, must have the same units as the overall system, i.e. the units of C/R.

Unity Feedback Systems

If the measured value, C_m, is considered to be the output rather than the actual controlled variable then the simpler *unity feedback system* of Fig. 3.23 results. This is equivalent to lumping the measurement system in with the process dynamics. In many control systems the measurement system will respond rapidly in comparison with the main process dynamics justifying the unity feedback approach. The unity feedback

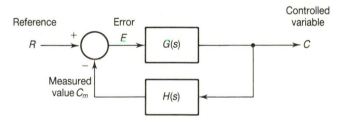

Figure 3.22   General feedback control system

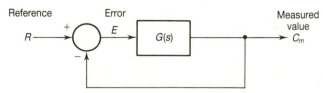

Figure 3.23   Equivalent unity feedback system

arrangement has certain advantages in that the reference, output and error signals all have the same units. In addition much of the theory used in the analysis and design of feedback control systems is simplified.

Combining the measurement system transfer function into the forward path, $G(s)$, the resulting unity feedback system has the transfer function

$$\frac{C}{R}(s) = \frac{G(s)}{1 + G(s)} \qquad (3.47)$$

For the sake of simplicity the suffix, m, has been dropped and the symbol 'C' used to represent the general output in both unity feedback and non-unity feedback systems.

The prime objective in feedback control systems is to minimize the difference between the output and the reference since this represents an error. The control system should quickly reduce this error to zero (or some acceptably small value) when there is either a disturbance or reference value change. The process dynamics cannot normally be modified but we are relatively free to alter the controller to give an acceptable overall performance within the constraints of the available control effort. *Proportional control* is one of the simplest feedback schemes where the control effort is simply proportional to the error.

Proportional Control of a First-order System

As an example of a simple proportional control system we will examine the dc motor speed control system shown in Fig. 3.24. Motor speed is measured by a dc tachogenerator which produces a voltage proportional to the motor speed, its sensitivity being K_t volt per rad/s. The tachogenerator signal is fed back to a unity gain differential amplifier where it is subtracted from a reference voltage derived from a potentiometer. The output from the differential amplifier drives the motor via a power amplifier with a gain of 'A' amps per volt. Considering the tachogenerator voltage as the output and the potentiometer voltage as the reference, Fig. 3.25 shows the block diagram of the system with the transfer function of the motor as derived in Chapter 2. The choice of measured

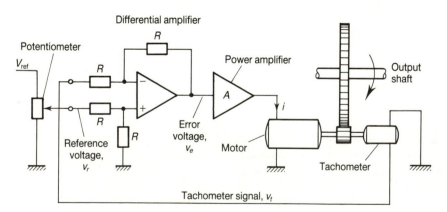

Figure 3.24 Proportional control of motor speed

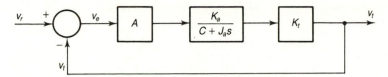

Figure 3.25 Speed control system block diagram

value (voltage) rather than actual speed as the output has resulted in a unity feedback system.

The block diagram reduces to give the forward path transfer function

$$G(s) = \frac{AK_aK_t}{(C + J_as)}$$

This is a first-order system with time constant

$$\tau = \frac{J_a}{C}$$

and steady-state gain

$$K = \frac{AK_aK_t}{C}$$

which is dimensionless. The forward path transfer function can be written as

$$G(s) = \frac{K}{1 + \tau s}$$

The overall closed-loop transfer function is

$$\frac{C}{R}(s) = \frac{\left[\dfrac{K}{1 + \tau s}\right]}{1 + \left[\dfrac{K}{1 + \tau s}\right]} = \frac{K}{1 + \tau s + K}$$

Now this is another first-order transfer function which can itself be put in standard form by dividing throughout by $(1 + K)$

$$\frac{C}{R}(s) = \frac{\left[\dfrac{K}{1 + K}\right]}{1 + \left[\dfrac{\tau}{1 + K}\right]s} \qquad (3.48)$$

The closed-loop system can be seen to have a time constant of

$$\tau' = \frac{\tau}{1 + K} \qquad (3.49)$$

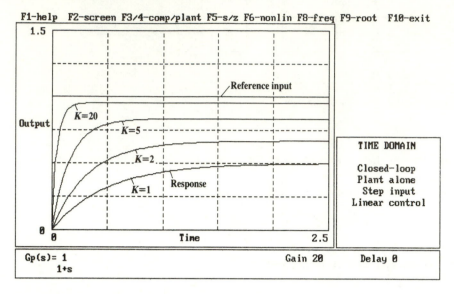

Figure 3.26   Step response of a first-order system with proportional control

and a steady-state gain of

$$K' = \frac{K}{1 + K} \tag{3.50}$$

The application of feedback has shortened the response time by reducing the time constant by a factor of $1/(1 + K)$, but the closed-loop steady-state gain is always less than unity. This means that the measured speed will always be less than the reference.

This steady-state error or *offset* is at first glance puzzling when it is considered that the purpose of the feedback was to remove errors. The offset can be explained by imagining a situation where there is no error; this could be achieved by suddenly altering the reference to be equal to the measured value. With the tachometer voltage equal to the reference voltage the input to the differential amplifier would be zero so producing zero motor current. The motor would therefore slow down until the error was sufficiently large to keep the motor running steadily.

As the overall gain, K, is increased (by perhaps increasing the amplifier gain) the response becomes faster and the offset becomes less. Figure 3.26 shows the step response of the system with a range of K values.

Exercise   Use CODAS to simulate the step response of a first-order system in closed-loop, with various values of gain, K. Use the cursor to find the steady-state response and the time to reach 63% and thus verify the values of the offset and time constant as calculated from Eqs (3.47) and (3.48).

Try the experiment of setting an initial condition which makes the error zero at $t = 0$, i.e. use a unit-step input, and an initial plant output of unity also. You should see the output drift away from the zero error initial condition to once again give an offset.

Steady-state Accuracy and Errors

The previous example showed that proportional control of a Type-0 system generally produces offset or steady-state error even when the reference is constant. The steady-state performance of open-loop systems was covered for a constant input in Section 3.4 and for ramp inputs in Section 3.9. Here we will examine the general steady-state performance of **unity feedback system** with various inputs.

For the general unity feedback system shown in Fig. 3.23 the transfer function relating the error, E, to the reference, R, is

$$\frac{E}{R}(s) = \frac{1}{1 + G(s)} \tag{3.51}$$

Constant references (positional inputs) When the reference input is a constant value

$$r(t) = A$$

the steady-state relationship between the error and reference input can be obtained by allowing 's' to tend toward zero in Eq. (3.51)

$$\frac{\bar{e}}{\bar{r}} = \frac{\bar{e}}{A} = \lim_{s \to 0} \frac{1}{1 + G(s)}$$

$$\frac{\bar{e}}{A} = \frac{1}{1 + \lim_{s \to 0} G(s)} = \frac{1}{1 + K_p} \tag{3.52}$$

$\bar{e}$ represents the steady-state error when the control system is subjected to a constant input of amplitude A. K_p is called the *positional error constant*.

$$K_p = \lim_{s \to 0} G(s) \tag{3.53}$$

Since the error has the same units as the amplitude, A, the positional error constant is dimensionless.

For a system in which the forward path transfer function is of Type-0, K_p is equal to the open-loop Bode gain, K_B (Eq. (3.5)). Thus Type-0 systems with a constant input will exhibit steady-state error which can be reduced by increasing K_p.

Type-1 systems and above have infinite positional error constants and so do not exhibit steady-state error with constant inputs. Therefore to achieve zero offset with a constant input the control system must have at least one free integrator in the forward path.

In a stable closed-loop system the input to any integrator must be zero in the steady-state. If this were not so then the integrator output would still be changing and the system would not be in steady-state. When the integrator input is proportional to the error then this scheme will force the steady-state error to be zero. We must be careful to make sure that the system is exactly as in Fig. 3.23. In practice the input to the integrator may not necessarily be proportional to the true error. One way in which this can occur is when the measured value is inaccurate, perhaps because of leaky signal transmission or inaccurate transducer calibration. Another less obvious cause is where

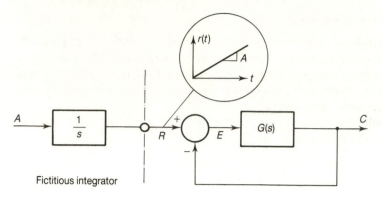

Figure 3.27 Use of a fictitious integrator to produce a ramp input

a disturbance enters the control loop before the integration stage. This second case is common in controlled processes which naturally have an integration as part of the process dynamics. Position and level control systems often fall into this category. This problem will be examined more thoroughly in Chapter 7.

Ramp references (velocity inputs) Many control systems must be able to cope with reference inputs which change at a steady rate. A ramp input is a suitable test signal for such systems. This type of input is also called a *velocity input*, since the reference changes at a constant velocity.

$$r(t) = At$$

where on this occasion 'A' is the velocity of the reference.

A ramp input may be produced by integrating a constant (the velocity, A). Figure 3.27 shows a *fictitious integrator* introduced at the system input to represent the relationship between the ramp input and its velocity.

The steady-state gain of the unity feedback system in cascade with the fictitious integrator thus gives the steady-state error in terms of the ramp slope.

$$\frac{\bar{e}}{A} = \lim_{s \to 0} \frac{1}{s(1 + G(s))} = \frac{1}{\lim_{s \to 0} sG(s)} = \frac{1}{K_v} \tag{3.54}$$

Here $\bar{e}$ is the steady-state error when the control system has a velocity input of rate A. K_v is called the *velocity error constant* and has units of $(\text{time})^{-1}$.

$$K_v = \lim_{s \to 0} sG(s) \tag{3.55}$$

The velocity error constant of a Type-0 system is zero and so the steady-state error is infinite with a ramp input.

For a Type-1 system the velocity error constant is the open-loop Bode gain. Type-1 systems therefore exhibit a finite steady-state error when subjected to a ramp input. The larger the velocity error constant the less the steady-state error.

Type-2 and higher systems have infinite velocity error constants and will therefore show no steady-state error with a velocity input.

Parabolic references (acceleration inputs) The above process can be extended to cater for systems with inputs which have a constant acceleration. Two fictitious integrators with a constant input of 'A' must be introduced to derive the reference, giving

$$r(t) = \iint A \, dt^2 = \int At \, dt = \frac{A}{2} t^2$$

The steady-state error is now

$$\frac{\bar{e}}{A} = \lim_{s \to 0} \frac{1}{s^2(1 + G(s))} = \frac{1}{\displaystyle\lim_{s \to 0} s^2 G(s)} = \frac{1}{K_a} \tag{3.56}$$

K_a is called the *acceleration error constant* and relates the steady-state error to the acceleration rate of the reference.

$$K_a = \lim_{s \to 0} s^2 G(s) \tag{3.57}$$

The acceleration error constant of both Type-0 and Type-1 systems is zero so leading to infinite steady-state error with accelerating inputs. Type-2 systems show a finite error and Type-3 and higher systems show zero steady-state error. Figure 3.28 summarizes the results.

Further fictitious integrators can be added to extend the ideas to higher-order inputs. However, for all practical purposes there is no need to go beyond acceleration inputs.

Example 3.12 Aerial positioning servo A servomechanism designed to position a rotating aerial consists of a geared motor and amplifier driven by the difference between a reference voltage and the feedback voltage from a position transducer. In an open-loop test (i.e. without feedback) a steady voltage of 1 volt applied to the amplifier input produced a steady-state aerial speed of 2 revolutions per minute. Given that the sensitivity of the position transducer is 250 mV per degree of error in position, find the closed-loop steady-state error when the servo demand input is changing at a rate of 2° per second.

SOLUTION Since the open-loop system settles to a constant velocity when a constant voltage is applied to the amplifier, the system must have a single free integrator in the forward path, i.e. it is a Type-1 system. The Bode gain is given by

$$K_B = \frac{2 \times 360}{60} \frac{\text{degree}}{\text{second} \cdot \text{volt}} \times 0.25 \frac{\text{volt}}{\text{degree}}$$

The velocity error constant is thus

$$K_v = 3 \text{ second}^{-1}$$

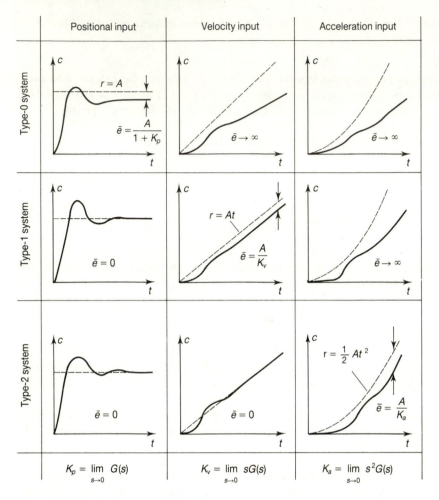

Figure 3.28 Steady-state error in closed-loop systems

In closed-loop the steady-state error is given by

$$\bar{e} = \frac{A}{K_v} = \frac{2}{3} \text{ degree}$$

PROBLEMS

3.1 The transfer function of a second-order system is given by

$$G(s) = \frac{4}{s^2 + 0.4s + 3}$$

Analytically determine the amplitude of the third overshoot of the response when the system is subjected to a unit step input. Calculate the time between the application of the step and the third overshoot. Be sure to check your answers on CODAS.

3.2 A system consists of two cascaded, non-interacting first-order lags (i.e. no loading effect) of time constants τ and $a\tau$. Express the system transfer function in standard second-order form and show that the system can never be underdamped. What value of 'a' gives the minimum damping ratio?

3.3 Show that the 10% to 90% rise time for a first-order system is about 2.2 time constants.

3.4 Using the results from Example 3.6, find the 63% rise time of a critically damped second-order system in terms of its natural frequency.

   Hint: the natural frequency of the system in Example 3.6 is $(1/\tau^2)$.

3.5 Figure P3.5 shows a positional servomechanism employing velocity feedback. Determine the values of K_c and K_t which produce a step response with 9.5% overshoot and a 0 to 100% rise time of 0.2 second.

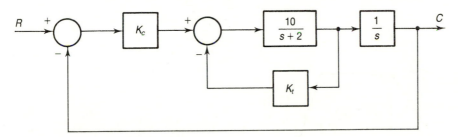

Figure P3.5   Position servo with velocity feedback

3.6 A spring return pneumatic actuator has a transfer function relating displacement, x, to applied pressure, p, given by

$$\frac{X}{P}(s) = \frac{86}{ms^2 + 2s + 1} \text{ mm/barg}$$

where m is the total mass (in kg) of the moving parts including the load.

   (a) Calculate the peak actuator displacement when a pressure of 3.0 barg is suddenly applied to the actuator with a total mass of 3.0 kg.

   (b) Obtain an expression for the damped natural frequency of the system in terms of the total mass, m. By differentiation of the expression find the value of 'm' which maximizes the damped natural frequency. What is the maximum damped natural frequency?

3.7 A system has a transfer function

$$G(s) = \frac{2.5 \times 10^4}{(s + 15)(s^2 + 10s + 425)(s^2 + 8s + 41)}$$

Determine the pole/zero pattern of the system and identify the dominant pole or poles. Estimate the salient performance parameters of the dominant pole(s) and hence sketch the approximate response to a unit step input. Simulate the open-loop unit step response of the system on CODAS to verify the accuracy of your sketch.

3.8 Write down the transfer function of a third-order system with unity steady-state gain which has a real pole at $s = -1$ and a pair of complex poles at $s = -1 \pm ja$. Simulate the open-loop step response of the system for 'a' values of 1, 2, 4 and 10, and compare the responses obtained with that of a first-order system alone.

3.9 Determine the dominant dynamic characteristics of the following transfer function. Estimate the system's 5% settling time and use CODAS to check your answer.

$$G(s) = \frac{1 + 2s}{(s + 2)(s^2 + 0.8s + 1)(1 + 3s)}$$

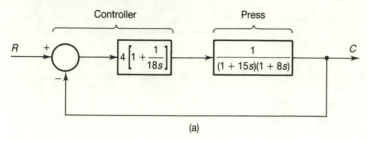

(a)

Figure P3.10(a) Pressure control system

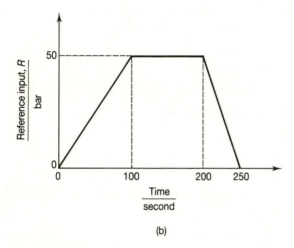

(b)

Figure P3.10(b) Reference input profile

3.10 Figure P3.10(a) shows the block diagram of a proportional plus integral controller controlling the hydraulic pressure supplied to a press. During press operation, the controller reference input is ramped up to the required pressure, held for a period and then ramped down again as shown in Fig. P3.10(b). Find the steady-state error during the three periods and sketch the expected response.

FOUR

FREQUENCY RESPONSE

4.1 INTRODUCTION

In Chapter 3 the transient and steady-state behaviour of open- and closed-loop systems to inputs such as steps and impulses was examined. In this chapter we will examine how systems respond to periodic input signals and in particular sinusoidal signals.

Reasons for Frequency Response Testing and Analysis

It is fair to ask why sinusoidal inputs are considered at all and why frequency domain methods are used in analysing and designing control systems. The s-plane/time-domain relationship is after all exact and furnishes essentially all the data on the behaviour of a system including its frequency response. A fundamental reason is that in order to use the s-plane one must have a transfer function model of the system, whereas frequency domain methods can be applied directly to systems for which only experimental data is available.

Another related reason is that by knowing how the sinusoidal response of a system can be predicted from its transfer function, one then can do the converse and fit transfer functions to experimentally obtained frequency response data. In other words frequency responses can be used to help produce models of systems or identify system transfer functions.

Certain types of control system are normally subject to sinusoidal inputs. For example a fatigue-test rig used in testing aircraft structures is designed to impose cyclical load changes and so its demand signal will be sinusoidal. Thus the frequency response of the system will feature strongly in the design specification of the control systems. Again other control systems, such as ship stabilizers, are designed to reject cyclical disturbances. Clearly if the ship stabilizer has a pronounced resonance around the frequency of ocean waves in mid-Atlantic there will be a lot of unhappy passengers complaining to the shipping company!

One last example of where frequency domain tools are also important is when dealing with arbitrary periodic inputs. For example, vibration absorbers may have to be designed to handle complex periodic waveforms. Here Fourier methods are often employed and so a knowledge of the frequency response of the system is again essential.

4.2 STEADY-STATE FREQUENCY RESPONSE OF LINEAR SYSTEMS

When a linear system is subjected to a sinusoidal input, the output will be sinusoidal too, at least after the initial transient has died away. The output signal will differ from the input signal in amplitude and phase, and furthermore, for a given system, the amplitude and phase relationship will depend on the frequency of the applied signal.

It is quite instructive to do a simple test with a friend to get the idea of a frequency response. Hold your hand out horizontally with your index finger extended. Get your friend to do likewise and point his/her index finger at yours. Tell your friend that you will start moving your finger up and down and that he/she should follow your movements by keeping his/her finger pointed at yours. Start off by cycling your arm up and down slowly and then gradually speed it up. Human beings are hardly linear systems, nevertheless you will have observed that your friend's output signal is similar to your excitation signal, but the output amplitude decreases with frequency and the phase lag increases as the frequency of your arm (sorry, input signal) increases.

Figure 4.1 schematically shows the principle of a frequency response test of a linear system. A sinusoidal signal, $x(t)$, of angular frequency ω rad/s and of amplitude, X, is applied to the system and the amplitude and phase of the output signal, $y(t)$ is observed.

Note: The formula relating the periodic time, T, of a sinusoidal signal and its angular frequency, ω, is:

$$\omega = \frac{2\pi}{T} \tag{4.1}$$

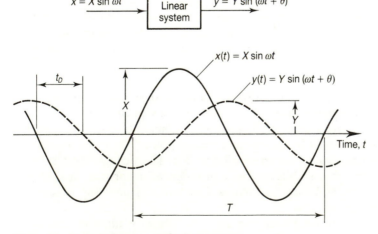

Figure 4.1 Sinusoidal response of a linear system

In Fig. 4.1 we have written

$$x(t) = X \sin(\omega t) \tag{4.2}$$

as the equation of the input signal, but $\cos(\omega t)$ could have been used just as well or any combination of $\cos(\ )$ and $\sin(\ )$. The diagram showing the input and output signals has deliberately been drawn without a time origin to emphasize the point. It is impossible to tell, merely by looking at the input waveform, whether it is a sine or a cosine or a combination of the two. In other words the phase of the input signal is indeterminate and in fact is quite immaterial. The output signal, $y(t)$, has the same frequency as the input signal but its amplitude is Y and its phase **relative** to the input signal is θ. The time delay between the two signals is t_D. The output signal has been drawn as **lagging** the input signal, and so the phase difference, θ, is

$$\theta = -\omega t_D = -2\pi \frac{t_D}{T} \tag{4.3}$$

Rather than depicting the input and output signals as waveforms it is easier to think of them as rotating vectors, i.e.

$$\mathbf{X}\,e^{j\omega t} \quad \text{and} \quad \mathbf{Y}\,e^{j\omega t} \tag{4.4}$$

as shown in Fig. 4.2. $\mathbf{X}$ and $\mathbf{Y}$ are two vectors each with a certain magnitude and phase. By multiplying each vector by $e^{j\omega t}$, the vectors are swept round at an angular frequency of ω rad/s. The **relative** magnitudes and phases of the two vectors are unaffected by the fact that they are rotating. The actual input and output time signals, $x(t)$ and $y(t)$, are the projections of the respective vectors on to the real axis.

As the two vectors $\mathbf{X}$ and $\mathbf{Y}$ are rotating counter-clockwise at ω rad/s, the diagram represents one 'still' from a 'movie'. The snap-shot as drawn in Fig. 4.2 is taken when the input vector, $\mathbf{X}$, is real or, as far as the time waveform is concerned, when the input signal is at its maximum value.

The output vector, $\mathbf{Y}$, as drawn in Fig. 4.2, can be represented by

$$\mathbf{Y} = Y\,e^{j\theta} \tag{4.5}$$

where Y is its magnitude or length (i.e. OB) and θ is its angle **with respect** to $\mathbf{X}$. It is

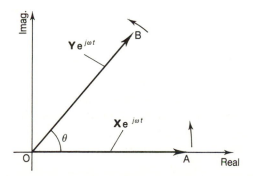

Figure 4.2   Vector representation of sinusoidal input and output signals

important to realize that as the 'movie' progresses the angle between the vectors will not change. For example, after a quarter of a period the input vector, **X**, will be pointing along the imaginary axis, but the output vector, **Y**, will still be θ radians ahead of it.

The amplitude ratio of the output and input signal is known as the *gain*. The gain of the system at a particular frequency is represented by the ratio of the length of the two vectors (Y/X). Again the magnitude ratio of the input and output vectors does not change as the two vectors rotate round the origin.

The total relationship between the output and input can be represented concisely by the vector ratio of **Y** and **X**. This ratio can be considered as a **complex gain**, **G**, i.e.

$$\mathbf{G} = \frac{\mathbf{Y}\,e^{j\omega t}}{\mathbf{X}\,e^{j\omega t}} = \frac{\mathbf{Y}}{\mathbf{X}} \tag{4.6}$$

The magnitude of **G** ($|\mathbf{G}|$) is the amplitude ratio of the output and input sinusoids, and its angle or argument ($\angle\,\mathbf{G}$) is the phase difference between the output and input signals. Note that **G** is **not** a rotating vector, the $e^{j\omega t}$ components having cancelled out in taking the quotient, and so is not a function of time.

As was stated before, the amplitude ratio and phase shift are dependent on the excitation frequency, ω, and so the ratio of output to input can be considered as a vector or complex function of frequency. This function is written as $G(j\omega)$ (the reasons for this nomenclature will become apparent in the next sections), and it is termed the *steady-state frequency response function*, or just the frequency response function.

4.3 FREQUENCY RESPONSE AND THE TRANSFER FUNCTION

The differential equation or transfer function of a system completely describes the behaviour of the system to any input both during the initial transient and when the system has reached steady-state. Thus one can derive from first principles how the system will respond to a sinusoidal input. We will do this here for a first-order system and hence see how to extend the result to systems of any order.

Frequency Response of a First-order System

The differential equation of a standard first-order system (i.e. unit steady-state gain) is:

$$y + \tau \frac{dy}{dt} = x \tag{4.7}$$

The previous section described the steady-state input and output signals as a pair of rotating vectors (Eq. (4.4)). Strictly speaking, because the time signals are the projections of the rotating vectors on to the real axis, the two time functions, $x(t)$ and $y(t)$, should be written as

$$x(t) = \text{Real}\{\mathbf{X}\,e^{j\omega t}\} \qquad \text{and} \qquad y(t) = \text{Real}\{\mathbf{Y}\,e^{j\omega t}\}$$

In frequency response analysis a general sinusoidal input is considered, i.e. the time origin is immaterial. Furthermore only the amplitude **ratio** and phase **difference** of the input and output signals are considered. Under these conditions, it is quite legitimate to forget about the projection of the vectors on to the real axis and to treat the input and output signals as complex quantities, namely

$$x(t) = \mathbf{X}\,e^{j\omega t} \quad \text{and} \quad y(t) = \mathbf{Y}\,e^{j\omega t} \tag{4.8}$$

This simplifies the treatment and gives correct results.

Substituting for x and y from Eq. (4.8) into Eq. (4.7) we obtain

$$\mathbf{Y}\,e^{j\omega t} + \tau \frac{d}{dt}\{\mathbf{Y}\,e^{j\omega t}\} = \mathbf{X}\,e^{j\omega t}$$

or

$$\mathbf{Y} + j\omega\tau\mathbf{Y} = \mathbf{X} \tag{4.9}$$

Hence, the frequency response function, $G(j\omega)$, for a first-order system is

$$G(j\omega) = \frac{\mathbf{Y}}{\mathbf{X}}(j\omega) = \frac{1}{1 + j\omega\tau} \tag{4.10}$$

The gain (amplitude ratio of output to input) of the system is Y/X which is the magnitude or modulus of the complex quantity $G(j\omega)$. Appendix A briefly reviews the rules for products and quotients of complex numbers. The important point to remember here is that the modulus of a complex quotient is obtained directly from the quotient of the magnitude of the numerator and the magnitude of the denominator. There is no need to rationalize the expression first, i.e.

$$\frac{Y}{X} = |G(j\omega)| = \frac{1}{\sqrt{(1 + (\omega\tau)^2)}} \tag{4.11}$$

The phase shift, θ, is the argument of $G(j\omega)$. Appendix A shows that the argument of a complex quotient is the difference of the numerator argument and the denominator argument. In this case it is simply

$$\theta = \angle(G(j\omega)) = -\tan^{-1}(\omega\tau) \tag{4.12}$$

The gain and phase can be plotted as functions of $\omega\tau$ (Fig. 4.3). $\omega\tau$ is a non-dimensional quantity which reflects the fact that the frequency response of a first-order system depends on the product of the excitation frequency and the time constant. That is, a slow system excited at a low frequency will behave similarly to a fast system excited at a high frequency.

From Fig. 4.3 you can see that the gain at low frequencies is unity and as the frequency increases the gain drops. Eventually at very high frequencies the gain drops towards zero. Similarly the phase shift is negligible for small values of $\omega\tau$, but increases to $-90°$ as $\omega\tau$ increases. At intermediate values of $\omega\tau$, the gain and phase shift can be found by substituting for $\omega\tau$ in Eqs (4.11) and (4.12) respectively. For example, when $\omega\tau$ is 1, the gain of the system is $1/\sqrt{2}$ and the phase shift is $-45°$.

The low frequency gain of a system corresponds to its steady-state gain. In this example the steady-state gain of the system is 1 which is the same as the low frequency gain. The high frequency gain corresponds to the instantaneous gain. In this example

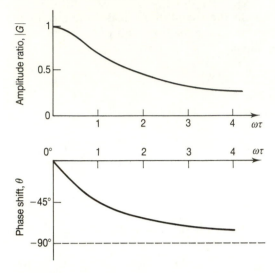

Figure 4.3 Frequency response of a first-order system

of a simple first-order system, there is no jump in the output when a step is applied and so its instantaneous gain is zero which is confirmed by a zero high frequency gain.

Exercise

(a) At what value of $\omega\tau$ is the amplitude ratio 0.5?
(b) At what value of $\omega\tau$ is the phase shift $-60°$?
(c) Confirm (a) and (b) using CODAS by applying a user defined input of $\sin(\omega t)$ to a first-order system of 1 second time constant. Use the cursor to make measurements. Don't forget to allow for the initial transient to die away.

4.4 GRAPHICAL REPRESENTATION OF FREQUENCY RESPONSE DATA

Frequency response data can either be calculated if the system transfer function is known or determined experimentally by doing a frequency response test on the system. The results can be presented graphically in a number of different ways. We have already seen one in Fig. 4.3. The way that the data was drawn there has a number of disadvantages and in fact is hardly ever used quite like that. There are four different ways in which frequency response data is usually presented graphically, namely polar plots (direct Nyquist and inverse Nyquist diagrams), logarithmic plots (Bode diagrams) and finally logarithmic gain versus phase plots (Nichols diagrams). Each of these has advantages and disadvantages from a practical and theoretical point of view. Figure 4.4 shows the four ways of presenting frequency responses for a first-order system. Let us look at each of these plots in turn.

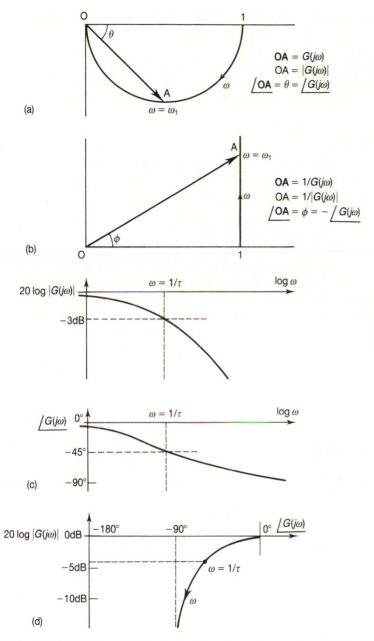

Figure 4.4 Different graphical representations of frequency responses. (a) Direct Nyquist diagram. (b) Inverse Nyquist diagram. (c) Bode diagram. (d) Nichols plot

Direct Nyquist Diagram

This is a direct polar plot of the frequency response, i.e. it is the locus of the tips of vectors whose lengths represent the amplitude ratio (gain) of the system and which are drawn at an angle corresponding to the system phase shift, θ. Figure 4.4(a) shows the direct Nyquist diagram of a standard first-order system.

The disadvantage of this plot is that frequency information is implicit and so the graph must include points labelled with the frequency value. A further disadvantage of the Nyquist diagram is that it is difficult to extract accurate data when the gain drops to a low value. It is however conceptually and theoretically very important.

Inverse Nyquist Diagram

As its name suggests this is a polar plot of the reciprocal of the frequency response, i.e. it is a polar plot of $1/G(j\omega)$. Referring to Appendix A, if a complex number z has a magnitude, r, and a phase angle, θ, then its reciprocal is

$$\frac{1}{z} = \frac{1}{re^{j\theta}} = \frac{1}{r}e^{-j\theta}$$

Thus in the inverse Nyquist plane each vector has a length which is equal to the reciprocal of the gain and is at an angle which is **minus** the system phase shift. Figure 4.4(b) shows the inverse Nyquist diagram of a first-order system.

The inverse plot suffers from a similar disadvantage to the direct Nyquist diagram in that frequency points are implicit. However, small gains are magnified and it has certain very important advantages from a theoretical point of view as we shall see in the next chapter.

Bode Diagrams

Bode diagrams are similar to the plots that were produced in Fig. 4.3, except that gain and frequency are plotted logarithmically. Bode diagrams are conventionally drawn on log–linear graph paper where the gain is usually plotted in decibels (dB) on a linear scale and the frequency is plotted on a logarithmic scale. A decibel strictly speaking is a logarithmic ratio of powers, i.e.

$$10 \log(P_2/P_1)$$

where P_2 is the output power and P_1 the input power to a system. Power is proportional to the square of the amplitude, other things being equal. For example, the average power dissipated by a sinusoidal voltage of amplitude V applied across a resistance R is $V^2R/2$. Provided that the resistance is the same, the power ratio associated with two voltages V_1 and V_2 is

$$10 \log[(V_2/V_1)^2] = 20 \log(V_2/V_1)$$

In control engineering the idea of a power ratio is lost, but nevertheless the gain of a system is often expressed in terms of decibels. In this context it simply means expressing the gain as $20 \log|G(j\omega)|$. As an example, a gain of 6 dB means an amplitude ratio of 2, a gain of 0.5 is -6 dB. The advantage of working in dBs is that gains are additive and that much larger dynamic ranges can be shown when drawing graphs with logarithmic (decibel) scales.

The advantage of Bode diagrams are that they contain explicit frequency information and the use of the logarithmic scale for the gain overcomes the problem of the dynamic range limitations of the previous plots. For example a range of 100 dB covers an amplitude ratio range of 10^5. Bode diagrams do, however, require two separate plots to show the gain and the phase shift of a system. Figure 4.4(c) shows the Bode diagram of a first-order system.

Bode diagrams were used extensively to produce approximate frequency response curves by short-cut methods. The underlying ideas in these short-cut methods (Bode asymptotes) are still useful (see Section 4.9), but with the advent of the computer they are no longer needed for drawing frequency response curves.

Nichols Plots

Nichols plots combine aspects of Nyquist and Bode by producing a single plot in which logarithmic gain (dB) is plotted against phase shift. Nichols plots are very useful vehicles for design as we shall see later. Figure 4.4(d) shows the Nichols plot for a standard first-order system.

4.5 FREQUENCY RESPONSE OF AN ARBITRARY LINEAR SYSTEM

The more astute reader will have noticed that the symbol, G, used for the frequency response function is the same as that used for the transfer function. Furthermore the frequency response function in Eq. (4.10) is identical to the transfer function of a first-order system except that the operator 's' has been replaced by '$j\omega$'.

Is this true in general? Indeed it is and the proof is quite straightforward. Every time you differentiate $e^{j\omega t}$, a '$j\omega$' drops down, i.e.

$$\frac{d}{dt} \{e^{j\omega t}\} = j\omega \, e^{j\omega t}$$

Hence every 's' in the transfer function (which represents differentiation) is replaced by '$j\omega$'.

Exercise Convince yourself that the above argument is true for a more general system by following the steps described in Section 4.3 for the first-order system for a system which is governed by the differential equation:

$$a_2 \frac{d^2 y}{dt^2} + a_1 \frac{dy}{dt} + a_0 = b_1 \frac{dx}{dt} + b_0$$

i.e. its transfer function is

$$G(s) = \frac{Y}{X}(s) = \frac{b_1 s + b_0}{a_2 s^2 + a_1 s + a_0}$$

Thus, in general, if the transfer function is known, the frequency response function can be obtained by replacing s with $j\omega$, i.e.

$$G(j\omega) = G(s)|_{s=j\omega} \tag{4.13}$$

In turn its frequency response characteristics can be drawn, i.e. how the gain and phase of the system vary with frequency.

Frequency Response of Typical Elements of a Transfer Function

Pure integrator, $G(s) = 1/s$ The frequency response of a pure integrator is given by

$$G(j\omega) = 1/j\omega; \qquad |G(j\omega)| = 1/\omega; \qquad \angle\, G(j\omega) = -\pi/2 \qquad (4.14)$$

Hence a pure integrator exhibits a constant phase lag of $90°$ and a gain which is very large at low frequencies and becomes smaller with increasing frequency.

Simple Lag, $G(s) = 1/(1 + \tau s)$ Simple first-order lags were discussed earlier in the chapter, but for completeness we will repeat the salient points here. Its phase characteristic starts at zero and increases to $-90°$ at high frequencies. Its gain goes from unity to zero as frequency increases. The phase shift is $-45°$ and the gain is $1/\sqrt{2}$ when $\omega = 1/\tau$.

Second-order systems, $G(s) = 1/[1 + 2\zeta s/\omega_n + (s/\omega_n)^2]$ This system is rather more complex than the previous examples and needs to be considered in more detail. The frequency response function is

$$G(j\omega) = \frac{1}{1 - (\omega/\omega_n)^2 + j2\zeta\omega/\omega_n} \qquad (4.15)$$

Let us consider the gain and phase characteristics in terms of the non-dimensional frequency u, where $u = \omega/\omega_n$.

$$|G(u)|^2 = \frac{1}{(1 - u^2)^2 + (2\zeta u)^2} \qquad (4.16)$$

$$\angle\, G(u) = -\tan^{-1}\left(\frac{2\zeta u}{1 - u^2}\right) \qquad (4.17)$$

In the time domain, the step responses of underdamped second-order systems exhibit an overshoot (i.e. when the damping ratio is below unity). In the frequency domain there is an analogous situation where the system exhibits *resonance* for damping ratios below $1/\sqrt{2}$, i.e. the gain reaches a peak value, M_p, greater than unity. The peak gain is referred to as the *peak magnification factor*, hence the use of the letter, M. The peak magnification factor is often used as a design specification for closed-loop systems as we shall see later.

The phase of the second-order system is zero at low frequencies. At the natural frequency ($\omega = \omega_n$, $u = 1$), the phase shift is $-90°$. At high frequencies the s^2 term in the denominator dominates the response, and so the high frequency phase shift tends to $-180°$.

Figure 4.5 shows the Nyquist diagram of a set of second-order systems with different damping ratios. As you can see the curves are essentially circular. Figure 4.6 shows a similar set of curves using Bode gain and phase representation. Here the peaking of the

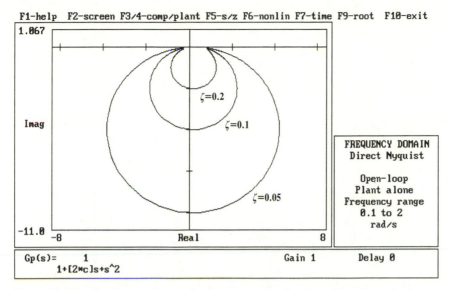

Figure 4.5 Nyquist diagram of second-order systems as a function of damping ratio, ζ

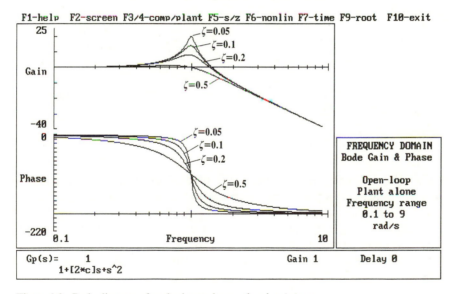

Figure 4.6 Bode diagram of underdamped second-order systems

gain curve and the associated rapid change of phase are the distinguishing features associated with underdamped second-order systems.

The frequency at which the gain peaks is the *resonant frequency*, ω_r, which can be found by differentiating the denominator of Eq. (4.16) and equating it to zero. If you do this you will find that the non-dimensional resonant frequency, u_r, is given by

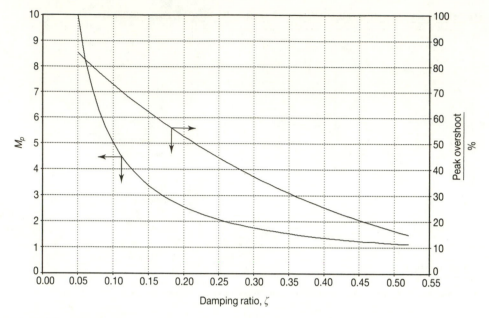

Figure 4.7 Peak magnification factor, M_p, and peak overshoot as a function of damping ratio, ζ

$$u_r = \frac{\omega_r}{\omega_n} = \sqrt{(1 - 2\zeta^2)} \qquad (4.18)$$

By substituting this value of u back into the gain equation, the peak magnification factor, M_p, is found to be,

$$M_p = \frac{1}{2\zeta\sqrt{(1 - \zeta^2)}} \qquad (4.19)$$

Figure 4.7 is a graph which shows how the peak magnification factor, M_p, changes with the damping ratio, ζ. Using this graph it is quite easy to obtain the damping ratio knowing the size of the resonance peak.

The natural frequency, ω_n, the damped natural frequency, ω_d, and the resonant frequency, ω_r, are all quite close together. In fact for lightly damped systems they can be considered identical. For more heavily damped systems there is a small but significant difference between them, i.e. $\omega_n > \omega_d > \omega_r$.

Exercise A system has a transfer function

$$G(s) = \frac{25}{s^2 + as + 25}$$

Calculate its resonant frequency and peak magnification factor for values of 'a' of 2, 4 and 8. Use CODAS to draw the Nyquist diagrams over the frequency range 1 to 10 radians/s and confirm your calculated results using the cursor.

4.6 GENERAL FREQUENCY DOMAIN PERFORMANCE SPECIFICATIONS

The ideas of resonance and peak magnification factor have been described above specifically in terms of a second-order system. In fact any system which has complex poles may exhibit resonance. However, simple analytical expressions relating ω_r and M_p to transfer function parameters are not available for general systems. Nevertheless both resonant frequency and peak magnification factor are used in describing and specifying frequency domain performance characteristics of arbitrary systems.

Bandwidth, ω_b

Another important measure of the frequency response of a system is its bandwidth, ω_b. Bandwidth is a measure of the frequency range over which a system can effectively operate. Bandwidth is an important requirement when specifying the performance of control systems. Sometimes a maximum bandwidth is specified where there is a danger of high frequency noise becoming a problem and sometimes a minimum bandwidth is specified to make sure that the control system will cope adequately with the frequency range of its demand signals.

There are various slightly different definitions of the term 'bandwidth'. In control systems it is generally taken to mean the frequency range over which the gain of the system exceeds $1/\sqrt{2}$, in other words, is greater than -3 dB. This definition is appropriate to the vast majority of control systems where the low frequency gain is unity. However, for certain Type-0 systems where the low frequency gain is small, this definition breaks down and clearly becomes meaningless if the gain never exceeds -3 dB. We will nevertheless adopt this definition for the sake of simplicity.

It is possible to derive an expression relating the bandwidth of the standard second-order system to damping ratio and natural frequency. The expression is rather complicated and the derivation is quite involved and so it won't be done here; do, however, try the next exercise.

Exercise Use CODAS to determine the -3 dB bandwidth of the system defined in the previous exercise for the cases $a = 1, 2, 4, 8, 16$. Plot a graph of normalized bandwidth (ω_b/ω_n) as a function of damping ratio.

4.7 SKETCHING THE FREQUENCY RESPONSES OF SYSTEMS

Even though the detailed calculations may be performed on a computer, it is important to be able to sketch frequency responses of systems so that insights into their behaviour are developed. Furthermore, through synthesizing frequency responses from transfer functions, the ability to propose reasonable transfer function models for systems from their experimental frequency response is enhanced.

To produce a rough sketch first consider the behaviour of the system at the extremes

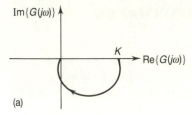

(a)

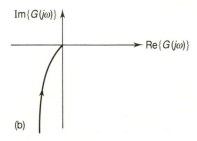

(b)

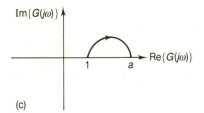

(c)

Figure 4.8 Nyquist diagrams of typical systems. (a) Nyquist diagram of $K/(1 + \tau_1 s)(1 + \tau_2 s)$. (b) Nyquist diagram of $K/s(1 + \tau s)$. (c) Nyquist diagram of $(1 + as)/(1 + s)$, $(a > 1)$.

of the frequency range, then consider the gain and phase contributions of the individual terms as the frequency is swept through the entire range. Here are some examples:

Double lag $G(s) = \dfrac{K}{(1 + \tau_1 s)(1 + \tau_2 s)}$ At low frequencies the denominator terms will contribute zero phase shift and their gain is unity. Hence the low frequency gain of the system is K and the phase shift is zero. At very high frequencies each denominator term introduces a phase lag of $90°$, and their magnitude is large, hence the high frequency system gain will tend to zero. In between the magnitude and phase of the denominator terms increase uniformly. The Nyquist diagram of this system is sketched in Fig. 4.8(a).

Type-1 system $G(s) = \dfrac{K}{s(1 + \tau s)}$ At low frequencies the response is dominated by the integrator, and so the system behaves like a pure integrator, i.e.

$$G(j\omega) \approx \frac{K}{j\omega}$$

Thus at very low frequencies, the system exhibits a phase lag of nearly 90° and has a very large gain. As the frequency increases, the gain drops and the phase lag increases by a further 90° (see Fig. 4.8(b)).

Lead network $G(s) = \dfrac{1 + as}{1 + s}$ A 'lag' is a simple pole, i.e. a denominator term which contributes a negative phase shift (phase lag). The analogous 'pure lead' would be a simple zero, i.e. a numerator term which contributes a positive phase shift (phase advance). As discussed in the previous chapter, a system consisting exclusively of a zero is not proper and hence not realizable. A physically realizable lead system will include at least one pole to make the system proper. Thus the above system termed a 'lead network' is the combination of a simple pole and a simple zero. Let us assume that 'a' in the above transfer function is greater than one. At very low frequencies the gain of the system is unity, and its phase shift is zero. As the frequency increases, the gain and phase contribution of the numerator will increase more quickly than the denominator. At very high frequencies the gain of the system will approach 'a'. Its phase shift will tend to zero as the 90° phase advance of the numerator term is cancelled by the 90° phase retard of the denominator. Figure 4.8(c) shows the overall Nyquist diagram of this system.

Exercise For the three example systems above, try sketching their frequency responses as inverse Nyquist, Nichols and Bode plots. Check your results using CODAS with the following numerical values for the parameters: $K = 2, \tau_1 = 1, \tau_2 = 0.2, \tau = 1$ and $a = 5$.

4.8 NUMERICAL CALCULATION OF FREQUENCY RESPONSE DATA

In general, transfer functions consist of a set of multiplicative terms in the numerator and another set in the denominator. Typically the terms will be of the form:

$$ s \quad \text{or} \quad a + bs \quad \text{or} \quad a + bs + cs^2 $$

and a multiplicative constant, K.

Suppose the transfer function is of the form

$$ G(s) = K \frac{N_1(s)N_2(s)}{D_1(s)D_2(s)D_3(s)} \tag{4.20} $$

where each of the terms N_1, N_2, D_1, D_2, D_3, etc., is one of the forms described above.

The overall gain of the system is

$$ |G(j\omega)| = K \frac{|N_1(j\omega)||N_2(j\omega)|}{|D_1(j\omega)||D_2(j\omega)||D_3(j\omega)|} \tag{4.21} $$

i.e. the **product** of the gains of the terms in the numerator **divided** by the **product** of the gains of the terms in the denominator.

The overall phase shift is

$$\angle G(j\omega) = \{\angle N_1(j\omega) + \angle N_2(j\omega)\} - \{\angle D_1(j\omega) + \angle D_2(j\omega) + \angle D_3(j\omega)\} \quad (4.22)$$

i.e. the **sum** of the phase shifts of each numerator term **less** the **sum** of the phase shifts of each denominator term.

Normally frequency response calculations are done using a computer program or package. However, it is important to understand how the calculations are done and to be able to do them yourself.

The method is to substitute $j\omega$ for s in the transfer function and to calculate the gain and phase of each term and then obtain the overall gain and phase in the manner described above. This is the method, of course, that is used in programs. In doing hand calculations great care must be taken with the phase calculation to get the right quadrant for higher-order terms.

Example 4.1 Numerical calculation of gain and phase Given the transfer function

$$G(s) = \frac{15(s+2)}{(s^2 + 2s + 5)(s + 4)}$$

find its gain and phase when $\omega = 3$ rad/s.

SOLUTION Substitute $s = j3$ in the transfer function and write down the magnitude of each term (in fact it is easier to work with the modulus squared), i.e.

$$|G(j3)|^2 = \frac{225(3^2 + 2^2)}{((5 - 3^2)^2 + 6^2)(3^2 + 4^2)} = \frac{225 \times 13}{52 \times 25} = 2.25$$

Hence the gain is 1.50.
The phase shift is

$$\angle G(j3) = \tan^{-1}\left(\frac{+3}{+2}\right) - \tan^{-1}\left(\frac{+6}{-4}\right) - \tan^{-1}\left(\frac{+3}{+4}\right)$$

$$= 56.31 - 123.7 - 36.87 = -104.26°$$

Exercise Given the transfer function

$$G(s) = \frac{640}{s(s + 4)(s + 16)}$$

Calculate the magnitude and phase of its frequency response at $\omega = 10$ rad/s. Check your answer with CODAS.

4.9 ASYMPTOTIC BODE DIAGRAMS

In Section 4.7 the direct polar frequency response curves of typical transfer function elements were drawn. Similar plots can be drawn in the other frequency domain views as was illustrated in the exercise at the end of Section 4.7. The drawing of the inverse

Nyquist and Nichols plots will not be treated separately, but Bode diagrams will be covered in more detail.

In the era before computer packages were readily available, a short-cut method for drawing frequency responses was developed using the logarithmic plot (Bode diagrams). This short-cut method used straight line approximations, known as Bode asymptotes, to enable rapid sketching of frequency responses given the system transfer function. Although it is no longer necessary to become highly skilled in drawing Bode asymptotes, a knowledge of the method is important in further developing an understanding of the frequency response of systems and its correlation with the system transfer function.

The basic idea of the method is to use a logarithmic representation of gain (decibels) so that once the gain curves of individual elements in a transfer function have been drawn, the overall gain curve can be obtained by **adding** the individual curves rather than multiplying the gains as is implied in Eq. (4.21). In addition, by using a logarithmic plot the asymptotic behaviour of typical terms can easily be deduced, so facilitating the production of straight line approximations to the actual frequency response curves.

The Principle of Constructing an Asymptotic Bode Diagram

The Bode representation of transfer functions was covered in the previous chapter, i.e.

$$G(s) = \frac{K_B G'(s)}{s^n} \tag{4.23}$$

Where $G'(s)$ is such that

$$\lim_{s \to 0} G'(s) = 1$$

and K_B is the Bode gain and n is the number of free integrators (i.e. the system Type).

Furthermore, $G'(s)$ is normally represented in the form

$$G'(s) = \frac{(1 + s/z_1)(1 + s/z_2) \ldots}{(1 + s/p_1)(1 + s/p_2)(1 + s/p_3) \ldots} \tag{4.24}$$

where $-z_1$, $-z_2$, etc. are the zeros of the system and $-p_1$, $-p_2$, etc. are the poles of the system. As was discussed in Chapter 3, the number of poles, P, is always greater than or equal to the number of zeros, Z, in a physically realizable system.

Gain Asymptotes of a Simple Pole/Zero

Consider a typical term of the form

$$F(s) = 1 + s/q$$

where $-q$ could be a pole or a zero. The gain is

$$|F(j\omega)| = \sqrt{(1 + \omega/q)^2} \tag{4.25}$$

or in decibels

$$20 \log(|F(j\omega)|) = 20 \log(\sqrt{(1 + (\omega/q)^2)}) \tag{4.26}$$

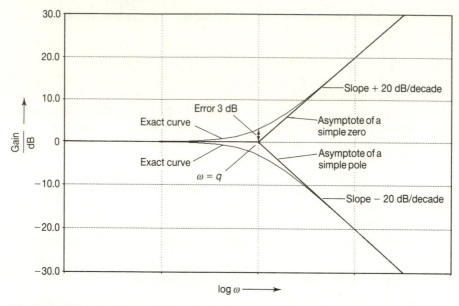

Figure 4.9 Gain asymptotes of a simple pole/zero

Now when $\omega \ll q$

$$20 \log(|F(j\omega)|) = 10 \log(1) = 0 \text{ dB}$$

which is the low frequency asymptote, and when $\omega \gg q$

$$20 \log(|F(j\omega)|) = 20 \log(\omega/q) \tag{4.27}$$

which is the equation of the high frequency asymptote.

If the gain in decibels is plotted as a function of the log of the frequency, the asymptotes are straight lines. The low frequency asymptote is a horizontal straight line and the high frequency asymptote is a line at a slope of 20 dB/decade, i.e. the gain changes by 20 dB as the frequency changes by a factor of 10. The two asymptotes intersect at a frequency q rad/s which is called the *corner frequency*.

If we have been dealing with a zero, (i.e. $q = z$) the gain asymptotes are exactly as derived above, and they are shown in Fig. 4.9. For a simple pole (i.e. $q = p$) the function, $F(s)$ would have been of the form

$$F(s) = \frac{1}{1 + s/p}$$

The low frequency asymptote is the same as for a zero (i.e. a straight line at 0 dB) whereas the high frequency asymptote is:

$$20 \log(|F(j\omega)|) = -20 \log(\omega/p)$$

When the decibel gain is plotted against the log of the frequency, ω, it yields another straight line, but whose slope is -20 dB/decade as shown in Fig. 4.9. Again the point of intersection of the two asymptotes is at a frequency of p rad/s.

The exact frequency responses are also drawn on the same diagram, and the error between the exact curve and the asymptotic lines is clearly a maximum at the corner frequency. The exact gain when $\omega = q$ from Eq. (4.26) is

$$20 \log(|F(j\omega)|) = \pm 20 \log(2) \approx \pm 3 \text{ dB}$$

In dealing with the frequency response of electronic amplifiers the slope of 20 dB/decade is often expressed approximately as 6 dB/octave, or in other words the gain changes by approximately 6 dB every time the frequency doubles.

Gain Asymptotes of Repeated Poles/Zeros

The derivation of asymptotes of terms of the type

$$F(s) = (1 + s/q)^n$$

follows closely what has been done in the previous section. It is left as an exercise to the interested reader to show that the gain asymptote is a line of slope $\pm 20n$ dB/decade intersecting the low frequency asymptote (i.e. the 0 dB axis) at the corner frequency.

Gain Asymptotes of Free Integrators

Terms of the type

$$F(s) = \frac{K_B}{s^n}$$

dominate the low frequency behaviour of the system. The gain of terms of this type is

$$20 \log(|F(j\omega)|) = 20 \log(K_B/\omega^n) = 20 \log(K_B) - 20n \log(\omega)$$

which is a straight line of slope $-20n$ dB/decade and which passes through the 0 dB line when $\omega^n = K_B$.

Summary of Asymptotic Bode Plots and Their Limitations

The above discussion shows that the gain asymptotes of a system consist of a set of straight line segments whose slopes are at integer multiples of ± 20 dB/decade. The slopes switch at the corner frequencies of the system.

Raw asymptotic Bode plots are a very good approximation when the system has well-separated simple poles/zeros. When poles or zeros lie close together the errors tend to build up and so it is essential to apply some correction to take account of the mismatch between the asymptotic lines and the exact values in the region close to the corner frequencies. The main weakness of the method is when dealing with underdamped systems or systems with complex pole or zeros, especially those exhibiting a pronounced resonance. In such cases one must resort to applying significant correction factors which is difficult and rather time-consuming to do.

4.10 MINIMUM-PHASE AND NONMINIMUM-PHASE SYSTEMS

All the systems studied so far in this chapter have been so-called minimum-phase systems. It is possible for two systems with different transfer functions to have identical gain characteristics but different phase characteristics. A transfer function whose phase characteristic is a minimum for its gain characteristic is termed *minimum-phase*.

Consider the two transfer functions,

$$G_1(s) = \frac{1 + 2s}{1 + 2s + 5s^2} \quad \text{and} \quad G_2(s) = \frac{1 - 2s}{1 + 2s + 5s^2}$$

$G_1(s)$ is minimum-phase whereas $G_2(s)$ is nonminimum-phase. The gain characteristic of the two transfer functions is identical, i.e.

$$|G_1(j\omega)| = |G_2(j\omega)|$$

The phase shift associated with $G_2(s)$ at high frequencies is bigger than that of $G_1(s)$, hence the term 'nonminimum-phase'. In fact

$$\angle G_1(j\omega) = -\tan^{-1}\left(\frac{2\omega}{1 - 5\omega^2}\right) + \tan^{-1}(2\omega)$$

whereas

$$\angle G_2(j\omega) = -\tan^{-1}\left(\frac{2\omega}{1 - 5\omega^2}\right) - \tan^{-1}(2\omega)$$

Figure 4.10 shows the gain and phase curves for the above two systems.

When there is no transport delay present, minimum-phase systems are systems

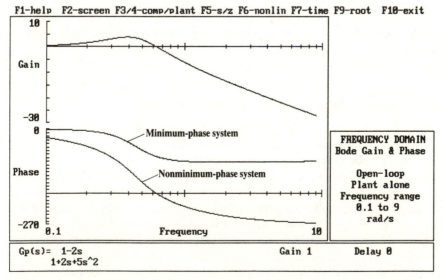

Figure 4.10 Gain and phase curves of minimum-phase and nonminimum-phase systems

whose zeros all lie in the left half s-plane. Hence nonminimum-phase systems, in the absence of transport delay, will have one or more zeros in the right half s-plane.

If it is known that a system is minimum-phase, then there is a unique relationship between its gain characteristic and its phase characteristic. This means that in principle, once its gain characteristic is known, its phase characteristic can be deduced and furthermore its transfer function can be determined. Generally minimum-phase systems are easier to control and their inverse function is stable, i.e. the inverse function has no poles in the right half s-plane which is important in certain control system design methods.

Frequence Response and Transfer Function of Transport Delays

Imagine a frequency response test being carried out on a plant consisting of a pure time delay of T_D second. The output waveform is a replica of the excitation signal, except that it is delayed by T_D seconds. In fact the input/output signals look exactly like the waveforms shown in Fig. 4.1. The 'experimenter' would conclude that the gain of the system is 1 and the phase lag is $-\omega T_D$ radians. Moreover, this observation would be true at all the test frequencies. If the gain/phase data is to be plotted on a Nyquist diagram, one would end up with a plot that circled the origin at a radius of 1. Now, the equation of a unit circle in polar coordinates is

$$r = e^{j\theta} \tag{4.28}$$

Hence the frequency response function of a transport delay is:

$$G(j\omega) = e^{-j\omega T_D} \tag{4.29}$$

In as much as $j\omega$ can be substituted for s to obtain the steady-state frequency response from the transfer function, the converse can also be done to replace s by $j\omega$ to obtain the transfer function knowing the steady-state frequency response function, i.e.

$$G(s) = e^{-s T_D} \tag{4.30}$$

Transport delay is another example of nonminimum-phase behaviour. The 'minimum-phase' version of a pure transport delay is a constant gain of 1!

Frequency Response of Systems with Transport Delay

The effect of a pure time delay on the overall frequency response of a system is best illustrated by considering a first-order system with a transport delay, i.e.

$$G(s) = \frac{e^{-s T_D}}{1 + s\tau} \tag{4.31}$$

Hence the gain and phase equations are respectively:

$$|G(j\omega)|^2 = \frac{1}{1 + (\omega\tau)^2} \tag{4.32}$$

$$\angle G(j\omega) = -\tan^{-1}(\omega\tau) - \omega T_D \tag{4.33}$$

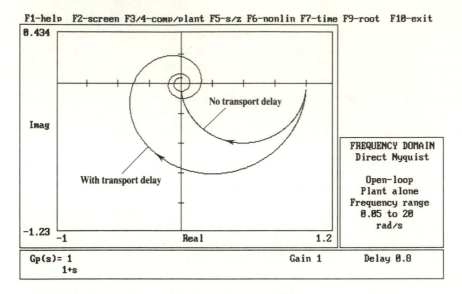

Figure 4.11 Effect of transport delay on the frequency response

The transport delay has no effect on the magnitude, but the phase lag associated with each vector is increased by ωT_D radians. As we saw in Fig. 4.4(a), the Nyquist diagram of a simple first-order system is semicircular; by adding the transport delay each vector is twisted by an additional angle which increases as the frequency increases. The net result is that the Nyquist diagram becomes distorted into a spiral as shown in Fig. 4.11. Clearly the presence of transport delay in a system renders it nonminimum-phase.

4.11 CLOSED-LOOP BEHAVIOUR: STABILITY

Up to now this chapter has dealt with open-loop systems. When a control loop is closed its behaviour is quite different from the open-loop case. It is therefore important to be able to predict how the system will behave in closed loop. For example the closed-loop frequency response characteristics of the system are of great interest. We will see that closed-loop frequency response behaviour of a system can easily be determined from its open-loop frequency response characteristics and furthermore one can predict whether or not the closed-loop system is stable.

Initially the question of stability of the general non-unity feedback system as shown in Fig. 4.12 will be considered. It will be assumed that the open-loop system is stable (i.e. no poles in the right half s-plane). As this is true of the vast majority of systems, the assumption is not too restrictive. Imagine that the switch, S, is in the position shown in Fig. 4.12, i.e. AB is open and at point A a sinusoidal signal is injected. If the system is linear, the signal emerging at point B is another sinusoid whose amplitude and phase

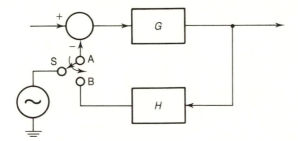

Figure 4.12 Stability of a general non-unity feedback system

relative to the input signal depends on the open-loop frequency response characteristics of the system.

Let us imagine that we adjust the frequency of the input sinewave (at A) until the output sinewave (at B) is exactly in phase with the input signal. Suppose, for the sake of argument, that the output signal is of lower amplitude than the excitation signal. The switch is flipped over closing AB and detaching the signal source so that the system is self-excited. Gradually, as the signal chases its tail round the loop, the output of the system will die away as each successive cycle of the signal is smaller than the previous cycle. This implies that the closed-loop system is *stable*.

Let us now imagine repeating the same experiment, but the gain of the system is adjusted so that the size of the signal at B is exactly the same as the excitation signal. Now when we flip the switch, the system will carry on oscillating at the excitation frequency indefinitely. In other words the system is *marginally stable* (also sometimes called critical stability).

Finally, consider the case where the system gain is increased so that the return signal is **bigger** than the excitation signal. Now when the switch is closed the output will grow and grow as each successive cycle augments the previous cycle. In other words the system is *unstable*.

The above experiments can easily be related to the open-loop frequency response of the system. The frequency which resulted in the output signal at B being in phase with the excitation signal at A, is equivalent to the frequency which produces $-180°$ phase shift in the open-loop system. The reason for this being that in the closed-loop system a further $180°$ phase lag is contributed when the controlled variable is subtracted from the reference value. The frequency at which the open-loop system contributes $180°$ phase lag is called the *phase-crossover frequency*, ω_{pc}.

Thus the critical point which separates stable from unstable systems is where the open-loop phase shift is $-180°$ and the open-loop gain is 1. This point corresponds to the -1 point on the Nyquist diagram as shown in Fig. 4.13. From the foregoing a stability criterion can be stated for frequency response loci in the direct Nyquist plane as follows:

Systems whose open-loop frequency response loci do not encircle the -1 point will be **stable** in closed loop, i.e. their open-loop gain is less than unity at the phase-crossover frequency.

Loci of systems which encircle the -1 point are **unstable** in closed loop.

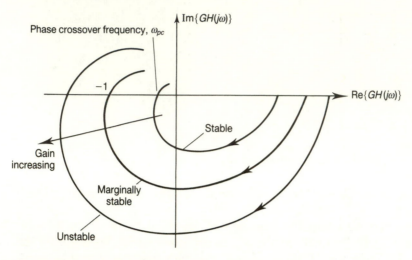

Figure 4.13 Stability in the Nyquist plane

Loci of systems which pass through the -1 point are **marginally stable** and will oscillate continuously at the phase-crossover frequency, ω_{pc}.

The stability criterion stated above is known as the simplified Nyquist stability criterion and is restricted to systems which are stable in open loop, i.e. no open-loop poles in the right half s-plane. The original contribution due to Nyquist is of a very mathematical nature and deals also with systems which are unstable in open loop. This topic is dealt with by many advanced texts, but it is more of theoretical interest than practical value.

The critical point concept can be extended to the other representations of frequency response. For example in the Bode gain and phase diagram, the critical point maps into a critical line where the phase shift is $-180°$ and the gain is 0 dB (see Fig. 4.14).

Exercise Draw diagrams similar to Figs 4.13 and 4.14 for a Nichols plot and an inverse Nyquist plot which show three frequency response loci: one at a low value of gain which is stable, one at a high value of gain which is unstable and one between which is marginally stable. Can you restate the simplified Nyquist stability criterion for these two frequency domains?

Relative Stability

In practice one is not merely concerned with whether a system is stable or unstable, one is also concerned with designing stable systems which produce an acceptable transient response (not too oscillatory) or which meet other design criteria such as maximum or minimum bandwidth, peak magnification factor, etc. If one is designing a control system to meet a time domain specificiation using empirical frequency response data, then it is necessary to be able to predict aproximately how the closed-loop system will behave in the time domain. Finally one is also interested in how *robust* the design

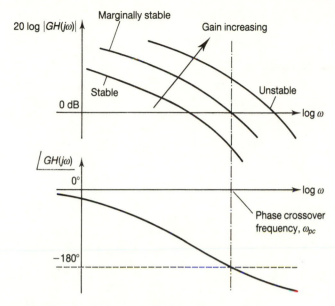

Figure 4.14 Stability using Bode plots

is in terms of the sensitivity of the control system performance to small changes in the characteristics of the controller or the system.

Gain Margin and Phase Margin

Suppose one has adjusted the system gain so that it is just on the point of instability. By reducing the gain slightly the system will be stabilized, but it will clearly still be very oscillatory. Furthermore a slight increase in the gain of the controller or the system will render the system unstable once more. The factor by which the gain can be increased before the system becomes marginally stable is termed the *gain margin, gm*. The gain margin of a system can easily be determined from its open-loop frequency response. It is merely the reciprocal of the open-loop gain at the phase-crossover frequency. Looking at a typical Nyquist diagram such as the one shown in Fig. 4.15(a) one can see that

$$gm = \frac{1}{\text{OA}} \qquad (4.34)$$

Often gain margins are expressed in decibels, i.e. the gain margin is

$$-20\log(\text{OA})\,\text{dB}$$

For a stable system OA is less than 1, hence the gain margin of a stable system is positive when expressed in dBs. This figure can be read straight off the Bode diagram as shown in Fig. 4.15(b).

A good gain margin is desirable from the point of view of robustness. However, one cannot select an arbitrarily large gain margin because the dynamic characteristics

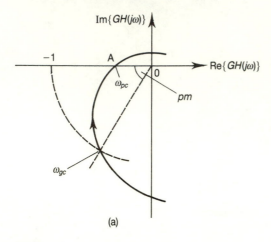

(a)

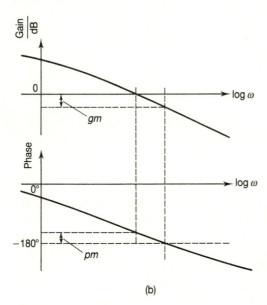

(b)

Figure 4.15 Stability margins. (a) Nyquist diagram. (b) Bode diagram

of the system are correlated with it and the response may become unacceptably overdamped and sluggish.

The gain margin **alone** is not enough to express the relative stability of a system. Consider the system shown in Fig. 4.16(a). This system has a very good gain margin, but it only requires a small additional phase lag to render the system unstable. Hence a further criterion is essential to cater for systems of this type. The criterion is termed the *phase margin*, *pm*, and is defined as the additional phase lag that will make the system marginally stable.

The phase margin can be determined easily from the open-loop frequency response.

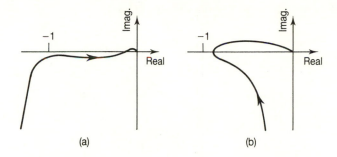

Figure 4.16 Systems satisfying one stability criterion. (a) Good gain margin, but poor phase margin. (b) Good phase margin, but poor gain margin

It is in fact the clockwise angle through which the unit vector must be rotated before it lies on the negative real axis. The frequency at which the open-loop system gain is unity is termed the *gain-crossover frequency*, ω_{gc}, i.e.

$$|GH(j\omega_{gc})| = 1 \qquad (4.35)$$

Hence the phase margin is the difference of the phase shift of the system and $-180°$ at the gain-crossover frequency, i.e.

$$pm = \angle GH(j\omega_{gc}) + 180 \qquad (4.36)$$

Situations are also possible where the system exhibits a good phase margin, but has a poor gain margin (Fig. 4.16(b)). Hence **both** criteria must be satisfied adequately to provide good overall stability.

Example 4.2 Simple design, changing gain A system is controlled by a simple proportional controller of gain, K. The open-loop transfer function is given by

$$G(s) = \frac{K\,e^{-0.1s}}{(1+s)^3}$$

Find the maximum gain K that will give the system a gain margin of at least 2 and a phase margin of at least 40°.

SOLUTION This is quite a complicated transfer function and performing the calculations by hand would take a considerable time; using CODAS we can obtain the frequency response rapidly (Fig. 4.17). When using CODAS a controller gain, K, of 1 is assumed. The resulting phase-crossover frequency is found to be 1.54 rad/s and at this frequency the system gain is 0.16 (i.e. $|G(j1.54)| = 0.16$). Hence to obtain a gain margin of 2, the gain must be increased by a factor 0.5/0.16, i.e. $K = 3.125$.

To determine the change in gain required for a phase margin of 40° one must first find the frequency where the phase shift of the system is $-140°$ (i.e. $pm - 180°$). This frequency will become the gain-crossover frequency when the gain has been adjusted. Using CODAS we find that the system exhibits a phase shift of $-140°$

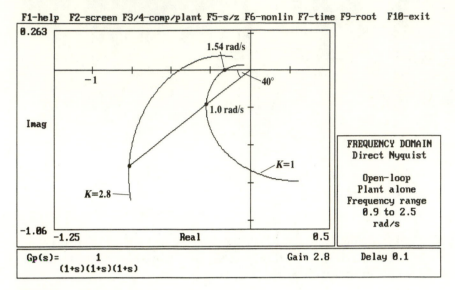

Figure 4.17 Simple gain design

when the frequency is about 1 rad/s, and at this frequency $|G|$ is 0.36. Hence for this frequency to become the gain-crossover frequency, $K = 1/0.36 = 2.8$.

Thus the maximum gain one can employ whilst satisfying **both** design criteria is 2.8.

Correlation with Time Response

Obviously systems with very small gain and phase margins will be very oscillatory, but is it possible to establish more quantitative correlations between open-loop frequency domain characteristics and closed-loop time domain behaviour? For example can one set target gain and phase margins to achieve a certain damping ratio or can one predict the settling time of a particular system from a knowledge of its open-loop frequency response? The answer to these questions is that one can make rough time-domain correlations from the frequency responses of systems. It is more difficult to make predictions in systems where there are pronounced resonances and where there is a significant mismatch between the gain and phase margins as for the systems shown in Fig. 4.16.

In process control systems a decay ratio of $1/4$ ($\zeta = 0.21$) is considered acceptable; this corresponds approximately to gain and phase margins of 2 and 40° respectively. In servo systems, however, a much less oscillatory response is required because of potential damage due to mechanical impingement and the requirement to minimize vibrations. Typically, peak overshoots of less than 20% are specified which correspond to gain margins of greater than 5 and phase margins of about 50°. The design must always take account of the worst case as it is most unlikely that both criteria are satisfied simultaneously.

The frequency at which the closed-loop system oscillates when it is marginally stable is the phase-crossover frequency. When the system is stable but underdamped, the oscillatory frequency is lower, and is closer to the closed-loop resonant frequency. We will see later how to estimate the oscillation frequency of the closed-loop system.

Exercise Use CODAS to display the closed-loop time response of the system used in Example 4.2 with $K = 2.8$ and measure the decay ratio. How close is the actual decay ratio to the target value of 1/4?

4.12 CLOSED-LOOP FREQUENCY RESPONSE

Whereas predicting the closed-loop time behaviour from open-loop frequency response data is a bit hit and miss, the closed-loop frequency response can be predicted exactly. Thus if the control system performance is specified in terms of frequency domain criteria such as the bandwidth, peak magnification factor, etc., complete designs can be made using empirical frequency response data.

Consider a unity feedback system whose open-loop transfer function is $G(s)$. The closed-loop frequency response is then

$$\frac{C}{R}(j\omega) = M\, e^{j\alpha} = \frac{G(j\omega)}{1 + G(j\omega)} \tag{4.37}$$

M is the closed-loop gain, and α is the closed-loop phase shift. Clearly both M and α are functions of frequency. The vector quotient on the right hand side of Eq. (4.37) can be interpreted geometrically by referring to Fig. 4.18(a). The vector **OA** represents G (i.e. $G(j\omega)$ at a particular frequency), and the vector **BA** which joins the -1 point to the tip of **OA** is equivalent in magnitude and direction to the vector $(1 + G)$. Hence the magnitude of the closed-loop frequency response, M, is merely the ratio of the distances OA and BA, i.e.

$$M = OA/BA$$

Thus given a Nyquist diagram it is very easy, by means of a geometrical construction, to find the closed-loop gain from the open-loop frequency response data.

The closed-loop phase shift, α, can also be found by a simple construction. Now, referring to Fig. 4.18(a),

$$\alpha = \angle G - \angle(1 + G) = \theta - \phi$$

From simple geometry this is the angle OAB. Using an inverse Nyquist diagram the results are even simpler and are summarized in Fig. 4.18(b). It is left as an exercise for the reader to confirm the results.

In a computer-aided design package such as CODAS, closed-loop data for specific frequency points is automatically displayed when using a cursor to extract detailed information from a frequency response curve. Furthermore it is quite simple to draw the entire closed-loop frequency response curve in any of the frequency domain representations discussed at the beginning of the chapter. Using hand methods simple geometrical constructions for obtaining closed-loop data are only available for the two Nyquist views.

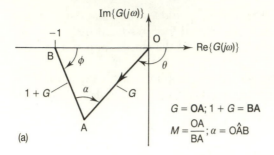

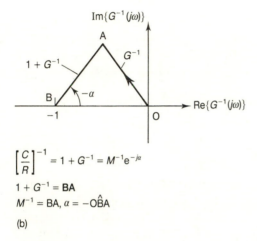

$$G = \mathbf{OA}; 1 + G = \mathbf{BA}$$

$$M = \frac{\mathbf{OA}}{\mathbf{BA}}; \alpha = \mathrm{O}\hat{\mathrm{A}}\mathrm{B}$$

(a)

$$\left[\frac{C}{R}\right]^{-1} = 1 + G^{-1} = M^{-1}\mathrm{e}^{-j\alpha}$$

$$1 + G^{-1} = \mathbf{BA}$$

$$M^{-1} = \mathbf{BA}, \alpha = -\mathrm{O}\hat{\mathrm{B}}\mathrm{A}$$

(b)

Figure 4.18 Finding closed-loop frequency response data from open-loop frequency data. (a) Direct Nyquist. (b) Inverse Nyquist

M-contours

The above technique for finding M values from the open-loop frequency response is fine for the purpose of analysis, i.e. one can obtain spot figures for the closed-loop performance and by trial and error one can find the peak magnification factor and the bandwidth of the closed-loop system. For design purposes, however, one requires loci of constant closed-loop gain which are termed M-contours. These loci are curves superimposed on the open-loop plot, on which the closed-loop magnification factor, M, is constant.

An M-contour in the direct Nyquist plane is the locus of points where the ratio OA/BA is constant. From simple geometry one can derive that the loci are a set of circles. In the inverse Nyquist plane the nature of the loci are self evident from the data in Fig. 4.18(b). The length of the vector $\mathbf{BA}$ is $1/M$, and since the point B is at -1, the loci in the inverse Nyquist plane are just a set of concentric circles centred at -1 and of radius $1/M$.

Although the geometry of the M-contours is very simple in the two Nyquist planes and well suited to hand methods of design, the Nichols diagram is a far more convenient computational tool for design using simple gain changes. The reason is that because

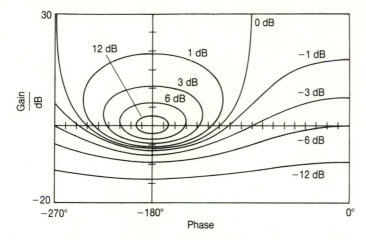

Figure 4.19 *M*-Contours in the Nichols diagram

the magnitude of the frequency response is plotted logarithmically on a Nichols plot, the frequency response curve merely moves up and down with changes of gain. The *M*-contours for a Nichols diagram are shown in Fig. 4.19. Note that *M*-contours cannot be drawn on a Bode diagram as they are functions of open-loop gain and phase and are **independent** of frequency.

It is interesting to observe from Fig. 4.19 that a phase margin of 60° and a gain margin of 6 dB corresponds to an *M* value of 0 dB. A system therefore that has these gain and phase margin values and lies close to the $M = 0$ dB contour will not exhibit any resonance peak, but will probably show a slight overshoot in the time domain. This helps to explain the rationale for the choice of desired gain and phase margin figures that were discussed in the previous section.

Example 4.3 Using *M*-contours for design using simple gain change Consider the same system as in Example 4.2, i.e.

$$G(s) = \frac{K\,e^{-0.1s}}{(1+s)^3}$$

(a) Find the maximum controller gain, K, that one can employ without exceeding a peak closed-loop magnification factor of 2 dB.
(b) What maximum controller gain will give the closed-loop system a bandwidth of 1.3 rad/s?

SOLUTION Initially make the gain, K, unity. It is generally best to start the design process with a gain of 1. Using CODAS, the Nichols view is selected and the open-loop frequency response curve is plotted followed by a 2 dB *M*-contour (Fig. 4.20). The problem is then to estimate what change in gain will make the frequency response curve tangential to this *M*-contour. This will ensure that the peak magnification factor is not greater than the design specification of 2 dB.

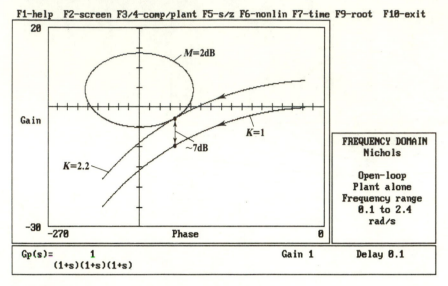

Figure 4.20   Design for a peak magnification factor

The frequency response curve will move up and down vertically when the gain is changed. By inspection a gain increase of about 7 dB will lift the frequency response curve sufficiently to make it tangential to the 2 dB M-contour, i.e. a gain, K, of 2.24. The free-wheeling cursor in CODAS is usefully employed here to measure the gain change required. With a little trial and error we can quickly refine the estimate and find that a gain of 2.2 is just about right. The final result is shown in Fig. 4.20.

To solve part (b), the frequency response curve for unity gain and the -3 dB M-contour are drawn (Fig. 4.21). The point on the original (i.e. $K = 1$) frequency response curve that corresponds to 1.3 rad/s is found and the distance this point must be moved to intersect the -3 dB contour is measured (see Fig. 4.21). The technique in CODAS is to use the normal cursor to mark the point on the open-loop frequency response curve where the frequency is 1.3 rad/s and then use the free-wheeling cursor in relative mode to measure the gain change required to move the marked point on to the -3 dB M-contour. It is found that an increase in gain of about 5 dB is required (i.e. a gain of 1.8).

Damped Natural Frequency and Damping Ratio Estimation

It was mentioned earlier that a good measure of the damped natural frequency of the closed-loop system is its resonant frequency, ω_r. This frequency can be found by locating the frequency at which the open-loop frequency response locus is tangential to the smallest M-contour. In Example 4.2 with K set to 2.2, the resonant frequency is found to be 1.06 rad/s, i.e. a period of about 6 seconds ($2\pi/1.06$).

If the effective damping ratio of a system could be obtained more accurately, the estimate of the settling time and peak overshoot could be improved.

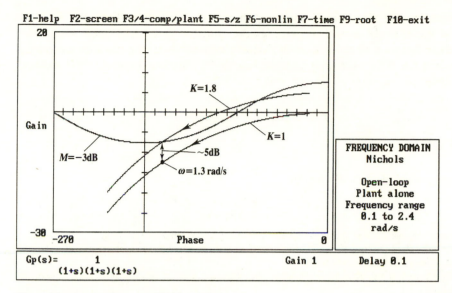

Gp(s)=   1   Gain 1   Delay 0.1
        (1+s)(1+s)(1+s)

Figure 4.21   Design for a specified bandwidth

Note: Strictly speaking damping ratio is a characteristic of second-order systems. Hence what one is doing is to model the response of a more complex system to an equivalent second-order system whose characteristics are similar to the actual system. This idea is similar to the dominant pole method that was discussed in Chapter 3.

The following technique uses the above principle to estimate the effective damping ratio of underdamped systems. The method is based on the fact that the gain of a second-order system at its natural frequency is $K_0/2\zeta$, where K_0 is its low-frequency gain (steady-state gain). This result can be deduced from Eq. (4.16) when $u = 1$. Don't forget that the material in that section was based on a **standard** second-order system with unit steady-state gain. The approach then is simply to determine the low frequency gain of the closed-loop system (K_0) and its gain at the effective natural frequency, i.e. when its phase shift is $-90°$. This latter gain can then be equated to $K_0/2\zeta$ to obtain an equivalent damping ratio, ζ. The resulting model is a second-order approximation to the actual system and is rather better that a simple dominant pole approach as the effects of the other poles are taken into account implicitly.

The controller gain, K, chosen for the system in Example 4.2 was 2.2. Hence, as the system is Type-0, its low frequency **closed-loop** gain, K_0, is $2.2/3.2 = 0.6875$. Using CODAS one finds that the closed-loop phase shift is $-90°$ at a frequency of about 1 rad/s and its closed-loop gain at this frequency is 1.23. Its effective damping ratio is, therefore, $0.6875/[(2)(1.23)] = 0.28$. Thus using this approximate second-order model we predict a peak overshoot of 40%.

One of the advantages of computer-aided control design and simulation programs is that one can easily verify how the system will respond in the time domain.

The dominant pole approach incidentally predicts a damping ratio of 0.26 and a damped natural frequency of 1.15 rad/s, so that in this case the differences are not pronounced. However you will have to wait till Chapter 6 to find out how to determine the dominant poles of a closed-loop system with transport delay.

Exercise Simulate the closed-loop step response of the system in Example 4.2 with $K = 2.2$ and measure the period of its oscillation and the percentage peak overshoot. Compare these figures with the 6 seconds period predicted using the resonant frequency correlation and the 40% overshoot predicted using the effective damping ratio method described above.

PROBLEMS

4.1 A system has an open-loop transfer function

$$G(s) = \frac{20s^2}{(1 + s)(1 + 2s)(1 + 4s)}$$

Sketch the direct Nyquist locus of the above system. Confirm your sketch with CODAS.

Using CODAS, find the frequencies for which the open-loop system exhibits a phase shift of $+90°$ and $0°$ respectively. Check the results analytically.

Use CODAS to find the frequencies for which the open-loop gain of the system is unity. Check the results analytically.

4.2 Draw the Bode gain asymptotes for the system in P4.1. Compare the values obtained for the frequencies obtained in P4.1 with the values predicted using the asymptotic plot.

4.3 A network used in a Wien bridge oscillator has a transfer function

$$G(s) = \frac{as}{1 + 3as + (as)^2}$$

where 'a' is time constant RC. Prove that the gain of the system is 1/3 when $\omega = 1/a$ and it has zero phase shift. Confirm the result using CODAS, and examine the time response of the system to an input signal of the form

$$x(t) = \cos\left(\frac{t}{a}\right)$$

4.4 Certain systems are *conditionally* stable, i.e. they are only stable for a range of gain values. Both below and above this range the systems are unstable. The system

$$G(s) = \frac{K(1 + 3s + 9s^2)}{s^3(1 + s)(1 + 0.2s)}$$

is conditionally stable. Sketch its frequency response, and then using CODAS draw its Nyquist diagram for $K = 0.05, 0.1$ and 0.5. For which of these plots is the closed-loop system stable? What are the limiting values for K for stability?

Use CODAS to draw the Nichols plot for each of the three cases considered and the associated closed-loop step responses.

Use the Nichols plot to determine what values of K give the best gain margin, the best phase margin and the best compromise. In each case examine the step response.

4.5 The open-loop transfer function of a position servo is

$$G(s) = \frac{K(1 + 0.1s)}{s(1 + 0.2s)(1 + 0.4s)}$$

Determine the value K which will satisfy the following design criteria:

(a) a phase margin better than $40°$;
(b) a gain margin better than 25 dB.
What is the resonant frequency of the resulting closed-loop system and what is its bandwidth?

4.6 Given the system whose open-loop transfer function is

$$G(s) = \frac{K(1.05 + s)\,e^{-0.3s}}{(1 + 0.2s)(1 + 0.5s)(1 + 0.6s)}$$

Find the gain K which will give the system a gain margin of 2.

What is the resulting peak magnification factor and resonant frequency. Using the techniques described in Section 4.12 to estimate the damped natural frequency and the effective damping ratio of the system, use the latter result to predict the peak overshoot of the system. Confirm your predictions by examining the closed-loop step response of the system.

FIVE

FREQUENCY DOMAIN COMPENSATOR DESIGN

5.1 INTRODUCTION

In the previous chapter we saw how to predict the closed-loop behaviour of a system from its open-loop frequency response and a certain amount of design was done using gain changes to meet design specifications. In general one is not free to change the gain of a system arbitrarily as the steady-state performance of the control system is determined by it. For example in a Type-0 system the Bode gain determines the steady-state accuracy of the loop to step changes in demand, and in a Type-1 system the Bode gain determines how accurately the system can follow a steadily changing demand signal (ramp input). Thus the designer is constrained on the choice of the overall system gain by steady-state design criteria. If the resulting system then meets all the dynamic requirements as well, the design is complete and can be implemented.

In most cases, however, the dynamic performance of the resulting control system will be unsatisfactory. The most obvious unsatisfactory situation that can arise is that the transient response is too oscillatory or at worst unstable. The control engineer must then find ways of improving the damping of the system whilst maintaining the target steady-state specifications.

Situations can occur where the system response is too slow. Here again the control engineer has at his disposal techniques to modify the control strategy, which will improve the speed of response of the system. Care must be taken, however, not to extend the bandwidth of the system excessively because the system may become too responsive. This can become a particular nuisance if noise is present in the demand signal as can happen with tracking servos.

Another example where the design of the control system may require further attention is when *measurement noise* is present. The term measurement noise refers to random effects such as mains pickup or other electromagnetic interference associated with sensors, commutator noise in tachogenerators, etc. Although the time domain behaviour of noise is indeterminate, in a particular application its frequency domain characteristics will be fairly well defined and the frequency range over which noise has significant strength can be determined by experiment. In the case of mains pickup, of

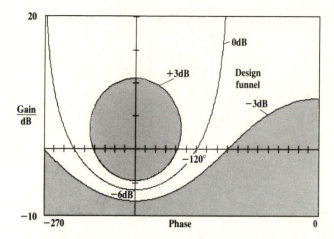

Figure 5.1 The design 'funnel'

course, there is only one significant frequency component. A control system behaves like a filter and generally speaking high frequencies are attenuated. If, however, the bandwidth of the control system is excessive relative to the noise bandwidth, then the control system must be redesigned to reduce its bandwidth and so make it less susceptible to measurement noise.

We saw in the previous chapter that the closed-loop dynamic behaviour of a control system was governed by the open-loop frequency response near the critical region on the Nyquist plot (i.e. the -1 point). The basic idea underlying all the techniques described in this chapter is to mould the frequency response of the system around the critical region to satisfy frequency domain stability and bandwidth criteria, but leave the low frequency response unaffected. On the Nichols plot this idea can be thought of as a 'design funnel' into which the frequency response must be channelled as shown in Fig. 5.1.

Control engineers have at their disposal a number of different tools for achieving the broad objectives outlined above. Most commonly they will involve placing a compensator (controller) in the forward path of the control system (cascade compensation) but sometimes additional internal feedback paths must be introduced. These techniques will be studied in detail in this chapter. It must be realized however that compensation techniques are a way of tailoring the control system behaviour. They are not a way of correcting poorly designed hardware or undersized actuators.

Finally all the methods presented here are only as good as the knowledge of the system frequency response or of its transfer function model. It is clearly a waste of time devising complex compensators subtly tuned to give superb results if the dynamics of the system under control are poorly defined or liable to change significantly in the short term. Process control systems are examples of systems where precise plant models are rarely known and where, furthermore, the plant behaviour varies with operating conditions. These types of control systems require a different approach and will be dealt with later in the book. Nevertheless the ideas and principles discussed in this chapter are very important in underpinning the material and approach used in process control.

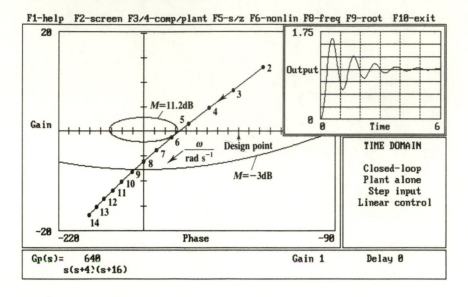

F1-help F2-screen F3/4-comp/plant F5-s/z F6-nonlin F8-freq F9-root F10-exit

Figure 5.2 Example system

Example problem Consider a position servo whose transfer function is

$$G_p(s) = \frac{K}{s(s+4)(s+16)} \tag{5.1}$$

Suppose the requirement is for a velocity error constant of 10 second^{-1}. This fixes K at 640 second^{-1}. Figure 5.2 shows the step response of the system and its associated Nichols plot. The response of the system is very oscillatory with a gain margin of 6 dB and a phase margin of only 17°. Its settling time is about 4 seconds. It has a resonance peak of 11.2 dB at 5.8 rad/s and its bandwidth is about 8.8 rad/s.

Let us suppose that a phase margin of 45° is required. We will see in the following sections how this requirement can be met using different types of compensator and what the consequences are of adopting one or another and how the design strategy affects the overall performance of the control system.

5.2 LAG COMPENSATION

If one were changing the gain to meet the above phase margin requirement, one would simply drop the system gain by about 8 dB (introduce an attenuation of about 2.5). The idea behind a lag compensator is to do just that but only in the region where the stability margin is defined and leave the system unaffected at low frequencies. In this way low frequency performance is maintained but the dynamic performance will be improved.

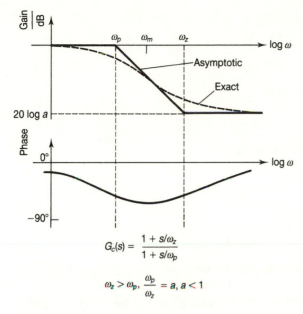

$$G_c(s) = \frac{1 + s/\omega_z}{1 + s/\omega_p}$$

$$\omega_z > \omega_p, \ \frac{\omega_p}{\omega_z} = a, \ a < 1$$

Figure 5.3 Bode diagram of lag compensator

We have already met a lag network or compensator in Chapter 3 where its transfer function was expressed as

$$G_c(s) = \frac{1 + a\tau s}{1 + \tau s}$$

As we are working in the frequency domain it is more convenient to express the transfer function of a lag compensator as

$$G_c(s) = \frac{1 + s/\omega_z}{1 + s/\omega_p} \tag{5.2}$$

where $\omega_z = 1/a\tau$ and $\omega_p = 1/\tau$ and the ratio of ω_p to ω_z is 'a', i.e.

$$\frac{\omega_p}{\omega_z} = a$$

The Bode diagram of a lag compensator is shown in Fig. 5.3. The lag compensator has unity low frequency gain and a gain of ω_p/ω_z at high frequencies, i.e. its high frequency gain is simply 'a'. If the numerator corner frequency, ω_z, is higher than the denominator corner frequency, ω_p, then the compensator will attenuate at high frequencies, i.e. $a < 1$.

By placing the compensator in cascade (in series) with the plant, the low frequency performance will be unaffected but the overall gain of the system will be reduced in the gain-crossover region to improve the stability margins. The amount of attenuation

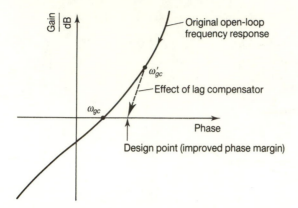

Figure 5.4 Compensation using a lag compensator

required to meet the design requirement determines the ratio of ω_p to ω_z and the only other decision that has to be made is the actual value (frequency) of either ω_p or ω_z.

As can be seen from Fig. 5.3, as well as introducing attenuation the lag compensator introduces a small phase lag at very low frequencies and at very high frequencies and a much larger phase lag between the two corner frequencies. Thus the introduction of a lag compensator in series with a system will not only reduce the overall gain but also increase the phase lag. On a Nichols plot this means mapping points on the uncompensated frequency response locus downwards and to the left. Hence the new gain-crossover frequency will be associated with a point on the original frequency response curve above and to the right of the design point. These points are illustrated in Fig. 5.4. The obvious effect is that the new gain-crossover frequency, ω'_{gc}, is lower than the original gain-crossover frequency, ω_{gc}.

By locating the corner frequencies of the compensator well away from (much lower than) the gain-crossover frequency, the amount of phase lag introduced, where attenuation is needed, is small. A typical design figure is for the compensator to introduce less than $5°$ phase lag at the design point. This figure translates approximately into placing the upper corner frequency, ω_z, of the compensator at a tenth of the frequency which will map into the design point.

We shall term the frequency associated with the design or specification point as, ω_s. In this example design frequency is the new gain-crossover frequency.

The validity that the one tenth rule ($\omega_s/\omega_z = 10$) introduces less than $5°$ phase lag can be verified as follows. The phase lag, ϕ, of the compensator is given by

$$\phi = \tan^{-1}(\omega_s/\omega_z) - \tan^{-1}(\omega_s/\omega_p) = \tan^{-1}(\omega_s/\omega_z) - \tan^{-1}(\omega_s/\omega_z/a)$$

$$= \tan^{-1}(10) - \tan^{-1}(10/a)$$

Consider a lag compensator with an 'a' value of 0.1; the phase lag introduced at the design frequency is thus

$$\phi = 84.29° - 89.43° = -5.14°$$

Larger values of 'a' will introduce less phase lag.

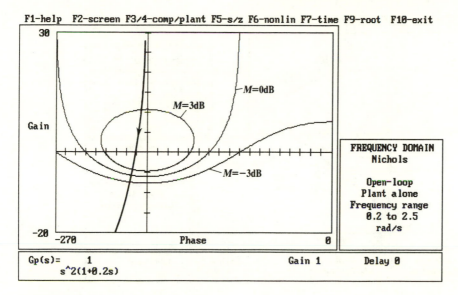

Figure 5.5 Nichols plot of a Type-2 system

There are systems where lag compensators are ineffective. Lag compensators, for example, cannot be used to compensate systems where there are two free integrators in the forward path transfer function, e.g.

$$G(s) = \frac{K}{s^2(1 + \tau s)}$$

The Nichols plot of this system is sketched in Fig. 5.5. In this particular example a lag compensator would have no effect at all as far as stabilization was concerned.

Exercise Use CODAS to draw the gain and phase characteristics of a lag compensator for 'a' values of 0.3, 0.2 and 0.1.

Example 5.1 Design of a lag compensator. Design a lag compensator for the example system (Eq. (5.1)) to satisfy a phase margin requirement of 45° whilst maintaining the K_v requirement of 10 second^{-1}.

SOLUTION For the example system in Fig. 5.2, the point where the original system exhibited a phase lag of 130° (i.e. $180° - pm - 5°$) will be mapped into the new gain-crossover point. This phase lag occurs at a frequency of 2.43 rad/s (see Fig. 5.6) and at that frequency the gain of the system is 3.4 (10.6 dB). Hence using the earlier rule of locating the upper corner frequency at one tenth of the design frequency, $\omega_z = 0.243$ rad/s and ω_p is $0.243/3.4 = 0.0715$ rad/s (i.e. ω_z/a). The resulting compensator is

$$G_c(s) = \frac{1 + s/0.243}{1 + s/0.0715}$$

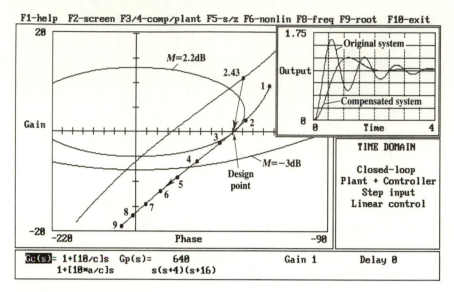

Figure 5.6 Lag compensation

Figure 5.6 shows the results of using the above compensator on a Nichols plot and the effect of the compensator on the transient response. It can be seen that the oscillation has been reduced and the peak magnification factor has been reduced to 2.2 dB. However the time to first peak has virtually been doubled. This is a consequence of reducing the gain-crossover frequency to 2.43 rad/s which is less than half its original value of 5.6 rad/s. All the other key frequencies will be lowered in a similar manner, hence reducing the overall bandwidth of the system to about 4.3 rad/s.

The method described here for designing lag compensators is essentially the method adopted long before there were hand calculators, never mind personal computers. Nevertheless it is quite a good method for most systems. However, a number of questions do arise about this method of design. Is it always best to aim for 5° phase lag; why not less and why not more; and what would the consequences be of adopting a different strategy? We will try to answer these questions later, but first let us consider a different type of compensator.

5.3 LEAD COMPENSATION

Lead compensation works on a different principle from lag compensation. Imagine that one could have a device that simply introduced phase advance. As far as the Nichols plot is concerned this would mean translating the frequency response locus to the right. In the case of the example system in Fig. 5.2, the desired phase margin would be achieved by incorporating a phase advance of 28°. The advantage of such a magic device over a gain reduction is that the stability margin would be improved without reducing the

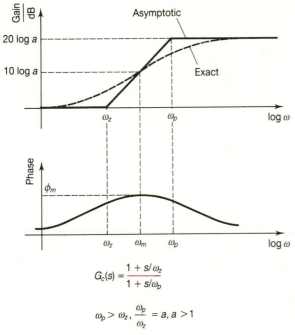

Figure 5.7 Bode diagram of a lead compensator

gain-crossover frequency and hence without sacrificing bandwidth and speed of response. Another advantage of a device that operated on this principle would be the possibility of stabilizing 'hopeless' cases such as systems with a double integrator which cannot be tackled with a lag compensator.

It is physically impossible to realize a compensator which produces a constant phase advance for all frequencies, nor is such a device necessary. All that is required is to produce a network that will introduce phase lead where it is needed, i.e. near the critical region. Practical phase advance networks or lead compensators, which produce phase lead over a frequency range, are easily realizable. However their gain characteristic is not constant; it also changes with frequency and this must be taken into account when using them as compensators.

The transfer function of a realizable lead compensator has exactly the same form as the lag compensator (Eq. (5.2)) except that the numerator corner frequency, ω_z, is **less** than the denominator corner frequency, ω_p. In other words this time the ratio, ω_p/ω_z, is greater than 1, i.e. $a > 1$. The Bode diagram of a lead compensator is shown in Fig. 5.7.

At low frequencies the phase lead contribution of the numerator term will dominate and at high frequencies the lag contribution of the denominator term will tend to cancel the effect of the lead. The maximum phase advance, ϕ_m, occurs at a frequency of ω_m, exactly half-way between the two corner frequencies as drawn on the logarithmic plot, i.e.

$$\log(\omega_m) = \frac{\log(\omega_z) + \log(\omega_p)}{2}$$

or

$$\omega_m = \sqrt{\omega_p \omega_z} \tag{5.3}$$

Since $\omega_p/\omega_z = a$, one can easily deduce that

$$\omega_p = \omega_m \sqrt{a} \tag{5.4(a)}$$

and

$$\omega_z = \omega_m / \sqrt{a} \tag{5.4(b)}$$

We will rewrite the basic form of the compensator of Eq. (5.2) in terms of ω_m by eliminating ω_p and ω_z using Eqs (5.4(a)) and (5.4(b)), i.e.

$$G_c(s) = \frac{1 + \sqrt{a}(s/\omega_m)}{1 + (1/\sqrt{a})(s/\omega_m)} \tag{5.5}$$

The square of the gain of the compensator is thus

$$|G_c(j\omega)|^2 = \frac{1 + a(\omega/\omega_m)^2}{1 + (1/a)(\omega/\omega_m)^2} \tag{5.6}$$

Now at its centre frequency ($\omega = \omega_m$) the gain of the compensator can be found by substituting for ω in Eq. (5.6) which results in:

$$|G_c(j\omega_m)|^2 = \frac{1 + a}{1 + 1/a} = a \tag{5.7}$$

i.e. the actual gain is $\sqrt{a}$ or $10 \log(a)$ in dB. The phase advance of the network at this frequency is a maximum and can be expressed as:

$$\phi_m = \tan^{-1}(\omega_m/\omega_z) - \tan^{-1}(\omega_m/\omega_p) \tag{5.8}$$

Taking the tan of both sides of Eq. (5.8) one obtains

$$\tan(\phi_m) = \frac{\omega_m/\omega_z - \omega_m/\omega_p}{1 + \omega_m^2/(\omega_p \omega_z)} \tag{5.9}$$

On substituting for ω_m from Eq. (5.3), the maximum phase lead can be expressed as

$$\tan(\phi_m) = \tfrac{1}{2}(\sqrt{\omega_p/\omega_z} - \sqrt{\omega_z/\omega_p}) = \tfrac{1}{2}\left(\sqrt{a} - \frac{1}{\sqrt{a}}\right) = \frac{\tfrac{1}{2}(a-1)}{\sqrt{a}} \tag{5.10}$$

An alternative representation that is found in most text books is

$$\sin(\phi_m) = \frac{a-1}{a+1}$$

This result follows directly from Eq. (5.10) and can be proved easily by drawing the associated triangle. Rearranging the above expression in terms of 'a' we have

$$a = \frac{1 + \sin(\phi_m)}{1 - \sin(\phi_m)} \tag{5.11}$$

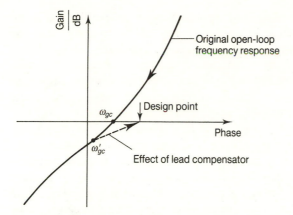

Figure 5.8 Compensation using a lead compensator

In a lag compensator, the amount of attenuation it offers is of primary interest and the phase lag it produces is discounted by centring the compensator well away from the design point. With a lead compensator, however, its phase characteristic is what the designer is after and its gain characteristic is a nuisance that he cannot wish away.

The effect of a lead compensator can be seen by referring again to a Nichols plot (Fig. 5.8). The introduction of the lead network in series with the plant maps points to the right (phase advance) and upwards (gain increase). Here it can be seen that the introduction of phase lead has the effect of mapping higher frequencies to the design region. The effect of the lead compensator is thus to increase the system bandwidth and hence the overall speed of response of the closed-loop system.

With a simple lead network one cannot have more than 90° phase lead and even this would require an infinite 'a' value (i.e. $\omega_p \rightarrow \infty$), but is there any other limitation on the value of 'a' and hence the amount of phase lead that one can introduce? One answer to this question is that, in general, it is undesirable to have too big a value of 'a' as it tends to increase the bandwidth of the system excessively, so making it very responsive and susceptible to any noise in the reference input or to measurement noise. However, one other important factor is the transient behaviour of the lead compensator to step inputs. If one applies a unit step to a lead compensator, the instantaneous response of the compensator is 'a' units (i.e. a sudden jump of 'a'). It is precisely this extra control effort that has the beneficial effects of improving the speed of response. However, the apparent improvements in response predicted by linear models when employing larger and larger 'a' values are simply not realized in practice. The reason is that large transient excursions of signals will be clipped to a smaller value by supply rails, and the achievable control effort will be constrained by actuator limits, etc. Hence, as a general rule, designs should restrict the amount of phase advance introduced by the lead compensator to less than 55° ($a \leqslant 10$).

Design of Lead Compensators

Designing a phase lead compensator is a little more tricky than a lag compensator.

Now the frequency that is going to be mapped to the design point or specification point is termed ω_s. In the previous discussion and examples this frequency was the gain-crossover frequency, but it could be any other frequency depending on the design criterion, e.g. phase-crossover frequency, resonant frequency, -3 dB bandwidth frequency, etc. The usual way of going about the design of a lead compensator is to exploit its phase advance fully, i.e. the specification frequency is made equal to the frequency where the lead compensator exhibits maximum phase advance ($\omega_m = \omega_s$).

The first step is to pick ω_s. How do you go about this? Basically it is a pure guess, but a little common sense helps and remember that a lead compensator introduces some gain as well as phase lead. Don't pick a point where the required phase advance is greater than about 55° for the reasons described earlier. But equally do not go straight for 55°, as generally it is preferable to design a compensator that introduces as little phase advance as possible. Suppose that you have chosen a point and you can see how much phase lead you require. Now you can use Eq. (5.11) to find 'a' and hence the gain of the compensator using Eq. (5.7). If this maps you on to your design point, then the design is finished, otherwise you must guess another point until the design criterion is satisfied.

> **Example 5.2 Design of a lead compensator** Design a lead compensator for the system in Fig. 5.2 to meet a 45° phase margin requirement using the above technique.

SOLUTION Looking at Fig. 5.2, it would seem that an initial guess of 7 rad/s for ω_s is reasonable as it lies to the left and below the target design point. The phase advance required is about 39° with a gain of less than 4 dB. Using Eq. (5.11) we find that an 'a' value of about 4.8 is required to obtain this amount of phase advance. The gain of the compensator is about 6.4 dB (i.e. $10 \log(a)$). As a result the point at 7 rad/s on the original locus would be mapped above the 0 dB line. Hence the new gain-crossover frequency will be higher than 7 rad/s using these design values. Since the phase lag of the original system is increasing uniformly and as the design method utilizes maximum phase advance, points at frequencies above 7 rad/s will be mapped to the left of the design point. Hence the compensated locus will pass inside the design point, and the design requirement will not be satisfied. The only way actually to meet the phase margin requirement using this first attempt would be to reduce the Bode gain of the system by 2.4 dB which would compromise the specification of the velocity error constant.

One can go on trying higher and higher frequencies, but as the design frequency, ω_s, goes up, the required phase advance goes up too and hence a higher 'a' value is needed. The problem is that the higher 'a' value results in a compensator with a higher gain at the design frequency, and so we start chasing our own tail to some extent. Table 5.1 shows the required phase advance, the maximum allowable gain of a lead compensator, the required 'a' value and the actual gain of the lead compensator as a function of frequency for the system in Fig. 5.2. As can be seen at 10 rad/s the required phase advance is more than 55° and the gain introduced by the compensator is still fractionally too much. However no-one can design to within 0.1 dB, and furthermore from a practical point of view a slight adjustment of gain will be needed anyway to trim for inaccuracies in the model and the

Table 5.1

Frequency (rad/s)	Required phase advance	Allowable gain	'a'	Compensator gain
8	45.0°	6.0 dB	5.83	7.65 dB
9	50.4°	8.1 dB	7.71	8.87 dB
10	55.2°	10.0 dB	10.31	10.13 dB

implementation of the compensator. Nevertheless, this example points out the difficulties in designing a lead compensator.

Finally we can calculate the transfer function of the actual compensator for $\omega_m = 10$ rad/s and $a = 10.31$ using the relationships of Eqs (5.4(a)) and (5.4(b)) to find that $\omega_p = 32.1$ rad/s and $\omega_z = 31.1$ rad/s, i.e.

$$G_c(s) = \frac{1 + s/3.1}{1 + s/32.1}$$

The effect of the compensator is illustrated in Fig. 5.9. The results are quite staggering compared to those obtained for the lag compensator in Fig. 5.6. The settling time is well under a second compared to about 2 seconds for the lag compensator. The bandwidth of the system, however, has been increased to over 17 rad/s compared to that of the original system at just under 9 rad/s and that of the lag compensated system at 4.3 rad/s. This very large increase in bandwidth makes the system much more sensitive to measurement noise.

However, there is one other point that makes this design less than totally

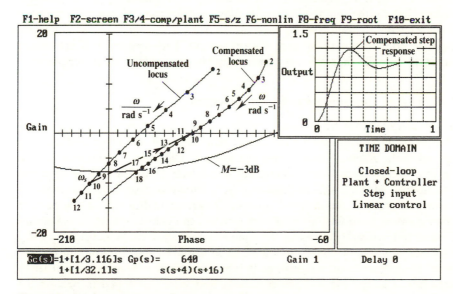

Figure 5.9 Lead compensator using maximum phase advance

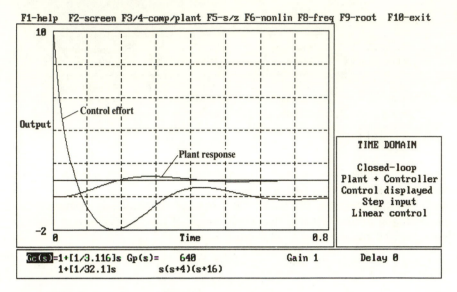

Figure 5.10 Control effort generated by a lead compensator

satisfactory. In order to achieve what appears to be a superb transient performance, one has generated a huge amount of control effort. This is shown in Fig. 5.10 but the vertical scale has had to be enlarged considerably to show it, and that is why the plant response looks so different from the previous figure.

In systems with no transport delays it is possible with an infinite amount of control effort to achieve a settling time of zero, but in reality the change one can achieve is dictated by the size of the signals that can be generated and ultimately by the amount of control effort that can be 'pumped' into the plant. Compensators can only massage the performance of the overall system, they cannot turn a system that has a natural response time of 1 hour into one that responds in 1 second. Thus any design that changes natural frequencies by more than a factor of two should be viewed with great scepticism.

Improved Method of Designing Lead Compensators

In the previous section the 'traditional' design method for lead compensators was covered. The problem with the method is that it relies on trial and error and leads to rather large amounts of phase lead being introduced. The basic assumption in the approach is that every ounce (sorry, degree) of phase lead going should be squeezed out of the compensator. This approach places the compensator so that it contributes maximum phase advance at the design frequency. Referring to Fig. 5.7, it can be seen that the phase curve is quite flat in the centre, whereas the gain curve has a maximum slope at its centre frequency. By offsetting the compensator centre frequency, i.e. $\omega_s < \omega_m$, virtually the same amount of phase advance can be achieved but the amount of gain introduced by the compensator is significantly less at the design frequency. In order to turn this idea into a design method the equations that were developed in Section 5.3 will have to be reviewed.

The gain and phase of the compensator can be expressed in terms of a non-dimensional frequency, b, which is the ratio of the actual frequency, ω, to the centre frequency, ω_m, i.e.

$$b = \omega/\omega_m \tag{5.12}$$

Using the transfer function expression of Eq. (5.5), and substituting $s = j\omega$ and introducing the parameter, b, the complex gain function of the compensator may be expressed as:

$$G_c(a, b) = \frac{1 + jb\sqrt{a}}{1 + jb/\sqrt{a}} \tag{5.13}$$

From which it is easy to obtain the gain and phase shift of the compensator in terms of a and b. In fact

$$|G_c|^2 = \frac{1 + b^2 a}{1 + b^2/a} \tag{5.14}$$

and

$$\tan(\phi) = \frac{b}{1 + b^2} (\sqrt{a} - 1/\sqrt{a}) \tag{5.15}$$

As can be seen the above equations reduce to the special cases of Eqs (5.7) and (5.10) when $b = 1$. These two equations represent a family of curves in the parameters a and b, and can be conveniently drawn on one graph using Nichols plot axes as shown in Fig. 5.11. These curves will be referred to as 'D-contours' as they resemble a set of D's. Superimposed on this plot is the data of Table 5.1 and some additional data representing required phase lead and maximum allowable gain to satisfy the 45° phase margin requirement of the example system.

It is now clear to see that the design adopted in the previous section ($b = 1$) is right on the limit of the allowable phase advance for a lead compensator. By choosing a lower design frequency and a lower value of 'b' (i.e. using the idea of the off-centred design frequency), we can design a lead compensator which will satisfy the design criterion and introduce far less phase advance (i.e. a lower value of 'a'). It can be seen from Fig. 5.11 that by choosing 'b' at about 0.6, the required 'a' value is about 6 and the design frequency, ω_s, is 7.5 rad/s. The centre frequency of the compensator (from Eq. (5.12)) is ω_s/b, i.e. at 12.5 rad/s.

Rewriting Eq. (5.5) in terms of ω_s we obtain

$$G_c(s) = \frac{1 + \sqrt{a}\,(b/\omega_s)s}{1 + (1/\sqrt{a})(b/\omega_s)s} \tag{5.16}$$

In CODAS, transfer functions can be expressed using parameters rather than numerical values. Hence in CODAS, the compensator can be defined using parameters. 'a' and 'b' can be left as they are, but 'c' will have to be used to represent the design frequency ω_s. Thus the equation as used in CODAS for the compensator becomes

$$G_c(s) = \frac{1 + [b/c * \text{sqr}(a)]s}{1 + [b/c/\text{sqr}(a)]s} \tag{5.17}$$

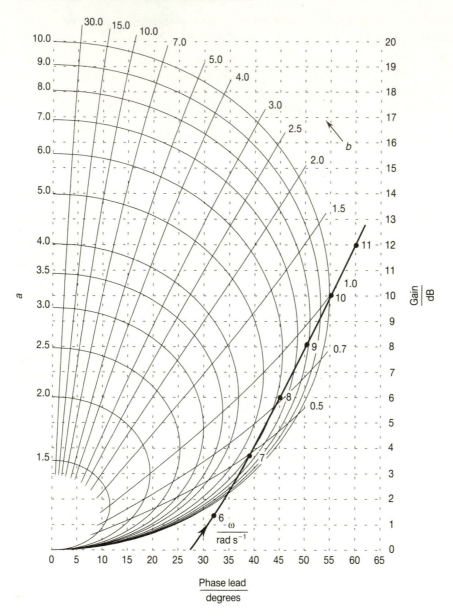

Figure 5.11 Characteristic curves of a lead compensator

The actual transfer function of the compensator is

$$G_c(s) = \frac{1 + s/5.1}{1 + s/30.6}$$

Figure 5.12 shows the results of applying a phase lead compensator using the values of 'a', 'b' and 'ω_s' determined above. The results should be compared with those of Fig.

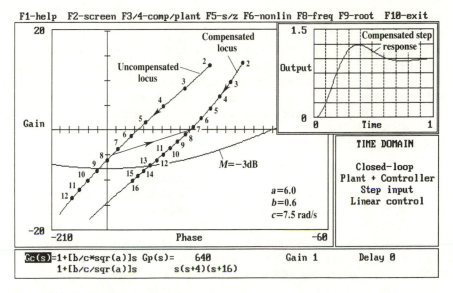

F1-help F2-screen F3/4-comp/plant F5-s/z F6-nonlin F8-freq F9-root F10-exit

Figure 5.12 Lead compensation using an 'off-centred' compensator

5.9. The bandwidth has been reduced considerably (from 17.5 rad/s to 13 rad/s), but the settling time is hardly any different. The response is in fact slightly slower, but it is better damped and so the settling time is approximately the same.

Exercise Using CODAS compare the compensated closed-loop Bode gain plots of the plant defined by Eq. (5.1) for each of the three compensators, i.e. the lag compensator of Example 5.1, the lead compensator of Section 5.3 and the improved compensator defined above. Verify the figures for the -3 dB bandwidths.

Designs of lead compensators using the technique described here can be facilitated by using an overlay. The Nichols plot of the open-loop system is drawn on graph paper and the curves of Fig. 5.11 should be reproduced to the same scale as the Nichols plot on a transparency. This transparency can be overlaid on the plot and slid along it and the values of 'a' and 'b' read off until a suitable combination is found. Using a computer makes it easier to superimpose the required gain and phase characteristics as was done in Fig. 5.11. In CODAS, however, the values of 'a' and 'b' are reported with the 'free-wheeling' cursor and so it is very easy to design a lead compensator with the package, based on the above technique.

5.4 A SECOND LOOK AT LAG COMPENSATORS

In Section 5.2 lag compensators were considered and a number of questions were left unanswered. The method for designing lag compensators used a rule of thumb that the amount of phase lag introduced by the compensator should be about 5° and this translated roughly into choosing the compensator upper corner frequency as one tenth of the design frequency ($\omega_z = \omega_s/10$).

Suppose the amount of phase lag chosen was less than 5°. This would mean that a higher design frequency could be utilized and so presumably would result in a system which had a higher bandwidth and a faster settling time. However the ratio ω_z/ω_s would be much smaller and the centre frequency of the compensator would be much lower which would introduce very low frequency dynamics into the system.

By making the centre frequency of the lag compensator higher, a lower design frequency will have to be chosen to allow for the greater phase lag introduced by the compensator. However the compensator dynamics will be faster and so the overall result on the speed of response and the settling time may be beneficial. Moving the compensator frequencies closer to the design frequencies creates a technical problem in design because both the gain and the phase of the compensator are changing more rapidly and one is entering a trial and error situation very similar to that described for the phase lead compensator. In fact with a little thought it is obvious that the contours of Fig. 5.11 developed for phase lead compensators could equally well be applied to a **lag** compensator, but instead of reading off the vertical axis as a gain it is interpreted as a **loss** (attenuation) and the horizontal axis is interpreted as a **lag** instead of a lead.

The only other problem is that if the form of Eqs (5.16) or (5.17) were adopted as the transfer function of the lag compensator, we would have to read off the 'a' values and take the reciprocal before using them. Rather than doing that, the transfer function of a **lag** compensator will be defined as

$$G_c(s) = \frac{1 + (1/\sqrt{a'})(b/\omega_s)s}{1 + \sqrt{a'}(b/\omega_s)s} \tag{5.18}$$

where 'a'' is greater than 1 and can be read off the D-contours directly. That is, 'a'' is the reciprocal of the 'a' used in Eq. (5.16) and so represents the ratio of ω_z to ω_p which is greater than unity for a lag compensator.

In order to test the consequences of different design strategies we will pick three design frequencies which will result in three quite different lag compensators for the example system that has been used throughout this chapter. At each of the three frequencies the attenuation and phase lag that is needed to meet the design specification is obtained. By plotting the required attenuation and lag on a set of D-contours, values of 'a'' and 'b' may be obtained. Table 5.2 summarizes the data. The rightmost column shows how the three strategies compare with the one-tenth rule discussed in Section 5.2. As you can see the traditional design approach falls between strategies 2 and 3.

Table 5.2

Design strategy	Design frequency rad/s	Required phase lag	Required attenuation	'a'	'b'	ω_s/ω_z
1	0.8	30.8°	21.76 dB	14	5.6	1.5
2	1.5	19.1°	15.87 dB	6.7	6.2	2.4
3	2.7	1.4°	9.60 dB	3.0	45'	26

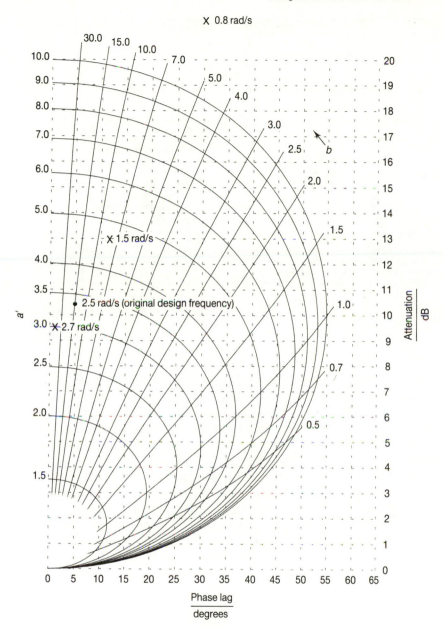

Figure 5.13 Alternative design strategies for a lag compensator

Figure 5.13 shows the points of Table 5.2 plotted on a D-contour as well as the point corresponding approximately to the original design ($\omega_s \approx 2.5$ rad/s, $a' = 3.4$ and $b = 18$). Design strategy 1 allows for some $31°$ of phase lag at the design frequency and so requires a considerable amount of attenuation ($a' = 14$). Strategy 3 on the other hand allows only $1.4°$ phase lag and therefore needs much less attenuation ($a' = 3$). The

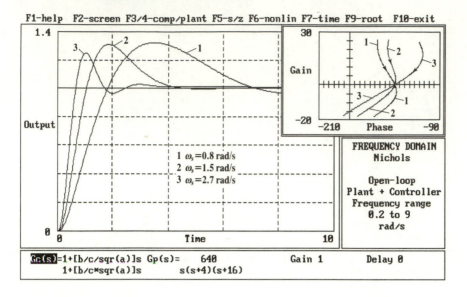

Figure 5.14 Effect of different lag compensators

transient responses and the associated Nichols plots of the overall system with these three compensators are shown in Fig. 5.14. It is very clear that strategy 1 results in a very sluggish response, and strategy 3 results in what seems a very nice response with a very fast settling time. The original design falls between 2 and 3 in terms of its response.

Closer examination of response 3 reveals that it has a very long 'tail', i.e. there is a very slow mode which does not allow the system to fully settle for a very long time. In this example the effects are of little practical significance and so it is safe to choose a compensator with a lower centre frequency than the one that is obtained using the traditional design approach. The reason that the slow mode is not observed is that the system is Type-1 and the response at low frequencies is so dominated by the term involving the integrator that the slow transient introduced by the compensator is not significant. In other cases and particularly with Type-0 systems the traditional design strategy produces this very long tail and a design which chooses a lower design frequency produces a better overall response.

Exercise Consider a system whose open-loop transfer function is

$$G_p(s) = \frac{(1 + 0.5s)\,e^{-0.3s}}{(1 + 0.2s)(1 + 0.5s)(1 + 0.6s)}$$

The specification for this system requires a steady-state loop gain of 4 and a phase margin of 45°.

(a) Is compensation necessary for this system and if so is it possible to use a lead compensator for this system?

(b) Design lag compensators which will contribute 5° and 20° phase lag at the new gain-crossover frequency.
(c) Which compensator gives the best 2% settling time?

5.5 LAG–LEAD COMPENSATION

Both lag and lead compensators affect the bandwidth of the system. By combining lag and lead compensators it is possible to design a compensator which has little or no effect on the overall bandwidth of the system. The design of lag–lead compensators was at one time considered a 'black art', but in fact it is quite a simple procedure which builds on the principles that have been established for the individual lead and lag compensators in the earlier sections.

To illustrate the methodology, consider designing a compensator for the example system that has been used throughout this chapter to meet a phase margin criterion of 45° as before, but leave the gain-crossover frequency **unaffected**. Clearly, using a lead compensator or a lag compensator on its own, it is impossible to satisfy both design criteria. However, using a lag–lead compensator both objectives can be fulfilled. The idea behind the design approach (Fig. 5.15) is to use a lag compensator to map the gain-crossover point down and to the left on the Nichols plot to some intermediate point, and then to use a lead to map that point to the right and up to satisfy the phase margin design criterion. The only question is where should the intermediate point be located? The simplest way is to have as the intermediate target a line which is about 5° lower than the phase shift of the system at the original gain-crossover frequency. Then work 'backwards' from the ultimate design point to find a lead compensator that will place it on this line and which results in an acceptable 'a' value.

Following these principles the figures in Table 5.3 are arrived at for the system whose frequency response is shown in Fig. 5.2. Figure 5.16 shows the design process. The lead element is designed first by finding the 'a' and 'b' values required to map the final design point back to a point which is 5° to the left of the original gain-crossover point. By trial and error a point is found where the 'a' value associated with the lead

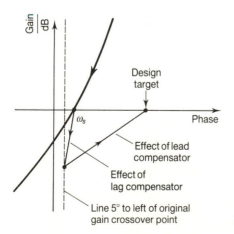

Figure 5.15 Design principle of a lag–lead compensator

Table 5.3

Compensator element	a	b	ω_s	ω_z	ω_p
lead	3.54	0.94	5.5	3.12	11.0
lag	1.89	5.87	5.5	1.29	0.682

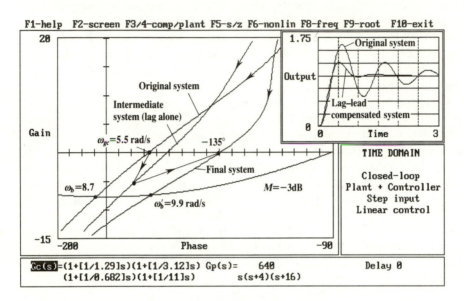

Figure 5.16 Lag–lead compensator design

element is minimal. Finally working back from that point to the original gain-crossover point the lag element is designed. Clearly the design frequency, ω_s, in this example is the gain-crossover frequency. The lag–lead compensator is thus

$$G_c(s) = \frac{(1 + s/1.29)(1 + s/3.12)}{(1 + s/0.682)(1 + s/11)}$$

Figure 5.16 also shows the time response of the overall compensated system contrasted with the original step response. It is very evident that the speed of the compensated system is very similar to that of the original system (though there has been a slight increase in bandwidth) but the damping has been considerably improved. The above compensator is not unique and other compensators which meet the required specification can be designed by taking a slightly different intermediate point.

As a final example we will consider a more complicated design.

Example 5.3 Design for peak magnification factor and bandwidth A system has an open-loop transfer function

$$G_p(s) = \frac{4}{s(1 + s)(1 + s/3)}$$

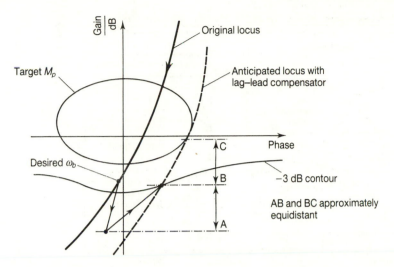

Figure 5.17 Design principle of a lag–lead compensator which must satisfy two criteria

Design a compensator to satisfy a requirement for a closed-loop system with a peak magnitication factor of 1.7 dB and a bandwidth of 2.6 rad/s.

SOLUTION There are different ways of approaching this problem and readers may like to think of alternative strategies. Here is one method that provides a feasible compensator quite quickly.

The design method is illustrated in Fig. 5.17. The first step is to introduce a lag compensator to map the point at 2.6 rad/s below the -3 dB contour in anticipation of the effect of the lead compensator. However, the bandwidth alone is not enough and so an eye has to be kept on the M_p criterion too. The assumption can be made that in the region of interest the gain and phase shift introduced by the lead compensator is more or less constant and so the shape of the compensated locus will be similar to that of the original system but merely slid to the right. Bearing these points in mind the lag compensator should map the point corresponding to 2.6 rad/s about the same distance below the -3 dB contour as the distance of the -3 dB contour from the target contour. Then when the lead compensator is introduced it will approximately meet both the bandwidth and peak magnification factor.

Proceeding in this manner for the actual system, a lag compensator is designed which maps the 2.6 rad/s point $5°$ to the left and to where the gain is about -16 dB (see Fig. 5.18). The details of the lag compensator are an 'a' value of 3 and a 'b' value of 14. A lead compensator is then designed to map the design point on to the -3 dB contour taking a conservative view of the amount of lead that is required, i.e. $a = 10$. The result is a compensator which exceeds the design requirement. Finally a slight adjustment is made to relax the amount of phase lead ($a = 7.7$, $b = 1.16$) and both design criteria are met. Figure 5.18 shows the final frequency response locus which meets both design specifications and in addition shows the transient response of the compensated system.

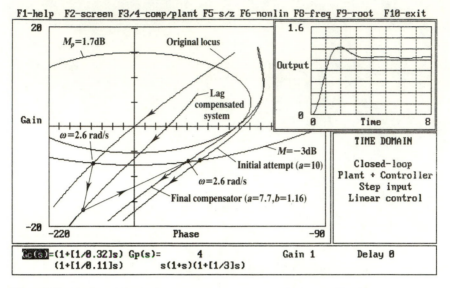

Figure 5.18 Design of lag–lead compensator meeting two design specifications

5.6 DOUBLE-LEAD COMPENSATOR

There are situations where neither a simple lead compensator nor a lag–lead compensator will stabilize a system adequately. One example of such a system is

$$G_p(s) = \frac{10}{s^2(1 + 0.2s)}$$

where the required phase margin is 35°.

A lag compensator is totally out of the question for a Type-2 system as the phase shift of the system is always less than $-180°$, so a lag compensator could not stabilize it, never mind meet the design specification. A lead compensator would stabilize it, but there is not enough phase advance available to satisfy the stability criterion without using absurd values of components ('a' values far in excess of 10). A lag–lead compensator is of no benefit because it could only degrade the modest stability margins achievable by a highly optimized lead compensator. One last desperate choice is two cascaded lead compensators, namely

$$G_c(s) = \frac{(1 + s/\omega_z)^2}{(1 + s/\omega_p)^2}$$

This compensator will give twice as much phase advance as a simple lead compensator, but the gain introduced by it will be the square of the gain of that introduced by the simple device (twice the gain on a dB scale). The technique for designing a compensator of this follows on from the previous methods in a straightforward manner. Basically it is a question of designing two simple lead compensators in cascade each of which contributes half the required phase advance and half the gain on a logarithmic scale.

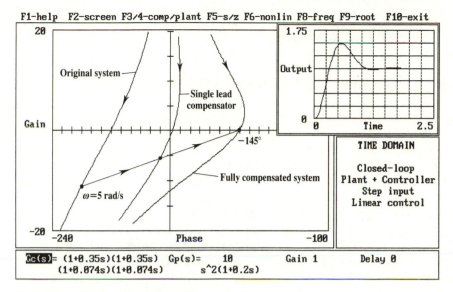

F1-help F2-screen F3/4-comp/plant F5-s/z F6-nonlin F8-freq F9-root F10-exit

Figure 5.19 Double-lead compensation

It is left as an exercise for the reader to complete a design for the above system, but Fig. 5.19 gives some clues as to how to approach it using CODAS.

In order to overcome the limitation of a single lead compensator, the duty has been 'shared' by two lead compensators to achieve a much greater phase advance. Nevertheless the points made earlier in discussing the limitations of lead compensators have even greater validity. The behaviour of the system in practice may be quite different from that predicted by simple linear models because the control effort will simply not be available for all sizes of disturbance and for all transient situations. The controller will work best when the system is near to equilibrium to start with and the disturbances or demand changes are small so that sufficient control effort can be generated to maintain an equilibrium situation in the way predicted by linear models.

Summary of Cascade Compensators

Lead compensators increase the system bandwidth and hence the speed of response of the closed-loop system. Care must be taken in designing them not to increase the bandwidth excessively and so make the system overly responsive and susceptible to reference input noise. Another factor that must be borne in mind is that the behaviour predicted by linear models may not be achieved in practice because of signal limiting. Lead compensators are the only effective way of stabilizing systems which are unstable or marginally stable in open loop. Lead compensators are not effective in systems whose phase lag is increasing rapidly with frequency such as in systems with transport delays.

Lag compensators reduce the speed of response of the system, and can introduce extremely slow modes of dynamic behaviour if care is not taken in their design.

Lag–lead compensators are used to compensate systems where there is a requirement to maintain the bandwidth of the system or to achieve a bandwidth that cannot easily be met by using a lag or a lead compensator on its own.

Double lead compensators are used on systems where lead compensators alone cannot meet the stability specification.

5.7 INTERNAL (PARALLEL) COMPENSATION

The compensators that have so far been discussed are placed in cascade (series) with the system to be controlled. Another method of compensation is to use internal feedback or parallel compensation in the manner shown in Fig. 5.20. The most common application of internal feedback is in servo systems in the form of velocity or acceleration feedback. The velocity signal is commonly generated by means of a tachogenerator connected to the load or by means of an integrated tacho within the servomotor. A feedback signal proportional to acceleration is usually synthesized by using a network that produces an approximate derivative of the velocity signal.

In order to examine the effect of internal feedback we will consider a simple position servo with rate (velocity) feedback. $G_p(s)$ represents the transfer function of the motor, its drive amplifier and the load. As an example consider the case where the servo is controlling a large inertia, J, with viscous friction, C, and where the motor dynamics are insignificant. The transfer function relating the position of the load, $C(s)$, and drive amplifier input signal, $E_2(s)$, will then be of the form

$$G_p(s) = \frac{K}{s(C + Js)}$$

or if the viscous friction is negligibly small (i.e. $C/J \to 0$), $G_p(s)$ becomes

$$G_p(s) = \frac{K_1}{s^2}$$

As was mentioned earlier the rate signal is usually obtained from a tachogenerator or derived explicitly by differentiating the position signal. In either event the feedback transfer function is of the form

$$H(s) = \beta s$$

The above transfer function is of course improper but the accompanying lags of a real

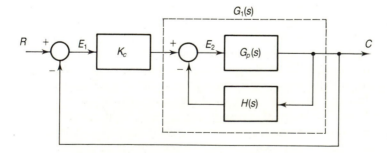

Figure 5.20   Internal feedback compensation

(proper) differentiator will make little difference to the arguments which follow. Let us first consider the system with no velocity feedback, i.e. $\beta = 0$. The open-loop transfer function of the system is

$$G(s) = \frac{K_c K_1}{s^2}$$

and hence the characteristic equation is

$$s^2 + K_c K_1 = 0$$

which is the equation of simple harmonic motion and the closed-loop system will oscillate sweetly at a frequency of $\sqrt{K_c K_1}$ rad/s. In other words the damping ratio of the system is 0.

Now velocity feedback is applied. The effective plant transfer function becomes

$$G_1(s) = \frac{K_1}{s(s + K_1\beta)}$$

and the characteristic equation becomes

$$s^2 + K_1\beta s + K_c K_1 = 0$$

which is the equation of a damped second-order system, and the damping term is due directly to the effect of the velocity feedback. The idea is thus very simple and seemingly effective. There is however a penalty to be paid, and that is that the system is now only Type-1, whereas initially without any compensation it was Type-2. This means that there will be a finite positional error when there are ramp demands to the servo. No doubt that seems a small sacrifice to pay for stabilizing the system. After all what is the use of a Type-2 system if it is unstable! This comment is fair enough, but we saw earlier that systems of this nature could be stabilized with a lead compensator without **any** sacrifice in the steady state performance of the system.

Before we take the discussion any further let us consider a slightly less troublesome system, namely

$$G_p(s) = \frac{K_1}{s(1 + \tau_m s)}$$

which might represent a servo system where the frictional component is significant. The effective plant transfer function with velocity feedback becomes

$$G_1(s) = \frac{K_1}{s(1 + K_1\beta + s\tau_m)}$$

This time there has been no change in the Type number, but there has been a reduction of the velocity error constant, K_v. The original K_v was $K_c K_1$, but now it is $K_c K_1/(1 + K_1\beta)$, a reduction directly attributable to the velocity feedback.

Use of velocity feedback certainly requires more auxiliary equipment and reduces the velocity error constant of the overall system. What then is the advantage of using it as a means of improving the dynamic characteristics of the system? Provided that the velocity error constant is not an overriding consideration, the main advantage of velocity feedback is that it improves the damping of the system without extending its bandwidth and in fact lowers it to some extent by virtue of the increased damping. This fact can be seen by examining the characteristic equation of the two systems that have been considered, where the natural frequency is unaffected by the introduction of velocity feedback.

Another more subtle reason for using tacho feedback directly round the motor is that any changes of motor speed are detected immediately and the resultant damping effect acts very quickly. With forward path compensation, the compensator also reacts to changes in speed because it acts on an error signal which includes a component proportional to the position of the load and hence indirectly to its speed. However, the signal that the cascade compensator operates on will, generally, have been subjected to further lags due to the dynamics of the load and the dynamics of the position transducer. Hence the speed of response of the control system may be slower than for systems stabilized by velocity feedback.

Design of Internal Compensators Using Inverse Nyquist Plots

For the simple example considered in the previous section, the amount of velocity feedback required to achieve a certain damping can be obtained by straightforward calculation. For more complex cases, or where the design criteria are less direct, or where only experimental frequency response data is available, the design of internal feedback compensators must be done graphically. The inverse Nyquist plot is the ideal vehicle for designing such compensators.

The plant transfer function and internal feedback transfer function can be lumped together and represented as an effective transfer function, $G_1(s)$, (Fig. 5.20) which is given by

$$G_1(s) = \frac{G_p(s)}{1 + G_p(s)H(s)}$$

or taking the reciprocal

$$G_1(s)^{-1} = G_p(s)^{-1} + H(s)$$

Thus in the inverse Nyquist plane the effect of the internal compensator simply adds to the inverse frequency response of the plant to obtain the inverse frequency response of the system as seen by the controller. In other words if the magnitude and phase of the plant frequency response are known, the resultant inverse behaviour of the effective plant transfer function, $G_1(s)$, is obtained by the **vector** addition of $H(j\omega)$ at each frequency (Fig. 5.21).

If one further takes into account the controller gain, K_c, the overall inverse forward path transfer function becomes

$$G(s)^{-1} = [K_c G_p(s)]^{-1} + H(s)/K_c$$

The design process is then essentially a simple matter. The frequency response $[K_c G_p]^{-1}$ is plotted and a decision is made regarding where to map a particular frequency point. The vector which performs the desired mapping is $H(j\omega)/K_c$, from which $H(j\omega)$ can be deduced.

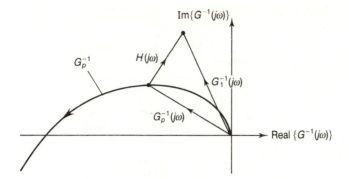

Figure 5.21 Effect of internal feedback using the inverse Nyquist plot

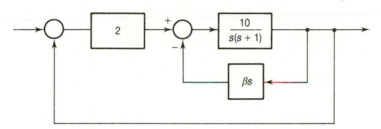

Figure 5.22 Example system for employing velocity feedback

To illustrate the design principle, we will determine the amount of velocity feedback for a peak magnification of 2 dB for the system shown in Fig. 5.22 without regard to the degradation in the velocity error constant. The compensator transfer function is

$$H(s) = \beta s$$

or

$$H(j\omega) = j\beta\omega$$

The effect of velocity feedback is thus merely to lift each point on the inverse Nyquist plot vertically by an amount proportional to the frequency. It is therefore obvious that the introduction of velocity feedback is to make the plot tangential to a larger M circle and so reduce the peak magnification and improve the damping.

Figure 5.23 shows the inverse Nyquist diagram of the original system and the target M_p circle. A guess has to be made where the point of tangency will lie and hence the frequency of the point that will be mapped on to the M_p circle (i.e. the new resonant frequency). As a first guess the point at $\omega = 4$ rad/s was chosen, and the vertical distance, r, to the M_p circle was measured as 0.56 units. Hence

$$\beta = K_c r/\omega = 2(0.56/4) = 0.28$$

Using CODAS it is easy to express the effective plant transfer function in terms of the parameter 'b' which represents the amount of velocity feedback, β, and so the new

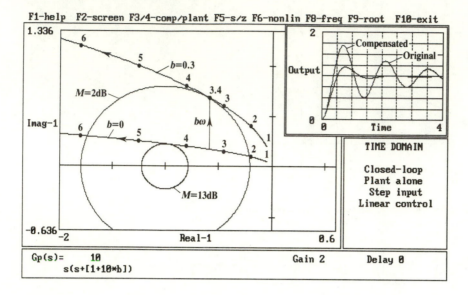

Figure 5.23 Design of internal compensation using inverse Nyquist plot

inverse Nyquist plot can be drawn immediately. On doing this it is found that the resulting curve intersects the target M_p circle slightly, and so with a little trial and error a value of 'b' of 0.3 is found to produce the desired performance. The units will depend on the input to the drive amplifier, but supposing it were volts, then the units of 'β' are volt/second.

Exercise Write down the closed-loop transfer function of the system in Fig. 5.22 and determine its damping ratio as a function of β. Use Eq. (4.19) to determine β analytically and compare with the result obtained graphically.

If the velocity error constant had to be maintained, then the gain of the controller would have to be increased. This would in turn require more velocity feedback to maintain the same damping. In the end the increase in gain leads to a system with a very much increased bandwidth, far more than would have resulted from a simple lead compensator. We shall see in the next section how internal feedback can be used to design a compensator to improve stability without loss of steady-state accuracy.

Exercise Design a lead compensator for the system in Fig. 5.22 to meet an M_p specification of 2 dB. Compare the transient response of the closed-loop lead-compensated system with that of the one which used velocity feedback.

Acceleration Feedback

Acceleration feedback implies that the internal feedback element has a transfer function of the form

$$H(s) = cs^2$$

Again the limitations of the 'properness' of this transfer function are ignored for the purpose of the following discussion. Assuming that $G_p(s)$ was a simple Type-1 transfer function as described earlier in this section, i.e.

$$G_p(s) = \frac{K_1}{s(1 + s\tau_m)}$$

the effective plant transfer function becomes

$$G_1(s) = \frac{K_1}{s(1 + s(\tau_m + K_1 c))}$$

and *voilà*, no degradation in K_v. This then is the clue of how to overcome the limitations of simple velocity feedback. However, there is a slight snag. The effect of acceleration feedback is to translate the inverse Nyquist plot horizontally to the left by an amount proportional to the square of the frequency, i.e. $H(j\omega) = c(j\omega)^2 = -c\omega^2$. The effect, therefore, of negative acceleration feedback in this case is to reduce the stability margins, decrease the damping and magnify the resonances. One possibility is to apply **positive** acceleration feedback, and indeed this is done sometimes, but again it will tend to increase the bandwidth of the system. The idea is to use internal feedback which is 'between' pure velocity feedback and pure acceleration feedback and which will map points to the left and **upwards**. The transfer function that will achieve the desired objective is acceleration feedback with a little lag or if you prefer velocity feedback with a little lead. Whatever the semantics, the transfer function of the feedback element is

$$H(s) = \frac{cs^2}{1 + as}$$

Example 5.4 Design of internal acceleration feedback compensator To see how to apply a compensator of this type we will consider the identical problem to the one examined in the previous section (depicted in Fig. 5.22). In order to arrive at a feasible compensator one must pick a point on the original locus that will be mapped on to the target M_p circle, that is to the right and below the target point. Figure 5.24 shows the design approach. A point on the original locus corresponding to 2 rad/s was chosen as the initial point and a point on the $M = 2$ dB circle was chosen as a reasonable point to which the original point is to be mapped. The translation can then be expressed as a distance, r, and an angle, ϕ. In this case $r = 0.5$ and $\phi = 64.5°$ (the free-wheeling cursor in CODAS is very helpful here).

The magnitude and phase of the compensator can be expressed as follows:

$$|H(j\omega)|/K_c = c\omega^2/\sqrt{(1 + (a\omega)^2)} = r$$

$$\angle H(j\omega) = -180° - \tan^{-1}(a\omega) = -180° - \phi$$

Hence from the second equation $a = 1.05$ and from the first equation $c = 0.58$. Again a parametric representation of the effective plant transfer function can be input into CODAS as

$$G_1(s) = \frac{10(1 + as)}{s(1 + [1 + a + 10*c]s + as^2)}$$

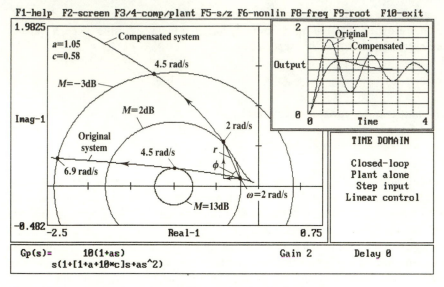

Figure 5.24 Compensation using modified acceleration feedback

It is quite simple using CODAS to verify the correctness of the design and to examine the resultant transient response. As can be seen the bandwidth of the system has been reduced and the resulting transient response is relatively slow but nicely damped.

Various design strategies can now be explored using the basic method outlined above to see what the consequences are of, say, picking a slightly different initial frequency, etc.

Exercise Examine how the performance of the system is affected by the use of a realizable feedback element of the form

$$H(s) = \frac{cs^2}{(1 + as)(1 + bs)}$$

where b takes on values from $0.5a$ to $0.05a$.

5.8 NON-UNITY FEEDBACK SYSTEMS

The systems that have so far been considered were overall unity feedback systems albeit in some cases there was a subsidiary internal feedback loop. The general feedback control system is always presented with a forward path transfer function, $G(s)$, and a feedback path transfer function, $H(s)$. Have we therefore been remiss in dealing only with unity feedback systems up to now? The answer to this question is a resounding NO, for most **control** systems are unity feedback as the main purpose of a **control** system is to make the controlled variable equal to the reference value. In other words the primary objective is to make the error zero or as small as is compatible with the design objectives of the control system.

If a transfer function, $H(s)$, is deliberately interposed between the controlled variable and the error box, then the signal fedback is a function of the output and the 'error' ceases to be a simple measure of the accuracy of the control system. With $H(s)$ present in the outer feedback path, the controlled variable becomes a complex dynamic function of the demand signal. This type of situation can arise in electronic amplifiers where, in the simplest case, the desired output of the amplifier is bigger than the input by a constant factor (the gain of the amplifier). Under these circumstances $H(s)$ is a fraction β. By employing combinations of resistance and capacitance elements in the feedback path of an operational amplifier complex transfer functions can be synthesized.

Is the study of non-unity feedback control systems then irrelevant? Many text books say that $H(s)$ represents the transducer dynamics. Certainly transducers may have appreciable dynamics and a transducer is an essential part of a control system. A controller can only act on a **signal** which represents the value of the actual process variable and in that sense it is not controlling the process variable (e.g. temperature), but the output of a transducer (e.g. a thermocouple). Thus the transducer is an integral part of the plant and hence is effectively in the forward path transfer function. If the transducer dynamics are known and one wanted to see how the process variable compared with the reference signal, then indeed the block diagram could be reorganized to place the transducer transfer function in the feedback loop, find the new overall closed-loop transfer function and deduce what was happening to the process variable. Alternatively the Laplace transform of the output signal can simply be divided by the transducer transfer function to obtain the process variable behaviour. Figure 5.25 summarizes these ideas.

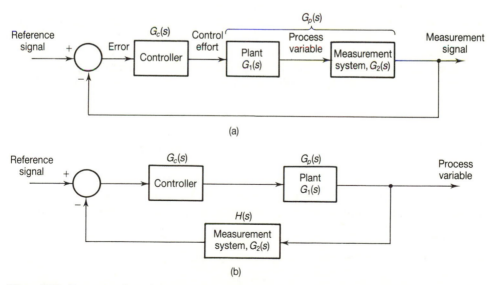

(a)

(b)

Figure 5.25 Representation of systems with measurement dynamics. (a) Measurement system in forward path. (b) Measurement system in feedback path

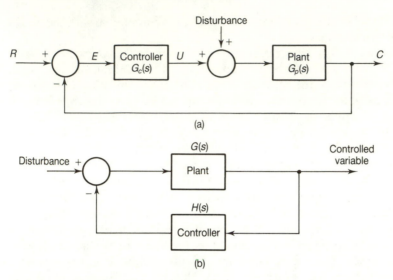

(a)

(b)

Figure 5.26 Control system as a regulator. (a) Normal block diagram of a control system. (b) Block diagram with respect to disturbances

But what if the transducer is inaccurate, nonlinear, too slow, too poorly damped, too far away, etc., so that the measured value is not close to the value of the true process variable? What can be done about it? Generally one is not going to 'doctor' the reference to correct the errors. The only sensible solution is to redesign the measurement system or select another transducer so that the output of the measurement system is closer to the true process quantity both statically and dynamically.

There is, however, one very important aspect of non-unity feedback that alone justifies its study and that is the performance of the control system as a regulator or in other words how well the control system deals with disturbances. Figure 5.26(a) shows the block diagram of a control system which is subject to disturbances. By rearranging the block diagram (Fig. 5.26(b)), the disturbance appears as a demand signal and the **controller** appears in the feedback path. In this case the output of the control system should **not** be equal to the demand signal! In fact ideally, the ratio of the controlled variable to the disturbance signal is **zero**, whereas normally the situation is that the output of the control system is equal to the demand signal or in other words the ratio is ideally **unity**. If it were not for the problems of stability both these objectives could be fulfilled in theory by having infinite controller gains.

In CODAS it is possible to investigate the performance of the loop as a regulator by placing the compensator or controller in the feedback path. The following exercise highlights the difference in considering the performance of a control system as a regulator or as a servo system.

Exercise A Type-0 plant has the following transfer function

$$G_p(s) = \frac{25}{(1 + 5s)(1 + s)(1 + 0.1s)}$$

Two compensators have been designed for this system to improve its stability margins, one a lag compensator,

$$G_{1c}(s) = \frac{1 + 6s}{1 + 22s}$$

and the other a lead compensator,

$$G_{2c}(s) = \frac{1 + 0.6s}{1 + 0.15s}$$

(a) Compare the closed-loop frequency response of the system as a servo for each compensator, i.e. the compensator in the forward path.
(b) Compare the closed-loop frequency response of the system as a regulator for each compensator, i.e. the compensator in the feedback path.
(c) Which compensator would you employ if there were significant disturbances to the system up to frequencies of 0.2 Hz?

Use of Inverse Nyquist Plots with Non-unity Feedback Systems

The use of inverse Nyquist plots to design systems with internal feedback loops was described earlier. Inverse Nyquist plots have similar application when dealing with non-unity feedback systems. The inverse relationship between the output, $C(s)$, and the reference input, $R(s)$, of the general non-unity feedback system is given by

$$[C(s)/R(s)]^{-1} = G^{-1}(s) + H(s)$$

In the frequency domain the above equation means that the closed-loop behaviour of the system can be found by **adding** the frequency response of the feedback element to the inverse frequency response of the forward path. Figure 5.27 shows the principle of the method. Note that the gains/phase shift of the closed-loop system is obtained by measurement from the **origin** of the plot and **not** the -1 point. The use of the -1 plot for predicting closed-loop behaviour is applicable **only** to unity feedback systems. In unity feedback systems '$H(s)$' is 1, and so the closed-loop frequency response is obtained by adding 1 to the open-loop frequency response, i.e. translating the inverse Nyquist locus to the right by 1. This is of course the same as taking the -1 point as the origin for closed-loop frequency response values. This point is quite logical, but is often confused by students.

> **Example 5.5 Simple system with sensor dynamics** Consider the system in Fig. 5.28. For convenience the plant and the controller are rolled into one block, $G_p(s)$, and the sensor transfer function is represented by $G_1(s)$. Obtain the frequency response of the true process variable with respect to the reference input and determine the peak magnification factor.
>
> SOLUTION To use CODAS to simulate the system, the controller transfer function, $G_c(s)$, is defined as $G_1(s)$ (the sensor transfer function). In the case of the unity feedback system it does not matter which way round the plant and compensator

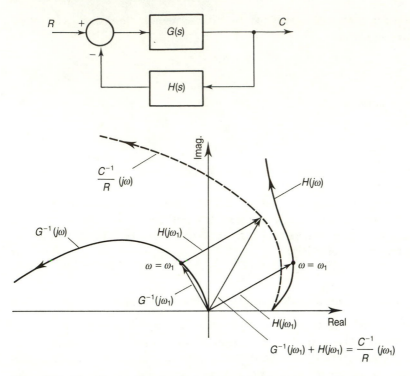

Figure 5.27   Use of inverse Nyquist plot for determining the frequency response of non-unity feedback systems

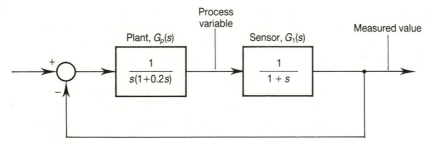

Figure 5.28   Plant with significant sensor dynamics

transfer functions are defined as long as one is only interested in the overall performance of the control system. Clearly the 'control' effort will be wrong if this is plotted with the two transfer functions interchanged.

Selecting the inverse Nyquist view, the peak magnification factor of the overall unity feedback system can be determined as about 3 dB (see Fig. 5.29). The overall step response can be shown readily using CODAS. The true plant output (the process variable) is different from the measured value. The true plant output can be displayed by switching to non-unity feedback mode. The peak magnification factor of the plant output with respect to the control system reference input can be

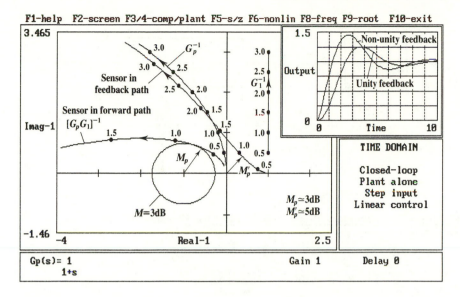

Figure 5.29 Performance of a control system with significant sensor dynamics

determined by drawing the vector sum of $G_1^{-1}(s)$ and $G_p(s)$ on the inverse Nyquist diagram, and finding the closest approach of that locus to the origin of the inverse Nyquist plane. This locus is also drawn in Fig. 5.29 and the value of this peak magnification factor is about 5 dB, which is significantly greater than the apparent closed-loop peak magnification factor as determined by using the output signal of the sensor.

Example 5.6 Complex system A final example illustrates the power of inverse Nyquist diagrams for handling quite complicated systems. Figure 5.30 shows a control position servo subject to disturbances. The servo system alone is very poorly damped and so a simple phase lead compensator of the form $G_c(s) = 1 + sT_d$ is employed to improve the gain margin. What value of T_d will give a gain margin of 3 and at what frequency will sinusoidal disturbances have the greatest effect on the output of the servo system?

SOLUTION In order to see how to do the first part, we can write that at the phase-crossover frequency of the compensated system for a gain margin of 3,

$$G_c G_p = -1/3$$

or

$$-3 - 3j\omega T_d = G_p^{-1}$$

Referring to Fig. 5.31 one can see from the open-loop frequency response of the plant alone that the real part of G_p^{-1} is -3 when $\omega = 4.7$ rad/s and the imaginary

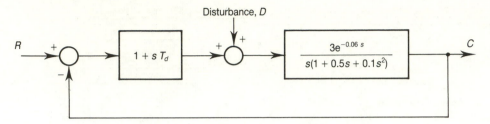

Figure 5.30 Control system subject to disturbances

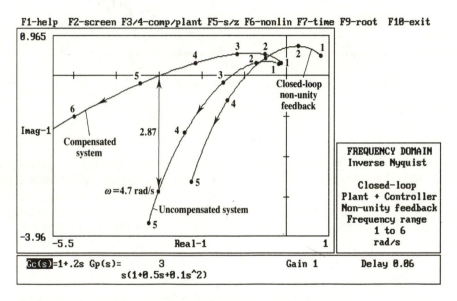

Figure 5.31 Frequency responses of lead compensated servo system

part is -2.87. Hence

$$3 \times 4.7T_d = 2.87$$

from which T_d is about 0.2 seconds. The open-loop frequency response of the compensated system with this value of T_d is plotted which confirms the desired gai.ı margin requirement. The closed-loop response of the system with respect to disturbances is plotted by adding the compensator frequency response to the inverse plant response. (In CODAS the non-unity feedback frequency response is plotted automatically if the non-unity feedback option has been selected.) The frequency at which the locus approaches the origin closest is the frequency at which disturbances will have their greatest effect on the output of the system. This frequency is about 2.8 rad/s and the ratio of output to disturbance is about 4 dB.

PROBLEMS

5.1 Design a lag compensator for the system

$$G_p(s) = \frac{30}{s(s+1)(s+6)}$$

to obtain a phase margin of 40° based on the methods described in Section 5.2. What is the bandwidth of the resulting system and its 2% settling time?

5.2 Design a lead compensator for the system in P5.1 using the methods described in Section 5.3 which results in the same phase margin. Compare the bandwidth and settling time with that obtained in P5.1.

5.3 Repeat 5.1 using the techniques described in Section 5.4 to see if a different design strategy will improve the settling time.

5.4 Repeat 5.2 using the improved techniques described in Section 5.3 to reduce the amount of phase lead introduced. Is there any significant change in the settling time?

5.5 A system has an open-loop transfer function

$$G_p(s) = \frac{1}{s^2(1+0.1s)}$$

The design specification is for a gain-crossover frequency of 1.4 rad/s and a phase margin of 45°. Design compensators which fulfil the design objectives and each of which in turn:

(a) minimizes the 'a' value employed in the compensator;
(b) employs an 'a' value of 10;
(c) satisfies the design criteria **without** reducing the Bode gain.

In each of the three cases carefully examine the closed-loop step response and compare the 5% settling times.

5.6 A system has an open-loop transfer function

$$G_p(s) = \frac{K}{s(s+1)}$$

The velocity error constant, K_v, is required to be 10 second^{-1}, and a phase margin of 45° is also required. Design:

(a) A lead compensator to meet the design requirements and which minimizes the amount of phase advance introduced.
(b) A lag compensator that introduces 5° phase lag at the new gain-crossover frequency.
Compare the closed-loop time domain performance of the system with each of the two compensators.

5.7 A system has an open-loop transfer function

$$G_p(s) = \frac{8}{(s+1)(1+s/3)^2}$$

The design requirement is for an overall phase margin of 45° and no change in the gain-crossover frequency of the original system.

Design lead compensators which fulfil the design objectives and each of which in turn:

(a) Minimizes the reduction in the positional error constant K_p.
(b) Minimizes the amount of phase advance introduced.
For each case what is the sacrifice in K_p?

5.8 A system has an open-loop transfer function

$$G_p(s) = \frac{K}{s(1+2s)(1+8s)}$$

What value of K will satisfy a gain margin requirement of 15 dB and a phase margin requirement of 45°? Design a lead compensator which maintains the gain margin and phase margin but maximizes the bandwidth of the closed-loop system. Compare the closed-loop frequency response of the original system with the compensated system. How do the 5% settling times compare?

5.9 Design a lag–lead compensator for the system in P5.1 which will reduce the peak magnification factor to 2 dB and which will give the system an overall bandwidth of about 1.25 rad/s. Use the following design strategies for the lag compensating element:

(a) The lag element should introduce 5° phase lag at the design frequency.
(b) The lag element should introduce 20° phase lag at the design frequency.

Compare the step response of the two compensators and comment on the results. Which design strategy should be adopted for this system?

5.10 Figure P5.10 shows the block diagram of a position control system. $H(s)$ represents a feedback compensator and $B(s)$ represents the dynamics of a position transducer. The design specification of the system is a peak magnification factor of 3 dB and a 10% settling time of 5 seconds.

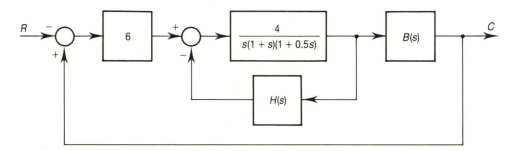

Figure P5.10

(1) $B(s) = 1$, negligible transducer dynamics.

(a) Ignoring the possibility of feedback compensation, how do you rate the chances of employing cascade compensation to meet the design specification?

(b) Design a feedback compensator of the form $H(s) = bs$ to satisfy the design specification. What is the resulting degradation in velocity error constant of the system? Observe the transient step response of the closed-loop system and comment on it.

(c) In order to maintain the velocity error constant design a feedback compensator of the modified acceleration type, i.e. $H(s) = cs^2/(1 + as)$ is used. Determine the values of 'c' and 'a' needed to meet the design specification and comment on the step response of the compensated system.

(2) $B(s) = 1/(1 + 0.25s)$, first-order response.

Repeat parts (b) and (c) above. Why is it easier to meet the design specification with a position transducer having a moderately slow response? What, if any, are the penalties?

5.11 Figure P5.11 is a block diagram of a plant whose temperature is controlled by means of a floating-rate

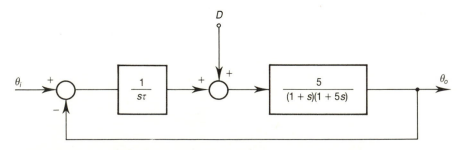

Figure P5.11 Floating-rate control of temperature

controller, $G_c(s) = 1/s\tau$ and which is subject to steam pressure disturbances D. What value of τ will give the system an overall gain margin of 3?

What is the period of sinusoidal steam pressure disturbances, D, which have the greatest effect on the output temperature, and what is the ratio of θ_o to D at this frequency?

ROOT LOCUS

6.1 INTRODUCTION

The concept of the s-plane was introduced in Chapter 3 and there it was shown how the positions of the poles determined the response of open-loop systems. The behaviour of closed-loop systems can similarly be predicted from a knowledge of the position of the closed-loop poles. The prediction of system behaviour from a knowledge of pole positions is essentially one of **analysis**. The s-plane can be a powerful tool for **synthesis** as well, i.e. the **design** of closed-loop systems.

The design of closed-loop systems was very much the theme in Chapter 5 using frequency domain methods. Why then study yet another method? One reason is that it is not always possible to obtain the frequency response of systems by experimental methods. The system may not yet exist physically, or it may not be accessible, or it may not be possible to determine its open-loop frequency response because the system is open-loop unstable. For many systems good mathematical models (transfer functions) are available and so s-plane techniques offer a more direct way of tackling the design problem.

There is of course a direct relationship between s-plane pole and zero positions and time domain behaviour. For simple cases where there is clear dominance of one or two poles without any complicating effects of zeros, good time domain correlations can be made based on first- or second-order models. Where there are several poles and zeros present of similar significance, it is difficult to predict how the time response will look without using a computer package.

The s-plane allows systems to be probed in some detail whereas frequency domain techniques are more 'broad-brush'. On some occasions detailed probing analysis is required and on other occasions a broader approach is more appropriate. In any event, it is always advantageous to have more than one design tool at one's disposal and to use one or other or all techniques to examine a problem and produce the best design.

The methods and concepts that will be discussed in this chapter should be considered as complementary to the frequency domain methods and not as replacements.

6.2 BASIC CONCEPTS OF ROOT LOCUS

Figures 6.1(a) and 6.1(b) show the block diagrams of two position control systems, the one in Fig. 6.1(b) having an extra lag. Both systems are Type-1 and so in open loop they will tend to drift away, and it would be impossible to control the position of the load without feedback. Thus position feedback is introduced and the error signal is passed through an amplifier of gain K before being applied to the servomotor input. As we have seen in earlier chapters the behaviour of the closed-loop system depends on the value of gain. The basic design objectives are to use a sufficiently high gain to satisfy steady-state accuracy requirements (K_p, K_v, K_a, etc.) within the constraints of an acceptable dynamic performance (ω_n, ζ, etc.).

First consider system (a). Remember that the roots of the characteristic equation are the poles of the **closed-loop system**. The characteristic equation of system (a) is

$$1 + G(s) = 1 + \frac{K}{s(s+2)} = 0$$

or

$$s^2 + 2s + K = 0 \tag{6.1}$$

It is simple to determine the roots (s_1 and s_2) of this quadratic equation for different values of K.

Table 6.1 shows how the roots vary for a range of gain values.

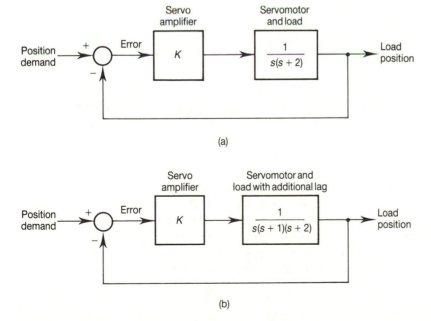

(a)

(b)

Figure 6.1   Two position control systems

Table 6.1

Gain, K	s_1	s_2	Comments
0	0.0	-2.0	Open-loop poles
0.5	-0.29	-1.71	Overdamped response
1.0	-1.0	-1.0	Critically damped
2.0	$-1 \pm j$		Underdamped response $\omega_n = 1.4$, $\zeta = 0.707$
4.0	$-1 \pm 1.7j$		Lightly damped response $\omega_n = 2.0$, $\zeta = 0.5$

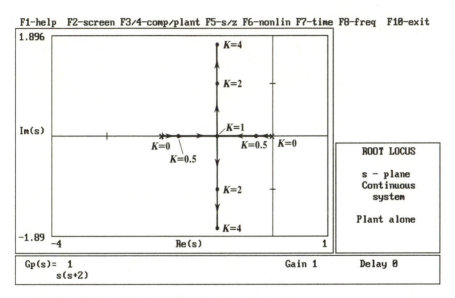

Figure 6.2 Migration of closed-loop poles for a simple servo

Rather than present the results in a table, the closed-loop pole positions can be drawn on an s-plane diagram as shown in Fig. 6.2. This figure shows very vividly how the closed-loop poles migrate as the gain is increased. The locus of these poles (i.e. the roots of the characteristic equation) is called the *root locus* and in this example is merely a set of straight lines.

The root locus diagram is very informative. For example, one can use it to find the range of gain that keeps the system damping ratio within certain bounds. This is done by first drawing the associated constant damping ratio lines. The range of gain that keeps the system damping ratio within these limits can then be read off the root locus.

The system in Fig. 6.1(b) has slightly more complicated dynamics. Following an analogous method to the previous example the characteristic equation of this system is

$$s^3 + 3s^2 + 2s + K = 0 \tag{6.2}$$

It is more difficult to find the roots of Eq. (6.2) as it is a cubic. However using trial and error or other methods the roots can be found and the resulting root locus can be

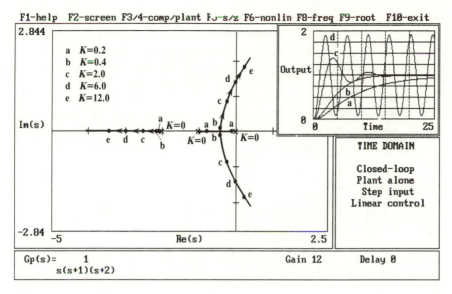

Figure 6.3 Root locus of a third-order system

plotted as shown in Fig. 6.3. (CODAS can be used to find the roots for a particular value of gain by using the 'C' command in the root-locus domain.) This time the root locus has more branches; a straight line branch along the negative real axis and a symmetrical pair of branches that eventually cross into the right-half plane. The point of incipient instability is where the locus crosses the imaginary axis ($K = 6$). In the inset the step response of the closed-loop system is drawn for various values of K.

6.3 GEOMETRICAL CONSTRUCTION OF ROOT LOCI

Introduction

The method used above for producing root loci relied on analytical techniques to find the roots of the characteristic equation, and other than for the very simplest systems such methods require the use of a computer. Indeed this is the method adopted by many computer programs for obtaining root loci. However this approach becomes very inefficient for higher-order systems and is simply not applicable to systems with transport delays because the characteristic equation involves a transcendental term (e^{-sT}) and cannot be solved by standard techniques.

The geometrical approach allows root loci to be sketched from a set of simple rules without explicitly finding the roots of the characteristic equation. This approach develops engineering intuition and promotes a feel for the way the system will behave in closed loop as the gain is varied. Furthermore, using the geometrical method, values of gain at any point on the root locus may be obtained by simple construction.

In the days before the era of personal computers, control engineers became highly skilled not only at sketching root loci but drawing them with some precision. In order to do this they used a great number of rules and were able to use a special-purpose drawing tool called a 'spirule'. The spirule has been replaced by the computer in conjunction with the graphics screen and there is no need to know every rule. It is desirable, however, to understand the fundamental concepts and to be able to draw approximate sketches of the loci using simple guidelines. In this way one does not become a slave of the computer and an engineering feel is developed. Hand sketching also helps with the recognition of obvious errors in the computer output that may have been caused by typographical errors, etc.

Angle and Magnitude Criteria

Let us write the open-loop transfer function of a general non-unity feedback system as

$$G(s)H(s) = KY(s)$$

where K is a parameter (usually the gain) and $Y(s)$ is a transfer function expressed in **pole/zero** form, i.e. the coefficient of s in each factor is 1. For example,

$$KY(s) = K\frac{(s + z_1)(s + z_2)\ldots(s + z_Z)}{(s + p_1)(s + p_2)(s + p_3)\ldots(s + p_P)} \tag{6.3}$$

The poles $-p_1$, $-p_2$, $-p_3$ and the zeros $-z_1$, $-z_2$ may be real or complex. If they are complex they will always appear as conjugate pairs for any real system. Furthermore for strictly proper systems the number of poles, P, will be greater than the number of zeros, Z.

The roots of the characteristic equation lie on the root locus and the characteristic equation is

$$1 + KY(s) = 0 \tag{6.4}$$

or

$$KY(s) = -1 \tag{6.5}$$

Now s is a complex variable and so the left-hand side of Eq. (6.5) is a complex number for a given value of s. This complex number must be identically equal in magnitude and sign to the right hand side of Eq. (6.5). In other words any value of s which lies on the root locus will make the angle of $KY(s)$ 180° (or any odd multiple of $\pm 180°$) and similarly this value of s makes the magnitude of $KY(s)$ equal to unity.

Thus Eq. (6.5) can be expressed in polar coordinates, i.e. an angle equation:

$$\angle Y(s) = \pm(2n + 1)\pi \qquad (n = 0, 1, 2, \ldots) \tag{6.6}$$

and a magnitude equation

$$|Y(s)| = 1/K \tag{6.7}$$

Where K is taken as an **unsigned** parameter. The variable s can lie anywhere in the complex plane but only values of s that simultaneously satisfy both the angle and magnitude equations (criteria) are roots of the characteristic equation.

The geometrical approach for constructing root loci consists essentially of finding

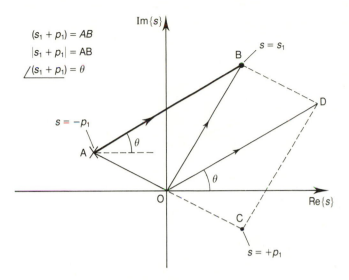

Figure 6.4 Angle and magnitude contribution of an isolated pole at $-p_1$

points on the s-plane where the angle criterion is satisfied, hence establishing points on the root locus. The corresponding value of K at those points on the root locus is determined by finding the value of K that satisfies the magnitude criterion at each point.

The Angle and Magnitude Contribution of a Pole/Zero

The angle and magnitude of $Y(s)$ can be obtained in an analogous manner to that used to obtain the frequency response of the overall open-loop system by adding/subtracting the phase contributions of individual terms in the transfer function and taking the products/quotients of the gain contributions. Similarly the overall angle and magnitude of the transfer function at particular point in the s-plane can be obtained by considering the contributions of each pole/zero separately.

In order to understand the geometrical interpretation of a term of the form $(s_1 + p_1)$, consider the situation drawn in Fig. 6.4 where an isolated pole has been drawn at $-p_1$. (The conjugate pole has not been drawn.) $+p_1$ is the vector **OC** and the vector **OB** represents s_1, and so by vector addition **OD** is $(s_1 + p_1)$. Normally only the pole/zero positions are shown on the s-plane (i.e. $-p_1$ **not** $+p_1$). Thus the vector **AB** is used to represent $(s_1 + p_1)$ which from simple geometry is equivalent to **OD** in magnitude and direction. The angle contribution, therefore, of any pole/zero can be obtained by finding the angle (with respect to the horizontal) of a line drawn from the pole/zero to the point on the s-plane under consideration. The magnitude contribution is the length of that line.

Zeros contribute positive angles and poles contribute negative angles to the overall angle of $Y(s)$ at a particular point on the s-plane. The overall magnitude of $Y(s)$ is obtained by taking the product of magnitude contributions of zeros and dividing by the product of the magnitude contributions of the poles. Figure 6.5 illustrates the principle. The system illustrated has poles at A, B, C and D and a zero at E.

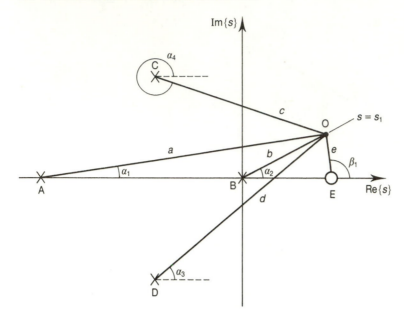

Figure 6.5 Angle and magnitude of an arbitrary system, $Y(s)$

$$|Y(s_1)| = \frac{e}{abcd}$$

and
$$\angle Y(s_1) = \beta_1 - (\alpha_1 + \alpha_2 + \alpha_3 + \alpha_4)$$

Exercise Enter the open loop transfer function

$$KY(s) = \frac{K}{s(s+1)(s+2)}$$

into CODAS. Select the root locus domain and change the X-range to -4 to 2 and make $+1.0$ the Y-centre. Use the cursor to determine points on the s-plane where the angle criterion is satisfied. Each time a point is found where the angle contribution is about 180°, press the 'M' key to produce a mark on the screen. In this way attempt to produce a locus similar to the one shown in Fig. 6.3.

Construction of a Root Locus Using the Angle Criterion

To illustrate the principles discussed above let us consider once more the simple system of Fig. 6.1(a). Its pole/zero diagram is shown in Fig. 6.6 and vectors have been drawn from the poles to an arbitrary point on the s-plane. The angle criterion of Eq. (6.6) requires that the angle of $Y(s)$ is an odd multiple of 180° on the root locus. For the

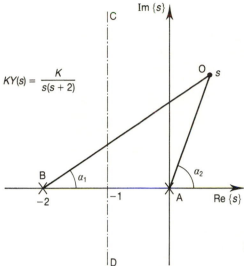

$$KY(s) = \frac{K}{s(s+2)}$$

Figure 6.6 Construction of the root locus using the angle criterion

system in Fig. 6.6

$$\angle\, Y(s) = -(\alpha_1 + \alpha_2)$$

The only way the angles α_1 and α_2 can add up to $\pm 180°$ is if s lies on a locus which is at the apex of an isoscles triangle whose base is AB (i.e. along the line CD), or if s lies on the line AB. This locus corresponds exactly to the one obtained by analytical methods in Fig. 6.2. The magnitude criterion (Eq. (6.7)) stipulates that the magnitude of $Y(s)$ is the reciprocal of the gain, K. For this system

$$|Y(s)| = \frac{1}{OA \times OB}$$

so we deduce from Eq. (6.7) that on the root locus for this system

$$K = OA \times OB$$

By way of example, the gain at the point where the locus leaves the real axis is 1, since at that point $OA = OB = 1$. This point on the locus corresponds to the critically damped case in Table 6.1.

More complex cases such as the one shown in Fig. 6.3 have to be drawn by searching the s-plane for points which satisfy the angle criterion. The gain at any point on the root locus can always be obtained using the formula

$$K = \frac{\text{Product of distances to the poles}}{\text{Product of distances to the zeros}} \tag{6.8}$$

Sketching a Root Locus

Before using a computer package to draw a root locus, it is important to be able to sketch some of its salient features by hand. In order to do this there are several guidelines that are helpful to know:

1. *Starting/end points and number of loci.* In a root locus diagram the terms 'start' and 'end' refer to the closed-loop pole positions when the value of gain, K, is respectively 0 and ∞. For strictly proper systems the loci start at the poles and finish at zeros or at infinity.

   Since K is the reciprocal of $|Y(s)|$ (Eq. (6.7)), the starting points occur when $|Y(s)|$ is infinity. $|Y(s)|$ is infinite at a pole and so the loci start at the open-loop poles of the system. The end points occur when $|Y(s)|$ is 0 (i.e. K is infinite). Clearly $|Y(s)|$ is 0 at a zero. Now for strictly proper systems, i.e. where the number of poles, P, is greater than the number of zeros, Z,

$$\lim_{s \to \infty} |Y(s)| = s^{-(P-Z)} \to 0$$

   Hence $|Y(s)|$ tends to 0 as s tends to ∞. Thus the loci terminate at a zero or at infinity.

   As the loci start at the poles in proper systems, the number of loci is equal to the number of poles. The root locus of Fig. 6.2 is an example of a strictly proper systems where there are 2 poles and no zeros. There are 2 separate loci all of which terminate at infinity.

   For systems that are not proper (more zeros than poles), all the loci terminate at zeros and so the number of loci is equal to the number of zeros. Now for systems that are not proper, $|Y(s)|$ will tend to ∞ when s tends to 0 or when s is at a pole. Hence some of the loci start at the open-loop poles of the system and the remaining loci start at infinity.

2. *The loci are symmetrical about the real axis.* This fact is self evident because complex poles occur in conjugate pairs in real systems.

3. *Asymptotes of the loci.* Loci which terminate at infinity do so along straight lines whose angles are:

$$\frac{\pm(2n+1)\pi}{(P-Z)} \tag{6.8(b)}$$

   Consider a point on the root locus very far away from all the poles and zeros. Taken sufficiently far away, the angle contributions of all the poles and zeros are the same, say θ. The angle criterion requires a total angle of $\pm(2n+1)\pi$. Thus for a surplus of $(P-Z)$ poles over zeros, the angle of $Y(s)$ when s is very large is $(P-Z)\theta$. Hence

$$\theta = \frac{\pm(2n+1)\pi}{(P-Z)}$$

   It can be shown furthermore that the asymptotes intersect on the real axis at a point,

$$\sigma = \frac{\text{Sum of the poles of } Y(s) - \text{Sum of the zeros of } Y(s)}{(P-Z)}$$

4. *Loci on the real axis.* Imagine a point very close to the real axis (Fig. 6.7). The angle contribution of real poles and real zeros to the left of the point is zero, whereas poles on the real axis to the right of the point each contribute $-\pi$ and zeros each

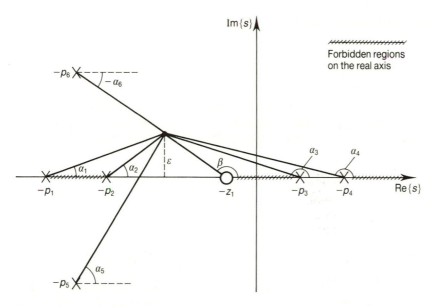

Figure 6.7 Loci on the real axis

contribute $+\pi$. Complex poles or complex zeros to the right or the left make a net angle contribution of zero, and so can be ignored as far as the angle contribution is concerned. Thus since the total angle must be an odd multiple of π, the number of poles and zeros to the right of a point on the real axis must be odd for that part of the real axis to be on the root locus.

In the situation shown in Fig. 6.7, as ε becomes infinitesimally small, α_1 and α_2 tend to zero, α_5 and $-\alpha_6$ add up to zero, whereas α_3 and α_4 each contribute $-\pi$ and the angle contribution, β, of the zero is $+\pi$. Thus the root locus will lie along the real axis between $-p_2$ and $-z_1$, between $-p_3$ and $-p_4$ and to the left of $-p_1$.

Example Root Loci

Example 6.1 A simple system has the transfer function

$$KY(s) = \frac{K}{s(s+1)(s+2)}$$

The root locus of this system was plotted in Section 6.2 using an analytical method to determine the roots of the characteristic equation for various values of gain. This time the guidelines that were developed above using the angle criterion will be used to justify the locus shown in Fig. 6.3.

There are three real poles $(0, -1$ and $-2)$ and no zeros. Hence there are three loci all terminating at infinity. Loci will appear on the real axis where there are an odd number of poles to the right of a given section, i.e. there are loci on the real axis between the pole at the origin and the pole at -1 and to the left of the pole at -2. As there are no zeros, $(P - Z)$ is three and hence the angles of the asymptotes are $\pm\pi/3$ and π which intersect at

$$\sigma = \frac{(-2) + (-1) + (0)}{3} = -1$$

on the real axis.

From the above points it is possible to produce a sketch along the lines of Fig. 6.3. An accurate plot in the region of interest could then be obtained by actually measuring angles to the poles and finding points which satisfy the angle in that area of the s-plane.

Example 6.2 A more complex system has the transfer function

$$KY(s) = \frac{K(s+2)}{s(s+4)(s+6)(s^2 + 6s + 12)}$$

This system has an open-loop zero at -2, open-loop real poles at 0, -4 and -6 and a pair of open-loop complex poles at $-3 \pm j\sqrt{3}$. Using the above guidelines one can deduce that there will be five separate loci as there are five open-loop poles. One of the loci terminates in the zero and the other four migrate asymptotically to infinity. The asymptotes are at angles of $\pm\pi/4$ and $\pm 3\pi/4$ and intersect the real axis at

$$\sigma = \frac{2 - (0 + 4 + 6 + 3 - j\sqrt{3} + 3 + j\sqrt{3})}{4} = \frac{-14}{4} = -3.5$$

The complete locus as drawn using CODAS is shown in Fig. 6.8.

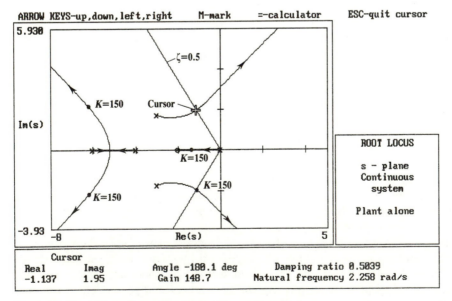

Figure 6.8 Root locus of example system with cursor enabled

The gain at any point on the locus is obtained by moving the cursor on to the locus and reading off the values in the bottom window (see Fig. 6.8). Suppose that it is desired to have the dominant conjugate poles exhibit a damping ratio of 0.5. To do this a damping ratio line is superimposed on the s-plane diagram and the cursor moved to the point where the damping ratio line crosses the locus. The gain at that point is about 150, and the conjugate closed-loop poles have a natural frequency of 2.26 rad/s.

Exercise Use CODAS to draw the root locus of the of the system in Example 6.2 above and confirm that with a gain of 150 the dominant conjugate poles have a damping ratio of 0.5. (The 'C' command can be used to actually mark all the closed-loop poles on the loci.)

Examine the unit closed-loop step response of the system and superimpose on it the **open-loop** response of a standard second-order system with the same damping ratio and natural frequency as the dominant conjugate poles of the example system. Why is there a significant discrepancy.

Hint: read Section 3.11 on dominant pole approximations.

6.4 NEGATIVE GAINS

There are situations where the root locus must be drawn for negative values of gain (e.g. where positive feedback is applied). Practical examples where negative gain values are needed will be seen later.

K is unsigned, hence for negative gain, Eq. (6.5) becomes

$$KY(s) = 1 \tag{6.9}$$

The fundamental consequence is that the angle criterion becomes

$$\angle Y(s) = \pm 2n\pi \qquad (n = 0, 1, 2 \ldots) \tag{6.10}$$

This modification to the angle criterion affects where the locus lies on the real axis and the angles of the asymptotes. The loci now lie on the real axis where there are an **even** number of poles/zeros or no poles/zeros to the right of a given section of the real axis. The angle of the asymptotes (Eq. (6.8(b))) now becomes

$$\theta = \frac{\pm 2n\pi}{(P - Z)} \tag{6.11}$$

CODAS will draw the root locus for negative gains by defining K as any negative number.

Example 6.3 Root locus with negative gains Sketch the root locus of

$$KY(s) = \frac{K(s + 2)}{(s + 3)(s^2 + 2s + 2)}$$

for negative values of K. What negative value of K will make the system marginally stable?

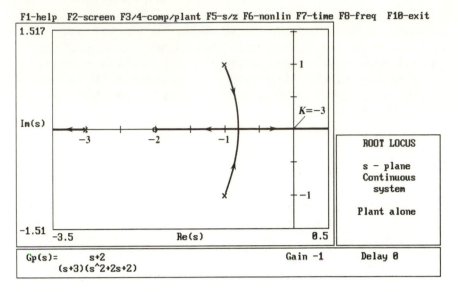

Figure 6.9 Root locus for negative gains

SOLUTION This system has one real zero at -2, a real pole at -3 and a pair of complex poles at $-1 \pm j\sqrt{1}$. Loci lie along the real axis to the right of the zero at -2 and to the left of the pole at -3. As there is an excess of two poles over zeros, the asymptotes lie at angles 0 and π, i.e. along the real axis. The locus for this example has been drawn in Fig. 6.9. The system becomes unstable when the locus crosses into the right-half plane. For negative gains the locus crosses the real axis at the origin of the s-plane. The negative value of gain which makes the system marginally stable can be obtained by applying the gain criterion (Eq. (6.8)) at the origin of the s-plane, i.e.

$$K = -\frac{3(\sqrt{2})(\sqrt{2})}{2} = -3$$

for marginal stability.

6.5 TRANSPORT DELAYS

The fundamental angle and gain criteria are valid whether or not transport delay is present in the system. The idea, however, that an open-loop pole/zero plot completely describes a system is not true when the system has transport delay. The presence of transport delay materially affects the root locus, yet in itself a transport delay will not affect the appearance of the open-loop pole/zero diagram.

Unfortunately there is no direct analytical way of solving the characteristic equation when transport delay is present and the only way to plot the root locus is by trial and error using the angle criterion. The algorithm in CODAS uses a trial and error approach and so there are no problems in drawing root loci for systems where transport delay is present.

Consider the following system:

$$KY(s) = KY'(s)\,e^{-sT}$$

where $Y'(s)$ is a rational transfer function (ratio of two polynomials). Let

$$s = \sigma + j\omega$$

Hence

$$e^{-sT} = e^{-\sigma T}\,e^{-j\omega T}$$

The angle criterion is thus

$$\angle\,Y'(s) = \pm(2n+1)\pi - \omega T \qquad (n = 0, 1, 2, \ldots) \tag{6.12}$$

and the gain on the root locus using the former criterion becomes modified to

$$K = \frac{e^{\sigma T}}{Y'(s)} \tag{6.13}$$

Since σ is negative for stable systems, the gain is lower than obtained with the standard rule of the pole distance product divided by the zero distance product (Eq. (6.8)).

The transport delay behaves as an infinite supply of poles and so the number of separate loci is infinite when there is transport delay in the system.

Example 6.4 Use CODAS to draw the root locus for the system below and determine the critical gain

$$KY(s) = \frac{K\,e^{-2s}}{(s+3)}$$

Figure 6.10 shows the result. This plot shows the effect of the transport delay and two set of branches are visible within the span of the plot. The critical gain is determined by the lowest branch crossing into the right-half plane.

On the whole systems with transport delay are better dealt with in the frequency domain. With transport delays much of the value and appeal of s-plane concepts is lost.

6.6 ROOT CONTOURS

Perhaps the most important aspect of root locus is the ability to plot how the roots of the characteristic equation change when parameters other than gain are varied. Such root loci are usually called root contours and the term root locus is generally confined to gain variations.

The principle for obtaining root contours is to recast the characteristic equation so that the parameter under examination appears as a multiplicative factor such that the overall form of the equation is identical to Eq. (6.5). This approach presumes that the

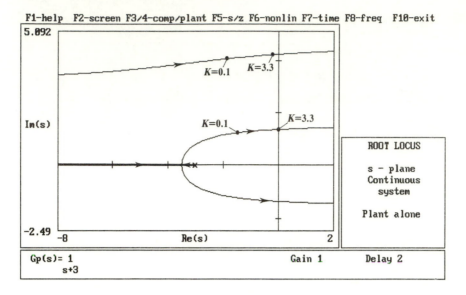

Figure 6.10 Root locus of system with transport delay

transfer function is rational, i.e. no transport delays and the parameter only appears in the transfer function as a linear term. Root contours are not applicable in a simple way for systems with transport delay.

Consider a position control servo system whose open-loop transfer function is of the form

$$KY(s) = \frac{K}{s(1 + 0.1s)(1 + s\tau_m)} \tag{6.14}$$

where K is the servo amplifier gain and τ_m is the motor time constant (notionally 0.4 seconds). The design objective is for the dominant closed-loop poles to have a damping ratio of 0.6 ± 0.1.

The root locus of the system is shown in Fig. 6.11(a) for variations in amplifier gain, K. The two limiting damping ratio lines have been superimposed on the plot. The value of amplifier gain that meets the design criterion for a notional motor time constant of 0.4 seconds is 1.3 per unit time, and the gain can drop to 1 or rise to 1.7 and keep within the design bounds.

Note: The open-loop transfer function presented above is **not** in pole/zero form. Great care must be taken in calculating the gain by hand methods from the root locus. The method derived using the product rule (Eq. (6.8)) was based on the assumption that the open-loop transfer function was in pole/zero format, i.e. the coefficient of the highest power of s in both numerator and denominator was assumed to be unity. Thus for the above system the open-loop transfer function should be thought of as

$$K'Y(s) = \frac{K'}{s(s + 10)(s + 1/\tau_m)}$$

where $K' = K10/\tau_m$.

(a)

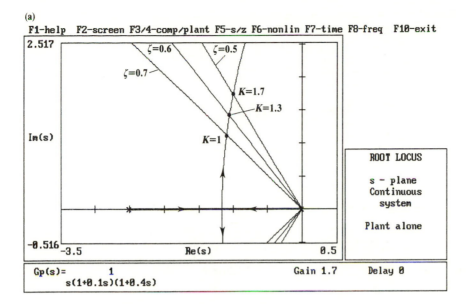

Figure 6.11(a)   Root locus of servo system with gain as a parameter

In other words if a gain is read off the root locus using the raw product rule the result must be divided by $10/\tau_m$ to get the correct amplifier gain. Using CODAS the correct value of gain is always reported irrespective of the form of the open-loop transfer function.

The characteristic equation of the system is

$$\frac{K}{s(1+0.1s)(1+s\tau_m)} = -1$$

The characteristic equation can be rearranged to make τ_m a multiplicative factor as follows:

$$K = -\tau_m s^2(1+0.1s) - s(1+0.1s)$$

or

$$K + s + 0.1s^2 = -\tau_m s^2(1+0.1s)$$

By dividing both sides of the previous equation by $(K + s + 0.1s^2)$ the desired result is achieved, namely

$$\tau_m \frac{s^2(1+0.1s)}{K + s + 0.1s^2} = -1 \qquad (6.15)$$

Which is exactly the same form as Eq. (6.5) except that the parameter τ_m has replaced the gain, K, and where

$$Y(s) = \frac{s^2(1+0.1s)}{K + s + 0.1s^2}$$

(b)

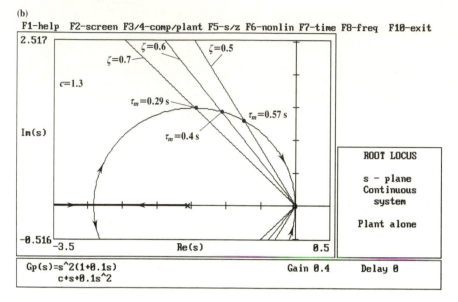

Figure 6.11(b)   Root contour with motor time constant as a parameter

Thus a root contour can be drawn using the same methods as before for root loci. The above system has a double open-loop zero at the origin and a zero at -10, poles at -1.54 and -8.46 for the nominal value of K $(K = 1.3)$. The root contour has been drawn in Fig. 6.11(b) with motor time constant as a parameter, i.e. K now is equivalent to τ_m. When τ_m is at its notional value of 0.4 the closed-loop poles coincide with the poles drawn in the root locus of Fig. 6.11(a) for $K = 1.7$ and $\tau_m = 0.4$. As τ_m is varied the poles migrate and the limiting values of damping ratio are reached when τ_m is 0.57 seconds and 0.29 seconds respectively.

Figure 6.11(c) superimposes the results of both plots on one diagram. This plot shows very graphically the sensitivity of the system both to parameter and gain changes.

The 'transfer function' on the left hand side of Eq. (6.15) is not proper; it has more zeros than poles. It is important to realize that this 'transfer function' has no physical meaning, it simply has arisen due to algebraic manipulation. The new zeros associated with this 'transfer function' are entirely fictitious, but the **poles** of the rearranged system are the poles of the real system since the characteristic equation of the rearranged system is the characteristic equation of the original system.

Note: The time response cannot be examined in CODAS once the transfer function has been rearranged to produce a root contour. Only the original system transfer functions can be used to display the time response of the system.

Exercise The open-loop transfer function of a system is

$$G(s) = \frac{K(1 + as)}{(1 + s)(1 + 2s)(1 + 6s)}$$

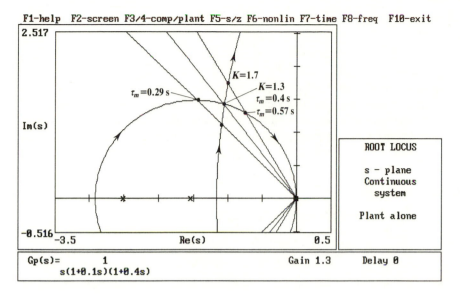

F1-help F2-screen F3/4-comp/plant F5-s/z F6-nonlin F7-time F8-freq F10-exit

Figure 6.11(c)   Combined root locus and root contour

The objective of this exercise is to find the value of 'a' that allows the gain, K, to be maximized whilst maintaining the damping ratio of the dominant closed-loop poles above 0.5.

Rearrange the characteristic equation so the 'a' appears as a multiplicative factor. Draw the root contour of the system with 'a' as the parameter with the gain, K, fixed at 5. Repeat for gain values of 6, 7 and 8. Draw the 0.5 damping ratio line and hence determine the best choice of K and 'a' that satisfy the design objectives. Determine the locations of the closed-loop poles with these values of K and 'a'. Draw the closed-loop step response of the system. Is the dominant pole approximation justified in this example?

6.7 COMPENSATOR DESIGN IN THE s-PLANE

The principles of lag and lead compensators were discussed at length in Chapter 5. Here we shall see how to design dynamic compensators in the s-plane and see the advantages and disadvantages over frequency domain methods.

The form of the transfer function that was used in Chapter 5 for lag and lead compensators was developed to allow a systematic design method in the frequency domain. Here we shall simply define a lag or a lead compensator as

$$G_c(s) = \frac{1 + a\tau s}{1 + \tau s} \tag{6.16}$$

where a is less than unity for a lag compensator and greater than unity for a lead compensator.

The same system that was used in Chapter 5 to illustrate the principles will be used here too; namely

$$G_p(s) = \frac{640}{s(s+4)(s+16)}$$

In the s-plane the dynamic requirements must be specified in terms of desired closed-loop pole locations, but more practically in terms of a time domain specification. In order to maintain broad consistency with the design requirements used in the previous chapter, we shall specify here a damping ratio of 0.45 or a peak overshoot of less than 25%.

Lag Compensator Design

The lag compensator that was designed in Section 5.2 was

$$G_c(s) = \frac{1 + s/0.243}{1 + s/0.0715} = \frac{1 + 4.1s}{1 + 14s}$$

i.e. 'a' is about 0.3 and τ is 14 seconds. The root loci of the compensated and uncompensated plants are shown in Fig. 6.12. (Because of the scale, the pole at -16 is not visible.) The compensator introduces a pole/zero pair very close to the origin of the s-plane, a pole at -0.0715 and a zero at -0.243. This pole/zero pair is often termed a 'dipole'. Compared to the poles of the plant at -4 and -16 the compensator pole/zero are insignificant and effectively cancel each other and so the shape of the compensated root locus is hardly different from that of the uncompensated plant.

However the presence of the dipole materially affects the gain read off at each point on the locus because of the effect discussed in Section 6.6 when terms are present in

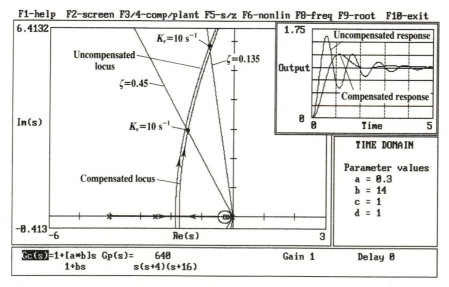

Figure 6.12 Root loci of a lag compensated system

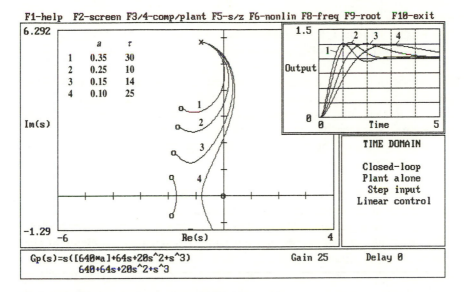

Figure 6.13 Root contours of a lag compensated system

the open-loop transfer function which are not in pole–zero form. In fact, because of the dipole, the gain at points on the compensated locus is about '$1/a$' times greater than the gain at similar points on the uncompensated locus. Hence the given Bode gain occurs at a earlier point on the locus where the conjugate poles are more heavily damped.

The above discussion can hardly be called design or synthesis and very little insight is offered by the root locus plot. In fact the raw root locus is not very helpful in designing a lag compensator. However by rearranging the characteristic equation to make τ a parameter, the effect of the choice of 'a' and τ on the design can be examined.

The characteristic equation of the compensated system is

$$\frac{(1 + a\tau s)}{(1 + \tau s)} \frac{640}{s(s + 4)(s + 16)} = -1$$

which can be rearranged to give

$$\tau s \frac{(640a + 64s + 20s^2 + s^3)}{640 + 64s + 20s^2 + s^3} = -1 \qquad (6.17)$$

The root contours of this system are plotted in Fig. 6.13 for different (a) values and the inset shows several step responses for those values of 'a' with τ chosen by trial and error to give similar peak overshoots for each response. Remember that the 'gain' read off the locus is actually the time constant, τ.

From the root contour one can see that as the 'a' value decreases the locus moves towards the real axis and consequently the responses tend to get slower. However because of the presence of a pole close to the origin when 'a' is small, it is difficult to make exact correlations and the choice of τ is a bit hit and miss and in the end relies on simulation. For bigger values of 'a' the conjugate poles are more dominant and better correlations are possible.

When 'a' is greater than 0.15, the closed-loop conjugate poles lie very close to the zeros which shows that there is little flexibility in the design and that the design criterion is rather too stringent for a simple lag compensator.

Note: CODAS can be used to draw root contours and show transient responses of the system on the same screen. The technique for accomplishing this is to store the root contour and system transfer functions in separate files. First the root contours are drawn and then the system description file for the system transfer functions is loaded but **not** the associated environment file. The transient responses can then be drawn in a separate window on the same screen.

The effort required to rearrange the equation to produce a root contour is not insignificant especially for higher-order systems and great care must be taken to avoid errors. The validity of the rearranged characteristic equation can always be checked by confirming that the poles of the rearranged equation are identical to those of the original system for the same parameter and gain values. Though some of the information that can be deduced from the root contour is interesting, it is not really very helpful in this case. On the whole a frequency domain approach, as described in Chapter 5, is better for designing simple lag compensators.

Lead Compensator Design

In Chapter 5 ('Improved method of designing lead compensators'), a lead compensator was designed with a transfer function

$$G_c(s) = \frac{1 + s/5.1}{1 + s/30.6} = \frac{1 + 0.196s}{1 + 0.0327s}$$

In terms of the prototype compensator (Eq. (6.16)), τ is 0.0327 seconds and 'a' is 6. The root locus of the system and compensator is drawn in Fig. 6.14. The lead compensator introduces a zero at -5.1, slightly to the left of the system pole at -4. The effect of the zero is to drag the locus to the left, so increasing the damping ratio for a given value of gain. As can be seen from Fig. 6.14, the closed-loop pole locations for the design value of velocity error constant ($K_v = 10 \text{ s}^{-1}$) lie very close to the 0.45 damping ratio line.

To design a lead compensator in the s-plane, the pole/zero locations have to be found that satisfy the design requirement. The best way of going about this is to produce a set of root contours for different values of τ for various values of 'a' in an analogous manner to that done for the lag compensator in the previous section.

Equation (6.17) can be used again for a lead compensator (values of 'a' greater than 1). The resulting root contour is shown in Fig. 6.15. Three contours have been drawn for 'a' values of 4, 5 and 6. As can be seen the design criterion cannot be met when 'a' is 4. With 'a' at 6, there are two points on the root contour that satisfy the design criterion. If a subsidiary design objective was to minimize the phase advance introduced, then clearly the smallest value of 'a' that will just satisfy the design criterion ($\zeta = 0.45$) is when 'a' is 5. The 'gain' on the root contour corresponds to the time

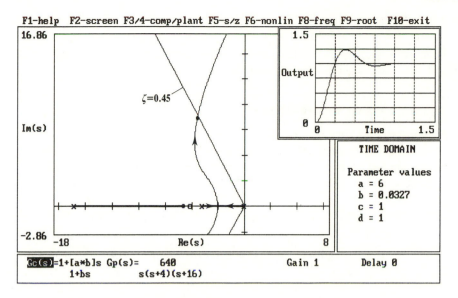

Figure 6.14 Root locus of a lead compensated system

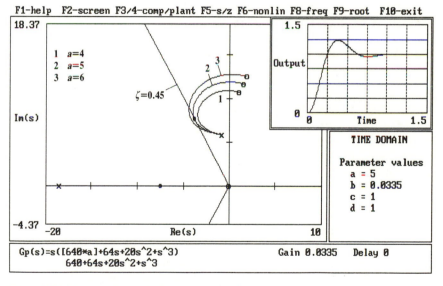

Figure 6.15 Root contours of a lead compensated system

constant, τ, and where the contour touches the damping ratio line it is 0.0335 seconds. The inset shows the step response of the system with these parameter values.

This example shows that root contours are a very good way of designing lead compensators and offer useful insights into the relationship between design parameters and control system performance.

6.8 CASE STUDIES

In order to expand the principles and techniques in this and the preceding chapter, two more extensive case studies will be considered.

Study No. 1: Ship Steering Control System

Statement of problem This case study deals with the design of an automatic steering system for a ship. The ship heading, ϕ, is measured using a radio compass and compared with the demand heading to produce an error signal. The error signal is processed by a controller whose output, u, is fed to an electrohydraulic rudder actuator which moves the rudder.

The relationship between the ship heading and the rudder angle, θ, can be modelled as:

$$\frac{\phi}{\theta}(s) = \frac{K_1}{s(1 + \tau_1 s)}$$

where K_1 is 0.017 second^{-1}, and τ_1 is 55 seconds.

The rudder actuator dynamics are essentially first order, namely

$$\frac{\theta}{U}(s) = \frac{1}{(1 + \tau_2 s)}$$

where τ_2 is nominally 5 seconds. The maximum rudder angle, θ, is limited to $\pm 35°$.

Performance specifications The performance requirements are:

(a) There should be no more than $3°$ steady-state error in heading when the demand heading is changing at the rate of $15°$/minute.
(b) When a small sudden change in demand heading is applied, the ship heading should not overshoot the demand heading by more than 20%. Furthermore the time taken to reach the demand heading should be as fast as possible.

Design evaluation Obtain the following data:

(a) the 5% settling time;
(b) the system peak magnification factor, resonant frequency and -3 dB bandwidth;
(c) the approximate minimum time to turn the ship about (i.e. through $180°$).

Design Part (a) of the design specification determines the overall velocity error constant of the system. A rate of $15°$/minute is equivalent to $0.25°$/second. Now the steady-state positional error is the rate of change of demand divided by the velocity error constant, K_v. Thus K_v is $0.25/3$ second^{-1} (i.e. 0.0833 second^{-1}). Assuming the controller gain is K then

$$K_v = KK_1$$

hence

$$K > K_v/K_1 = 0.0833/0.017 = 4.9 \text{ second}^{-1}$$

Rounding up, we have that K must be greater than 5 second^{-1}. It is quickly established using CODAS that the response using a simple proportional controller is very oscillatory with an overshoot of about 50% and so some form of dynamic compensation is necessary. The type of compensator that is required can be deduced from part (b) of the design specification. The requirement to reach the demand heading as quickly as possible is equivalent to having a minimum rise time. Hence in order to minimize the rise time a phase lead compensator will be used.

The form of the lead compensator is given in Eq. (6.16). The open-loop transfer function of the system is thus the product of the compensator, rudder and ship transfer functions, i.e.

$$G(s) = K \frac{(1 + a\tau s)}{(1 + \tau s)} \frac{1}{(1 + \tau_2 s)} \frac{K_1}{s(1 + \tau_1 s)}$$

As we saw in Chapter 5, generally the higher the 'a' value, the greater the system bandwidth. As there is a correlation between bandwidth and rise time (short rise times require wide bandwidths) a lead compensator will be used with the largest acceptable 'a' value (i.e. 10). The question then is to choose τ, if possible, so that the overshoot criterion is satisfied. Assuming that the response of the system will be dominated by a pair of conjugate poles, the 20% overshoot criterion is equivalent to designing for a damping ratio of about 0.45. The assumption of this second-order correlation takes no account of the presence of other poles or, perhaps more significantly, of zeros. Nevertheless this assumption will serve as a starting point for the design.

The characteristic equation $(G(s) = -1)$ can be rearranged to make τ a multiplicative factor so that a root contour can be drawn, i.e.

$$\tau s \frac{(0.017Ka + s + [55 + \tau_2]s^2 + 55\tau_2 s^3)}{0.017K + s + [55 + \tau_2]s^2 + 55\tau_2 s^3} = -1$$

Figure 6.16 shows the root contour for the above system. The parameter 'c' represents the gain of the controller, K, and parameter 'd' represents the rudder time constant, τ_2. The root contour crosses the 0.45 damping ratio line at two places. To maximize the rise time the higher point is chosen, i.e. at the higher damped natural frequency. The corresponding τ value is 5.3 seconds. It just so happens that with 'a' set at 10, this value of τ introduces a compensator zero that nearly cancels the pole due to the vessel dynamics. Figure 6.17 shows the root locus of the system with τ set at 5.3 seconds and the inset shows the step response with controller gain, K, of 5. Due to the approximate cancellation of the vessel pole and the compensator zero and the relatively large attenuation of the other system pole, the dominant conjugate pole assumption is justified, as can be seen from the time response which has an overshoot of about 20%.

This design however allows virtually no flexibility because the gain is nearly at its minimum allowable value. Can a compensator be designed that allows a gain of 6 without contravening any of the design specifications? If it can, then the system will be more robust as the gain can be dropped by up to 20% to compensate for uncertainties in the system model. By repeating the root contour exercise for a gain of 6, a value for τ of 4.6 seconds is found that produces conjugate poles with a damping ratio of 0.45. On simulation it is found that the overshoot is greater than 20% because the zero is

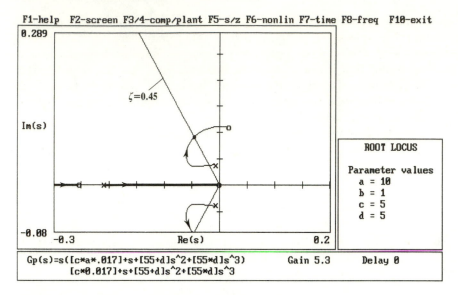

Figure 6.16 Root contour of ship control system with $K = 5$

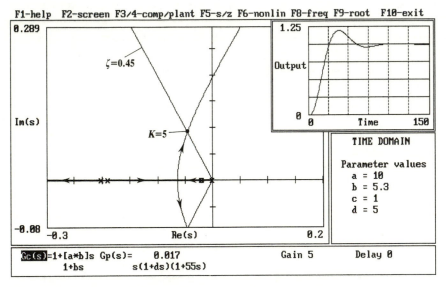

Figure 6.17 Root locus of ship steering system and its step response

no longer cancelled and tends to give the step response an extra 'kick'. Thus a slightly higher damping ratio must be chosen, i.e. a lower value of τ. By trial and error a time constant, τ, of 4 seconds is found suitable. The root locus and step response of the modified system is shown in Fig. 6.18. The rise is hardly any different from the previous case but the system is less oscillatory and settles more quickly. The root locus shown in Fig. 6.18 is unusual as it shows a break away from the real axis which is not at 90°.

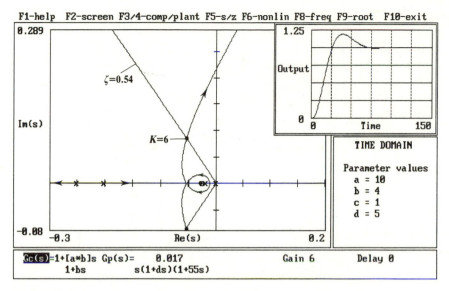

Figure 6.18 Root locus and step response of final ship steering control system

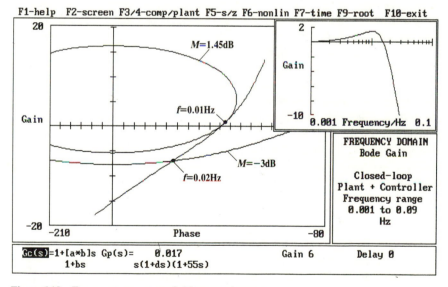

Figure 6.19 Frequency response of ship control system

Normally loci break away from the real axis at 90° because two poles come together, but on this occasion there is a 'collision' of three poles.

The 5% settling time for small step changes in demand is about 60 seconds. The frequency response data is shown in Fig. 6.19, where the resonance peak is measured at 1.45 dB at a frequency of 0.01 Hz and the −3 dB bandwidth is 0.02 Hz.

The final requirement is to estimate the time to turn the ship about. If the demand

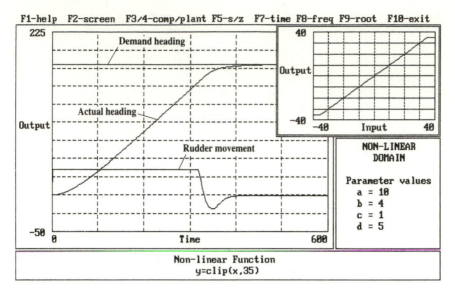

Figure 6.20 Turn about simulation of ship with rudder angle limitation

heading is suddenly changed by 180° the rudder will go hard over by the maximum amount of 35° and will stay like that until the error in heading is small enough for the rudder to come out of saturation. Thus for the major part of the turn-about manoeuvre θ is constant at 35°. Neglecting the initial and final transients, the heading will be changing at the rate of (35)(0.017) degree/second i.e. about 0.6 degree/second. Thus as a first approximation the 180° turn will take about 300 seconds (180/0.6).

A rather better approximation can be made if one realizes that during this manoeuvre the control system is effectively in open loop and the behaviour of the ship can be modelled as a ramp of 0.6°/s applied to a first-order lag of 55 seconds (i.e. the time constant τ_1 associated with the ship dynamics). As described in Chapter 3, the output will lag behind by 55 seconds. Thus the turn about time will be 355 seconds neglecting the final transient.

When the rudder comes out of saturation the rate of change of heading reduces and eventually the ship settles down and points towards the desired heading. This final transient contributes a further delay making the actual turn about time significantly greater.

CODAS-II can be used to simulate the whole situation exactly by introducing a clipping nonlinearity on the rudder angle. The use of CODAS-II to simulate nonlinearities will be discussed in Chapter 10. However the results are included here for interest in Fig. 6.20. This simulation shows that the time taken to complete this manoeuvre is actually about 367 seconds to bring the heading within 5° of the steady-state value.

Study No. 2: Control of an Inverted Pendulum

This second study deals with balancing an inverted pendulum. The system is more artificial than the one covered in the ship steering study. Nevertheless the problem of balancing systems similar to the simple one being considered here do occur in practice in the areas of missile stabilization and robotics. Furthermore the problem brings out some ideas and concepts not covered before.

Statement of problem The object of this study is to design a control system to balance an inverted pendulum. The 'pendulum' is connected by a simple pivot to a carriage. The effective length, l, of the pendulum is 250 mm. A servomotor is available to control the position of the carriage.

Performance specifications This is a very difficult system to control and so the main objective here is quite simply to balance the inverted pendulum. However the general guidelines are to design for a well damped system ($\zeta \approx 0.7$) which settles quickly.

Design evaluation What is the 5% settling time of the final system? What is the effect on the final design of a 20% variation in the effective length of the pendulum?

Design and instrumentation of the rig Although the position of the top of the pendulum cannot be measured directly it can be inferred from a knowledge of the angle of the pendulum and the position of the carriage. The angle of the pendulum, θ, can be obtained by attaching a potentiometer to the pivot, or indeed making the potentiometer the pivot. The position of the carriage can be measured by attaching a cable to the carriage which in turn is connected to a potentiometer to give a signal proportional to the carriage displacement, x. Figure 6.21 shows a schematic diagram of the rig with the associated instrumentation.

From simple geometry the position of the top of the pendulum, y, is related to x

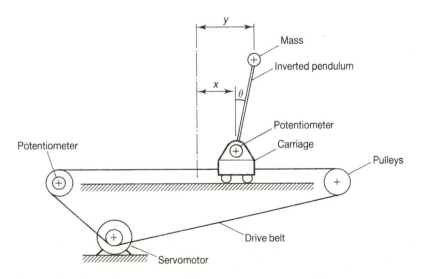

Figure 6.21 Schematic diagram of inverted pendulum rig

and θ by

$$y = l\sin(\theta) + x \approx l\theta + x$$

assuming small angles. Since, hopefully, the pendulum will be balanced or nearly so, the small angle approximation is valid.

At this point a strategic decision is required whether or not to use a position servo to position the carriage or to apply feedback around the entire loop to derive the control signal to the servomotor. A subsidiary position control loop offers a number of advantages. The response of the carriage position control system will be faster. Furthermore by putting feedback around the servomotor the overall stabilization problem is eased as one is not introducing an additional integrator into the forward path. The servo subsystem can be designed independently to behave like a well-damped second-order system with no positional errors. Step response and frequency response tests can be carried out on the subsystem without introducing the complicating factor of the inverted pendulum.

Tests on the servo system show that it can be modelled as a second-order system with a natural frequency of about 10 Hz and a damping ratio of 0.7.

Design: servo dynamics neglected Initially the servo dynamics will be neglected. Very often it is good practice to simplify the problem and produce an initial design neglecting complicating but second-order effects.

The transfer function relating the position of the pendulum mass to the position of the pivot (carriage) is

$$\frac{Y}{X}(s) = \frac{1}{1 - d^2 s^2}$$

where

$$d^2 = l/g$$

Hence for a pendulum whose effective length is 250 mm, d is about 0.16 rad/s and so the transfer function becomes:

$$\frac{Y}{X}(s) = \frac{1}{1 - 0.0255 s^2}$$

The form of the above transfer function is a little misleading when considering the s-plane because the coefficient s^2 is negative. In order to avoid any ambiguities the transfer function is better written as

$$\frac{Y}{X}(s) = \frac{-1}{(ds - 1)(ds + 1)} = \frac{-1}{\left(\dfrac{s}{6.26} - 1\right)\left(\dfrac{s}{6.26} + 1\right)}$$

i.e. there are a pair of real poles located at ± 6.26 and there is an inherent negative gain associated with this system. With simple proportional control using negative feedback, the two poles will simply move along the real axis: the positive pole will migrate to the right (towards to $+\infty$) and the negative pole to the left (towards to $-\infty$).

Stabilization can only be effected by inducing the pole in the right-half plane to move to the left. The only way of achieving this left migration is to use negative gain in the controller (i.e. positive feedback) which cancels the inherent negative system gain. In this case negative gain makes sense too from a practical and intuitive point of view. If the pendulum starts dropping to the right, the carriage must move to the **right** as well to catch up with the pendulum and balance the system. Normal negative feedback produces a control effort that is in the opposite sense to the change in measured value. In this case the control effort must be in the **same** sense as the change in measured value; hence the use of negative gain.

From the point of view of behaviour in the s-plane, negative gain will cause the poles to move towards each other and then break away along the imaginary axis. Thus negative gain alone is insufficient to completely stabilize the system, but at least it is moving the offending pole in the right direction.

In order to move the pole into the left-half plane, a zero is required in the left-half plane. One simple way of achieving this is to introduce a lead compensator of the form

$$G_c(s) = \frac{1 + a\tau s}{1 + \tau s}$$

The question is where to place the pole and the zero due to the compensator. One simple approach is to place the compensator zero so as to cancel the pendulum pole at -6.26; i.e. the form of the compensator becomes

$$G_c(s) = \frac{1 + ds}{1 + \tau s}$$

The compensator introduces a pole at $-1/\tau$ and so the system reduces to having two real poles, one due to the system at $+1/d$ and the other due to the compensator at $-1/\tau$. The lead compensator pole is to the left of the zero and so the root locus is essentially identical to the one dealt with at the beginning of the chapter in Fig. 6.2 except that the whole locus has been shifted to the right.

It is interesting to consider cancelling the system pole in the right-half plane, i.e. introduce a compensator zero at $s = -1/d$. The system can never be stabilized in this way for the reasons discussed in Section 3.11. The interested reader may like to try it out and see what happens!

The requirement is for a damping ratio of 0.7 and a fast settling time. The bigger the 'a' value, the further to the left the compensator pole lies and the higher up the 0.7 damping ratio line intersects the root locus. Thus using the maximum allowable 'a' value of 10 maximizes the speed of response and places the compensator pole at $-10/d$ (i.e. $\tau = d/a$). The root locus can now be drawn and a gain value determined that will place the dominant conjugate poles on the 0.7 damping ratio line. Figure 6.22 shows the root locus and the resulting transient response. The desired value of gain is about -5. It can be seen that the system settles with very little overshoot in about 0.2 seconds. There is a large static offset which is due to the low loop gain ($K_p = -5$), but in this case no requirement was stipulated for steady-state accuracy. All this offset means in practice is that the carriage is a little further along than desired.

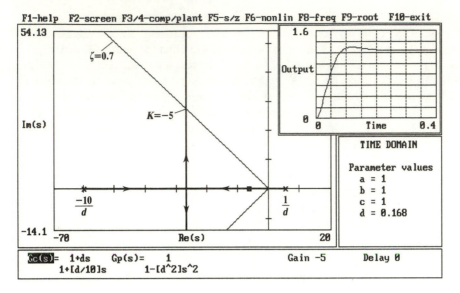

Figure 6.22 Root locus and step response of inverted pendulum with cancelling phase lead compensator and ignoring servo dynamics

Note: As the positional error constant is negative and greater than unity, the steady-state positional offset is negative, i.e. the steady-state position of the pendulum is greater than the demand position.

Design: servo dynamics included The foregoing design neglected the servo dynamics. The servo dynamics are essentially second order, i.e.,

$$\frac{X}{U}(s) = \frac{1}{(1 + a_1s + a_2s^2)}$$

Where

$$a_2 = 1/\omega_n^2$$

and

$$a_1 = 2\zeta/\omega_n$$

From the tests carried out on the servo system, its natural frequency and damping ratio are 10 Hz and 0.7 respectively, thus a_1 is 0.02 and a_2 is 0.000 25.

The overall relationship between control effort and the position of the top of the pendulum is hence of the form

$$\frac{Y}{U}(s) = \frac{1}{(1 + a_1s + a_2s^2)(1 - d^2s^2)}$$

The root locus of the system incorporating the servo dynamics and the pole cancelling phase lead compensator are drawn in Fig. 6.23. The allowable gain has now been

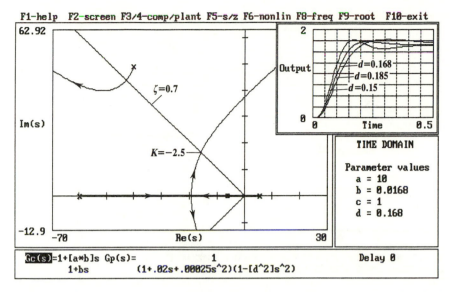

F1-help F2-screen F3/4-comp/plant F5-s/z F6-nonlin F8-freq F9-root F10-exit

Figure 6.23 Inverted pendulum including servo dynamics and variation in the effective length of the pendulum

reduced to -2.5 but the overall step response is still good. However, the steady-state offset is now rather large. Nevertheless as explained earlier this just means that the pendulum ends up a bit further over than desired.

A change in the length of the pendulum of 20% causes approximately a 10% variation in the parameter, d, of the inverted pendulum and so the pole cancellation will not be perfect. The effect on the root locus is not significant. This is evidenced by the responses drawn in the inset of Fig. 6.23. Parameters have been used extensively in CODAS to produce the results of Fig. 6.23. The parameter 'b' represents the time constant τ and 'd' is used to represent the natural frequency of the pendulum.

The final question that deserves some consideration is whether the pole-cancelling strategy is the best one. In order to investigate this a general lead compensator is considered and the characteristic equation of the system rearranged to make τ a multiplicative factor, i.e.

$$\tau s \frac{(1 + Ka) + a_1 s + (\omega_n^2 - a_2)s^2 - a_1\omega_n^2 s^3 - a^2\omega_n^2 s^4}{(1 + K) + a_1 s + (\omega_n^2 - a_2)s^2 - a_1\omega_n^2 s^3 - a_2\omega_n^2 s^4} = -1$$

The root contour of this system can be drawn for various values of gain, K. After a little bit of trial and error the root contour of Fig. 6.24 is obtained for a gain of -3.5. It is not entirely clear what value of τ to choose, but when τ is 0.0133 seconds at least one pair of conjugate roots lie on the 0.7 damping ratio line. The transient response shows a large overshoot, but the steady-state offset has been reduced as the gain has been increased from -2.5 to -3.5. The settling time is still good. The overshoot is due to the effect of the compensator zero which now no longer cancels the pendulum pole. The compensator zero is more dominant now as it lies to the right of the pendulum pole closer to the origin of the s-plane.

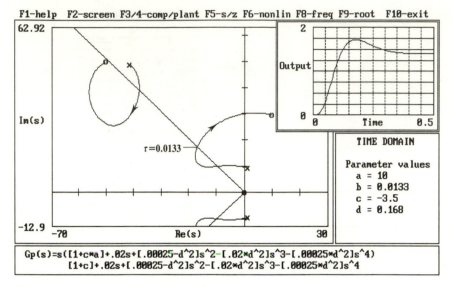

Figure 6.24 Root contour, inverted pendulum

At this point it is time to stop the design study and implement the system with a lead compensator with components nominally set to give an 'a' factor of 10 and a time constant, τ, of 15 ms. The compensator implementation should allow both parameters and the gain to be varied slightly *in situ*. Time spent in design and simulation is time well spent as it forms a sound basis for implementation. However, system nonlinearities and modelling errors make it unwise to spend an excessive time fine tuning a design.

PROBLEMS

6.1 The open-loop transfer function of a system is

$$KY(s) = \frac{K(s+1)(s+4)}{s(s+2)(s^2+6s+10)}$$

(a) Draw the open-loop pole/zero locations on graph paper.

(b) Establish the regions of the real axis where root loci will appear for positive values of gain, K.

(c) Determine the angle and intercept of the asymptotes of the root loci.

(d) Make a **sketch** of the root locus.

(e) Mark the point on the root locus where the imaginary value is 3 rad/s. Check that the location of this point is approximately correct by measuring the total angle contribution of the open-loop poles and zeros to it. If necessary move the point horizontally until the angle is correct to within $\pm 2°$.

(f) Using the point determined in (e), find by **measurement** the value of gain that gives the closed-loop complex poles a damped natural frequency of 3 rad/s. Where are the other closed-loop poles located at this value of gain?

Use CODAS to confirm the results you predicted.

6.2 A robot control system with unity feedback has an open-loop transfer function of

$$G(s) = \frac{K(s+2)}{(s^2+6s+10)(s^2+4s+6)}$$

Plot the root locus. With the controller gain, K, set at 20, determine the position of the dominant conjugate closed-loop poles. Plot the step response of the system. Superimpose on the trace the open-loop step response of a pure second-order system whose damping ratio and natural frequency are the same as the dominant poles and whose open-loop gain is the same as the steady-state gain of the original system. Comment on the accuracy of the dominant pole approximation.

6.3 A hydraulic servomechanism has an open-loop transfer function

$$G(s) = \frac{K(s^2 + 2s + 3)(s + 2)}{s^5 + 40s^4 + s^3 + 20s^2 + 10s + 6}$$

At what value of K do both pairs of conjugate poles have the same damping ratio?

Hint: because of the complexity of the system increase the number of points used in CODAS to 200.

With this value of K plot the closed-loop step response and determine the peak overshoot and 5% settling time of the system.

6.4 An exothermic chemical reaction can be modelled by the transfer function

$$G_2(s) = \frac{20}{s^2 - s + 2}$$

as shown in Fig. P6.4. In order to stabilize the plant a feedback compensator, $H_2(s)$, is employed.

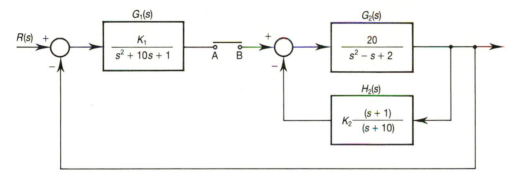

Figure P6.4 Stabilization using internal feedback

Consider the inner loop alone (i.e. link AB open) and plot the root locus as K_2 is varied and hence show that the damping ratio of the conjugate poles of this loop is maximized when K_2 is about 2.25. Obtain the closed-loop transfer function of the inner loop with this value of K_2.

Now consider the overall system with the main controller, $G_1(s)$ included (i.e. link AB closed). Plot the root locus as K_1 is varied and find a value of K_1 that gives the dominant conjugate poles a damping ratio of 0.7. Calculate the steady-state offset to a unit step demand.

Examine the step response of the overall system and verify the offset. Compare the peak overshoot of the actual system with that predicted by the dominant complex poles alone.

6.5 The open-loop transfer function of a unity feedback control system is given by

$$G(s) = \frac{K}{s(s^2 + 6s + a)}$$

(a) Given that the nominal value of 'a' is 16, obtain the position of **all** the closed-loop poles when K is 10 s^{-1}. From their positions decide on which pole(s) will dominate the response. Hence estimate the nature of the corresponding closed-loop step response and its salient characteristics. Confirm your predictions using CODAS. Repeat for K equal to 20 s^{-1}.

(b) Maintaining K at 20 s^{-1}, draw a root contour with 'a' as a parameter and obtain a value of 'a' for the three closed-loop poles to be equally attenuated. Plot the step response of the closed-loop system and find the 5% settling time.

6.6 The open-loop transfer function of an open-loop unstable position control system is given by

$$G(s) = \frac{K(1 + 0.5s)}{(s^2 - 1)(1 + \tau s)}$$

where the time constant, τ, is nominally 0.25 seconds and the desired positional accuracy requires that the gain, K, is at least 10.

(a) Plot the root locus of the system and locate the closed-loop poles with, K, set at 10. Examine the closed-step response. What is the settling time and steady-state offset?

(b) Draw a root contour with τ as the parameter and hence determine the value of time constant that (i) guarantees a damping ratio of the closed-loop conjugate poles of 0.7 and (ii) maximizes the damping ratio of the closed-loop conjugate poles.

Obtain the corresponding time responses for the cases (i) and (ii) and compare the performance with the results obtained in (a).

6.7 The open-loop transfer function of a system is given by

$$G_1(s) = \frac{30}{s^2 - 2s + 2}$$

and it is to be stabilized by a lead network of the form

$$G_c(s) = \frac{s + \alpha}{s + 10\alpha}$$

placed in cascade with the system. Determine the range of α values for which the closed-loop system will be stable and find the value of α which will maximize the attenuation of the conjugate poles. Obtain the step response of the closed-loop system and comment on the performance of this system and how it correlates to the closed-loop s-plane pole/zero positions.

6.8 The open-loop transfer function of a Type-2 system is given by

$$G_1(s) = \frac{1}{s^2(1 + 0.1s)}$$

The control system is to be stabilized by means of a lead compensator of the form

$$G_c(s) = \frac{1 + a\tau s}{1 + \tau s}$$

Design a compensator which will give the dominant conjugate pole a damping ratio of 0.7 and introduce as little phase advance as possible.

6.9 The open-loop transfer function of a liquid level control system is

$$G(s) = \frac{K}{(1 + \alpha s)(s + 2)(s + 3)}$$

The parameter α depends on the diameter of the vessel in which the liquid is contained and the system is designed for an α value of 1 minute. What value of K will position the closed-loop poles so that they exhibit a damped natural period of 5 minutes. What is their corresponding damping ratio?

What are the allowable variations in α that constrain the damping ratio of the conjugate poles to the range 0.6 to 0.7?

6.10 The open-loop transfer function of a unity feedback system with transport delay is

$$G(s) = \frac{K\,e^{-0.25s}}{s(1 + 0.2s)(1 + 0.05s)}$$

At what value of gain, K, do the conjugate poles exhibit a damping ratio of 0.5? Examine the corresponding closed-loop step response and compare with that of a pure second-order system having the same damping ratio and natural frequency as the conjugate poles of the actual closed-loop system.

SEVEN
PROCESS CONTROL

7.1 INTRODUCTION

Process control deals with the control of equipment and plant in the process industries. The terms 'process industry' and 'process plant' are rather vague and more or less cover every industrial process that exists. All industrial plant needs control, but the most important industries where process control systems are used and process control engineers are employed are in the fields of petrochemicals, food, steel, glass, paper and energy.

A process control engineer must understand the processes and equipment involved in his industry to great depth. He (or she) will need to understand how changes in operating conditions affect product quality and how variables within the plant interact, how the plant is integrated with other plants and how they affect each other. The process control engineer will be familiar with different types of transducers and instrumentation systems and their associated costs and performance characteristics.

Safety considerations are always paramount. Before any thought is given to conventional control systems, safeguarding and monitoring systems must be considered. During the design phase of the control systems, questions must be posed and answered of the type 'what happens if this equipment fails?'. Very often thinking of the right question is more difficult than getting the right answer. The requirements of the control systems can often have repercussions on the design of the processes and the selection and cost of associated equipment.

It is clear that no book or course can make a process control engineer. This chapter therefore provides a little glimpse into the subject of process control and deals with one main topic in some depth, namely three-term controllers.

7.2 PROCESS CONTROL TERMINOLOGY

The early development of the disciplines of process control and servo systems occurred side by side with very little cross-fertilization. During the Second World War the greatest advances were made in the field of control engineering but security prevented open publication and debate on developments. As a result the language and terminology used by process control engineers differed from that used by the servo designer despite the fact that both disciplines were striving for broadly similar objectives.

One rather basic difference between a process control system and a servo system is that generally the emphasis in process control is on the performance of the loop as a regulator, i.e. disturbance rejection, whereas in servo systems the emphasis is on how well the control system can follow changes in the reference or demand signal. That is not to say that process control systems are not subject to changes in reference conditions nor are servo systems never subject to disturbances. What is true is that in a typical process control system the reference value will not change frequently. For example, the required temperature of a particular product or stream may be constant for days. In fact during reference value changes the product is often recycled or dumped until the plant has stabilized, and so productivity and profitability require long runs with hardly any change or disturbance of any type. Disturbances will, nevertheless, occur and the most common and severe form of disturbance in the process industry is that due to a change in throughput or load brought on by changes in customer demand, lack of storage, shortage of feedstock, etc.

One important point that influences terminology in the process industry is that quantities are often treated **non-dimensionally**. For example a valve may be 50% open, a flow is 80% of design, etc. There are several reasons for adopting this approach, but one factor is the huge diversity of variables that have to be controlled even in a small plant. It is very difficult for operators to remember every absolute pressure, flow, temperature, etc., but by using non-dimensional representation of quantities the task of keeping track of what is going on is eased. Another factor is a firm awareness that everything within the plant is constrained to lie within certain limits. For example a valve can only operate from fully closed to fully open, a temperature sensor is designed to give correct readings over a certain range. Thus it is natural to express quantities as fractions or percentages.

The term *set-point* is used to represent the reference input to the process control system. The rationale for this terminology is that in process control systems the reference represents a desired constant operating point. The set-point can be determined by turning a knob, typing in a value or it can be transmitted to the controller from a computer or another controller.

The term *measured value* has been discussed previously. The measured value represents the output of the measurement system (transmitter). The measurement system produces a signal over a particular operating range which is a function (usually linear) of the actual value of the process variable that is being controlled. Some common measuring elements, however, produce signals which are nonlinear functions of the process variable, for example orifice plates used in flow measurement produce a differential pressure which is proportional to the square of the flowrate. The signal may be electronic (voltage/current, analogue or digital), pneumatic or even mechanical in very simple systems.

The set-point is compared to the measured value to produce a *deviation* which is simply the difference between the two. Clearly in order that these two quantities can be compared and a difference taken between them, the units of these two quantities must be the same. Thus the deviation could be quoted in engineering units (e.g. °C), but very often it is quoted as *fractional* or *percentage deviation* with respect to the measurement span.

In order to understand the idea of fractional deviation, let us consider a temperature control system where the measurement system has a measurement range of 50°C to 100°C, i.e. the measurement span is 50°C. Note that a **range** requires two numbers, whereas a **span** is a single figure. The figure of 50°C corresponds to a 0% measured value reading, and the 100°C corresponds to a 100% measured value.

Suppose that the set-point is 65°C and the measured value is 62°C. The absolute error is 3°C, but the percentage deviation is 6%, i.e.

$$\text{Percentage deviation} = \frac{\text{Error}}{\text{Measurement span}} \times 100 = \frac{3 \times 100}{50} = 6$$

The advantage of the non-dimensional approach is that it is more closely tied to the engineering reality. In itself an error of 3°C has very little meaning. Is 3°C a big error or a small error? Unless one is familiar with the process and the operating constraints, the question is meaningless. However a 6% deviation has greater meaning, for if the process had to be controlled to a much higher tolerance (say ±0.5°C), a temperature transmitter would have been used with a much lower span (say 10°C rather than 50°C) and then with the same **absolute** deviation the fractional deviation would have been much higher signalling that the process was totally out of control.

Another factor that supports the use of non-dimensional quantities is that percentage deviations greater than 100% or less than 0% are impossible. Once the process variable is outside the measurement range of the instrument, it is impossible to tell whether the deviation is 200% or 2000%. A measurement system produces a signal of finite range corresponding to a finite range of the process variable.

A non-dimensional approach is also adopted for describing the output of the controller. The control effort is expressed as a fraction or percentage of the controller output span. As an example, suppose a controller produces an output ranging from 4 mA to 20 mA and the control effort changes by 2 mA. The percentage change in control effort is thus (2/16)(100), i.e. 12.5%.

7.3 PRACTICAL CONTROLLERS

In the previous chapters target design specifications were met either by changing the system gain or by designing specific compensators for a particular system whose transfer function was known. In process control it is very uncommon to design a fixed compensator because the dynamics of the plant are uncertain and dependent on operating conditions. Thus a general purpose controller is normally utilized which has a number of variable parameters which can be changed or tuned to meet the static and dynamic requirements of the control system. In order to introduce the ideas behind the general purpose controller, a simple proportional controller will be considered first.

The concept of proportional control has been used many times previously, i.e. the control effort, u, is directly proportional to the error, e, i.e.

$$u = Ke \tag{7.1}$$

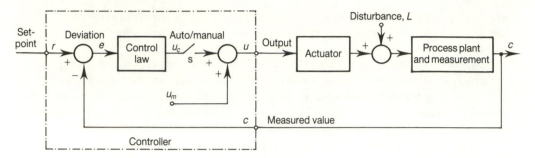

Figure 7.1 Block diagram of a process control system

where K is the gain or sensitivity of the controller. If one interprets Eq. (7.1) in a practical sense, say in terms again of a temperature control system, the control effort, u, would represent the heat input into the process, the error, e, is the difference between the set temperature and the measured temperature and the gain, K, tells us how much heat is put into the process for every °C error. The absurd conclusion that one reaches from this equation, however, is that when there is no error (deviation) there is no heat input into the process! In practice the output of a controller is generally **not** zero when there is no deviation but when a deviation does occur the controller **augments** (or diminishes) the control effort by an amount proportional to the deviation. Thus Eq. (7.1) should be correctly written as

$$u = Ke + u_m \qquad (7.2)$$

where u_m represents a base level or quiescent output which is constant. Hence the Eq. (7.1) that has been used for proportional control is correct provided it is interpreted as meaning a **change** in control effort rather than an absolute value.

Figure 7.1 schematically shows the elements of a practical controller in a process control system. In a practical controller the output can be adjusted by the operator directly. This is termed *manual* operation, i.e. the controller output can be adjusted directly by turning a knob or keying in a value and will not be affected by the deviation. This is the value u_m in Eq. (7.2). When the controller is switched to *automatic* (or more usually *auto* for short), an additional component, u_c, is added to the manual control effort, u_m, to produce the total control effort, u, i.e.

$$u = u_c + u_m \qquad (7.3)$$

The variable component, u_c, is the outcome of a control law which in the case of a simple proportional controller is simply Ke.

Note: Once the controller is in automatic, u_m is held constant at the value it was at the instant of change over.

Normal practice is to have the controller in manual, make the set-point equal to the desired value and adjust the controller output manually until the deviation is zero, i.e. the plant is operating steadily at the desired operating point. The operator can then switch to auto. Since at the instant of switching over to automatic control there is no deviation, the output is simply the manual output, i.e. $u = u_m$.

A problem can arise, however, in the switch over. Suppose the operator decides to change the operating point of the plant manually to a new value. On this occasion the operator forgets to change the set-point and leaves it at its previous value and simply observes the measured value and not the deviation as the controller output is adjusted. When the plant has settled to the new operating condition, the controller is switched to auto. At this instant a deviation exists because the old set-point does not match the new measured value and the control effort will suddenly change causing a significant disturbance or 'bump' so driving the plant away from the desired operating point.

A similar bump can occur when switching from auto to manual. In older controllers a separate dial was visible showing the manual control effort, u_m. Once the controller is in auto, changing the manual control effort has no effect on the overall controller output. Thus to bring one of these controllers back to manual, the procedure was to adjust the manual control effort to be equal to the controller output and then to switch to manual. In this way *bumpless transfer* could be achieved.

Modern controllers incorporate automatic bumpless transfer. That is not to say that the measures described above for avoiding bumps should not be followed. Indeed they should because they represent sound engineering practice. With modern controllers, even if a large deviation exists, no bump will occur when changing from manual to auto. However, the anomalous situation will exist of a plant operating steadily at its desired operating condition with a large deviation and an incorrect set-point! Various mechanisms exist for achieving bumpless transfer, but the details will not be discussed here.

7.4 PROPORTIONAL BAND

The idea of proportional control was discussed above in terms of a controller gain or sensitivity, K. Generally speaking in industrial controllers the gain of the controller is described in terms of its *proportional band* (*PB*) or its percentage proportional band (%*PB*). The proportional band represents the fractional deviation (or fractional change in measured value) that will generate 100% change in control effort. It is best described by means of a diagram (see Fig. 7.2). Line 'a' in Fig. 7.2 represents a controller with 100% *PB*. When the proportional band is narrowed a more sensitive controller is produced. Line 'b' shows a controller with a 50% *PB*. By widening the proportional band one gets a less sensitive controller such as line 'c', where the proportional band is 200%. Clearly one cannot get 200% changes in measured value; the 200% refers to a notional change in measured value that would elicit a 100% change in control effort. Equally with a narrow proportional band, changes in measured value outside that proportional band will have no effect on the controller output, i.e. it will be *saturated* at 0% or 100%.

The equation of proportional band, *PB*, is thus

$$PB = \frac{\text{Fractional deviation}}{\text{Fractional change in control effort}} \tag{7.4}$$

Now the fractional deviation is the error, e, divided by the measurement span, and the fractional change in control effort is the change in control effort, u_c, divided by the

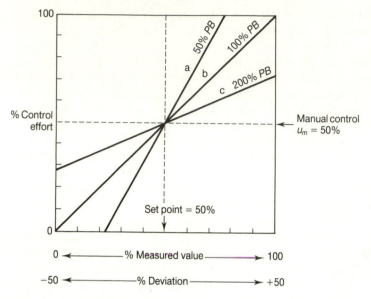

Figure 7.2   Proportional band

output span of the controller. Hence Eq. (7.4) can be rewritten as:

$$PB = \frac{e}{\text{Measurement span}} \frac{\text{Controller output span}}{u_c} \qquad (7.5)$$

Since the controller gain, K, is u_c/e, Eq. (7.5) can be written to show the relationship between proportional band or more commonly percentage proportional band and controller gain, namely

$$\%PB = \frac{1}{K} \frac{\text{Controller output span}}{\text{Measurement span}} 100 \qquad (7.6)$$

Equation (7.6) shows that the proportional band and the gain of a controller are related reciprocally. The higher the gain the narrower is the proportional band, and the lower the gain the wider is the proportional band. It is incorrect to state that the proportional band is simply the reciprocal of the gain. Generally controller gain is a **dimensioned** variable whereas proportional band is always **dimensionless**.

Example 7.1 A temperature controller has a 60% *PB* and its input range is 50°C to 100°C and its output range is 4 mA to 20 mA. What is its gain?

SOLUTION Equation (7.6) can be rearranged in to express gain in terms of proportional band. Furthermore the measurement span is equivalent to the controller input span, hence

$$K = \frac{100}{\%PB} \frac{\text{Output span}}{\text{Input span}} = \frac{100}{60} \frac{16}{50} = 0.53 \text{ mA/°C}$$

The above gain figure is pretty meaningless and can take on any value depending on the engineering units that are being used. The percentage proportional band, on the other hand, has some meaning even if one does not know anything about the plant. It is most unusual to employ proportional bands narrower than about 10% and wider than 500%. If one gets figures of proportional band outside this nominal range questions should be asked about the selection of instrumentation ranges and sizing of actuators on the plant.

7.5 DYNAMIC MODELS OF PROCESS PLANT

The dynamic behaviour of process plants is very varied. They can be unstable in open-loop (exothermic reactions) and can exhibit significant non-minimum phase behaviour (e.g. level in boiler drums caused by the 'swell' phenomenon). However the vast majority of process plants can be modelled adequately as first- or second-order (overdamped) systems with transport delay.

The simplest model of a process plant consists of a single lag plus a transport delay, i.e.

$$\frac{C}{U} = K_1 \frac{e^{-sT_D}}{1 + s\tau} \tag{7.7}$$

The constant K_1 is the steady-state gain of the plant. The step response of this model shows a sharp and unrealistic discontinuity after the transport delay has elapsed. A more realistic model is obtained by adding a second lag, i.e.

$$\frac{C}{U} = \frac{K_1 e^{-sT_D}}{(1 + s\tau)(1 + s\tau_2)} \tag{7.8}$$

PCS: Process Control Simulation Package

CODAS is a very general purpose program and it can be used to study process control systems, but a special purpose program for the study of process control systems is available called 'PCS'. The structure of the model used in PCS is shown in Fig. 7.3. The advantage of PCS is that it is tailored for process control and so it is quicker and simpler to use for this purpose. For example, load disturbances can be introduced easily in PCS for studying the regulator performance of the loop.

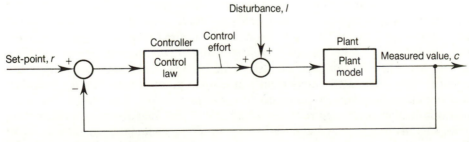

Figure 7.3 Model structure used in PCS

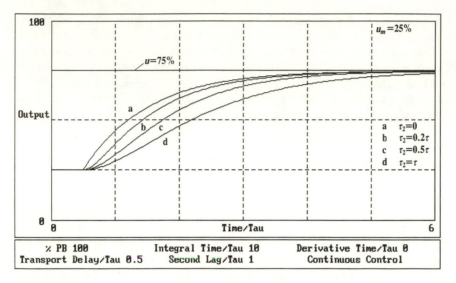

Figure 7.4   Open-loop responses of PCS prototype systems

PCS uses fixed forms of the plant model with variable parameters, whereas with CODAS the nature of the plant transfer function is determined by the user. The models in PCS are first- or second-order models with optional transport delay and so cater for typical process plants. For more complex plants with non-standard dynamic behaviour CODAS can be used as a simulation and design tool.

The transport delay is always expressed in terms of τ, e.g. the default plant has a transport delay of 0.5τ. The steady-state gain of the plant is fixed arbitrarily at unity. This does not matter as the actual plant gain can be 'transferred' to the controller when doing a simulation.

Figure 7.4 shows the open-loop step responses of the prototype models in PCS with various combinations of the two lags. The initial (manual) control effort is 25% and a step increase in control effort of 50% is applied.

7.6 CONTROL ACTIONS, THREE-TERM CONTROL

In servo system compensator design we used gain adjustment, lead compensation, lag compensation or a combination of all three corrective elements. The idea behind a general purpose process controller is to employ the three elements in a flexible manner so that they can be adjusted *in situ* to produce the desired response. The three most common elements or actions are called *proportional*, *integral* and *derivative* action.

In this text there is no detailed discussion of how three-term controllers are implemented; they are treated purely from a functional point of view. Briefly, however, the most popular technology for many decades was pneumatics. The reasons for using pneumatic controllers were their robustness and safety. The final actuator in the process industries was invariably pneumatic too, so by adopting pneumatics in the controller additional and different power sources were not needed. One other aspect about

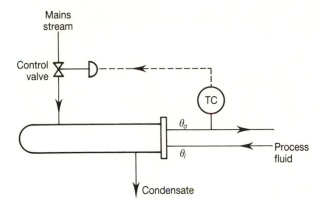

Figure 7.5 Conventional temperature control of a heat exchanger

pneumatics that made it very popular was that equipment maintenance and trouble shooting is relatively simple. The main drawback with pneumatics is the lack of flexibility and the cost compared to electronic controllers. Today digital microprocessor-based controllers have to all intents become standard because of their low cost, great reliability and sophisticated features.

Proportional control Proportional control has been discussed at some length in the previous chapters and so the treatment here is brief. One point that is important to remember is that most process plant is Type-0, and so with proportional control a steady-state error will exist after a set-point change or a load disturbance. In the process industries this phenomenon is termed *offset*.

To understand why offset occurs, consider the following practical situation. A certain plant is making a product which has to be kept at constant temperature by means of a steam heater (Fig. 7.5). The flow of steam is controlled by a pneumatically actuated valve.

The dynamic sequence of events described here is illustrated in Fig. 7.6. The temperature controller is in automatic and there is no deviation. After some time there is an increased demand for the product and so the product flowrate is increased (point A in Fig. 7.6).

The increase in flowrate places an increased load on the heater and so after the transport delay has elapsed the temperature of the product drops (point B). At that point a deviation is produced and so the controller increases the control effort (steam flow) but the temperature continues to fall for a while despite the increased heat input because of the dynamics of the plant (B–C). At point C the deviation starts falling and there is less heat applied but the product temperature rises for a while (C–D). Eventually the system settles down with a finite and constant deviation, i.e. the product temperature is below the set-point temperature. This constant deviation is the offset referred to earlier.

The only way that the offset can be removed is by increasing the steam flow to cater for the extra product. With proportional action alone, the controller can only produce extra control effort if there is a deviation. Thus there must be some offset. The

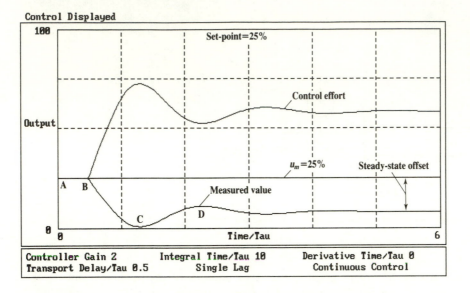

Figure 7.6   Response of a proportional control system to a load disturbance

amount of offset can be reduced by increasing the controller gain, but increasing the gain can make the response too oscillatory and eventually unstable.

The offset due to a lead increase, L, can easily be calculated as follows. With proportional control, the control law is

$$u_c = Ke = K(r - c) = -Kc \qquad (7.9)$$

As the control law is based on **changes** in the variables, the variable, r, representing the change in set-point is zero. The change in measured value, c, assuming a steady-state plant gain of K_1 is

$$c = K_1(u_c + L) \qquad (7.10)$$

Substituting for u_c from Eq. (7.9) into Eq. (7.10) and simplifying, gives

$$c = \frac{K_1 L}{1 + KK_1} \qquad (7.11)$$

The graph of Fig. 7.6 was produced using PCS with a controller gain of 2 and applying a load disturbance of -50%. As the plant gain in the PCS model is unity, the offset is $-50\%/(1 + 2)$, i.e. -16.67%. The only way of correcting the situation with a proportional controller is to restore it to manual (bumplessly of course!) and reset the control effort manually to accommodate the new operating conditions. This seems a bit pathetic and hardly worth the trouble of using a so-called automatic controller.

Proportional plus integral control In order to overcome the problems of offset encountered with proportional controllers, a control action is required which produces a finite control effort with zero deviation. This is achieved by augmenting the purely proportional term with an additional component which is proportional to the integral of the error. The introduction of an integrator into the forward path produces additional control effort which will continue to change until the error is zero, hence removing offset entirely. The term 'automatic reset' is used to describe controllers with integral action. The integral component is often referred to as 'reset action' to distinguish it from the proportional contribution (also called 'proportional action'). The amount of reset action is weighted by a constant, T_r, called the *reset time*. The variable control effort generated by a proportional plus integral controller (PI controller for short) is given by

$$u_c = K\left[e + \frac{1}{T_r} \int e\, dt \right] = u_p + u_r \qquad (7.12)$$

where the proportional contribution or proportional term, u_p, is

$$u_p = Ke$$

and the integral contribution or reset term, u_r, is

$$u_r = \frac{K}{T_r} \int e\, dt$$

The effect of the reset action can be seen more clearly by considering the open-loop response of a PI controller to a transient deviation as shown in Fig. 7.7. The inset schematically shows the make up of a PI controller and the two contributions to the

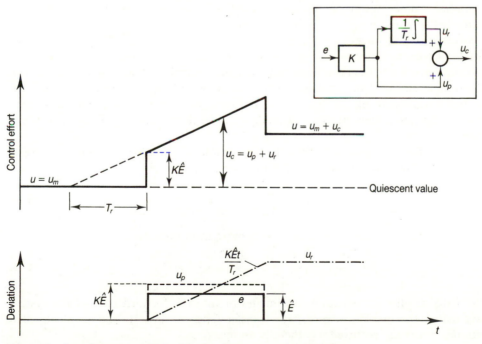

Figure 7.7   Open-loop response of a PI controller to a transient deviation

control effort. Initially the deviation is zero and the controller output is the manual value, u_m. Subsequently a step change of deviation of magnitude $\hat{E}$ is applied which remains constant for a while and then returns to zero. The initial jump in the control effort is due entirely to the proportional contribution, u_p, i.e.

$$u_p = K\hat{E} \qquad (7.13)$$

This contribution is constant provided the value of the deviation stays constant at $\hat{E}$.

The time integral of a constant is a linear function of time and hence the reset contribution, u_r, is

$$u_r = \frac{K\hat{E}}{T_r} t \qquad (7.14)$$

which is a ramp of slope of $K\hat{E}/T_r$ as shown in Fig. 7.7.

The smaller the reset time, T_r, the steeper the ramp and the greater is the integral contribution in a given time. The ramp persists until the deviation is zero, at which time the reset contribution stops changing and the proportional contribution jumps down by an amount $K\hat{E}$. Thus the controller output has changed to a new value even though the deviation is zero. This is precisely the effect that was required to remove offset.

It is interesting the note that after T_r seconds the reset contribution, u_r, is equal to the proportional contribution, u_p. (Substitute $t = T_r$ in Eq. (7.13) and compare u_r with the value of u_p in Eq. (7.14).) Referring to Fig. 7.7, the reset time can be found by projecting back the sloping part of the total control effort to where it intersects the manual control effort. The time from this point to where the step change of deviation was applied is the reset time, T_r. This fact is useful in checking the reset time of controllers independently of the proportional band setting.

When a PI controller is incorporated in a closed-loop system it will cause the control effort to change until the deviation is zero. However, as the system is in closed-loop, the deviation will not be constant and the way the control effort changes with time will be quite different from the open-loop situation shown in Fig. 7.7. Integral action will only have a significant effect if there is a deviation present for some time. That is, integral action introduces a slow dynamic element into the time response of the closed-loop system.

The transfer function of a PI controller can be found by applying Laplace transforms to Eq. (7.12). The transfer function, $G_c(s)$, of the controller is

$$G_c(s) = \frac{U_c(s)}{E(s)} = K\left(1 + \frac{1}{sT_r}\right) \qquad (7.15)$$

This transfer function can be written a little more meaningfully as

$$G_c(s) = \frac{K}{s}\left(s + \frac{1}{T_r}\right) = KG_c'(s) \qquad (7.16)$$

A PI controller introduces a free integrator into the forward path of the system, which makes the overall forward path transfer function Type-1, and so steady-state positional errors are removed, i.e. there is no offset.

Figure 7.8 shows the transient response of a process control system to a set-point

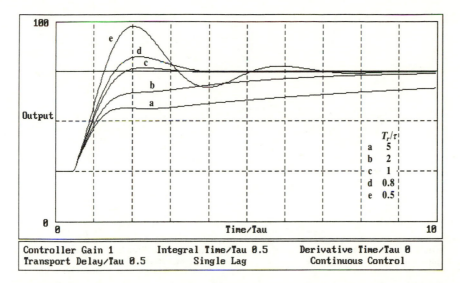

	T_r/τ
a	5
b	2
c	1
d	0.8
e	0.5

Controller Gain 1	Integral Time/Tau 0.5	Derivative Time/Tau 0
Transport Delay/Tau 0.5	Single Lag	Continuous Control

Figure 7.8 Response of a control system to a set-point change with varying amounts of reset action

change with varying amounts of reset time incorporated into the controller. The offset removing effect of reset action is clearly visible. Trace 'a', where the reset time is much too long shows the offset removal taking place far too slowly. Trace 'e' is at the opposite extreme. Here a reset time of 0.5τ has been used and the result is an oscillatory response which takes a long time to settle. The slowing down effect of reset action is apparent in the trace as there is an oscillation present with a period of about 4τ. Trace 'c' (reset time of 0.8τ) seems just about ideal. However before rubbing one's hands in glee, the performance of the loop as a regulator should be considered.

Figure 7.9 shows a set of responses when the loop is subjected to a step change in load. This time trace 'd' ($T_r = 0.8\tau$) is better because the disturbance is removed more quickly. In fact, even a more lightly damped response such as 'e' may be preferred as, provided there is a sufficiently wide tolerance band on product specification, it gets rid of disturbance more quickly. This is one reason why, on the whole, lower damping ratios and stability margins are more acceptable in process control systems than in servo systems.

Frequency domain interpretation of PI controller behaviour In order to understand the dynamic behaviour of the loop more clearly, the transfer function of the PI controller must be examined more closely. The transfer function in Eq. (7.16) has been written as $KG_c'(s)$ to contrast it with a proportional controller whose transfer function is simply K. The $G_c'(s)$ part represents the dynamic part of the transfer function and is present because of the integral action. In the frequency domain the dynamic part of the PI controller becomes

$$G_c'(j\omega) = \frac{j\omega + 1/T_r}{j\omega}$$

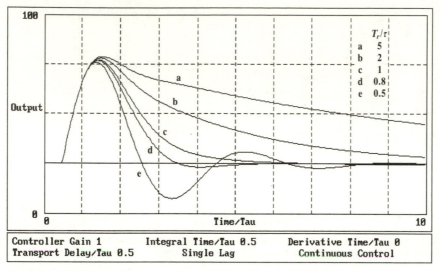

Figure 7.9 Regulator performance of a control system with varying amounts of reset action

i.e. its magnitude and phase contributions can be expressed as

$$|G'_c(j\omega)|^2 = \frac{\omega^2 + 1/T_r^2}{\omega^2} \qquad \angle G'_c(j\omega) = -\pi/2 - \tan^{-1} \omega T_r$$

which is very much like a lag compensator as it introduces a phase lag that increases with frequency.

The effect on a typical Nyquist plot of introducing reset action ($T_r = 0.8\tau$) into a control loop is shown in Fig. 7.10. Reset action introduces amplification which is very large at low frequencies. At higher frequencies the amplification factor reduces, dropping in the limit to a value of unity. Thus the effect in the direct Nyquist plane of reset action is to elongate vectors of the original plot and to twist them clockwise. The net result is that there is a slight degradation in stability margins and a marked lowering of the phase-crossover frequency. In the example in Fig. 7.10, the phase-crossover frequency has been lowered by about 25% which explains why the introduction of reset action slows down the dynamic behaviour of the loop.

Note: The Nyquist plot in Fig. 7.10 was obtained using CODAS. However be careful in comparing the time responses obtained in CODAS with those presented here using PCS. In CODAS there is no simple equivalent to the manual control effort that is available in PCS. To get the equivalent time domain responses from CODAS use zero initial conditions on control effort and plant output and apply a step change. The response relative to the **zero** line will then be the same as the time traces drawn in this chapter.

Regulator performance There is one point that is often misunderstood about the presence of an integrating element in a loop and the effect of disturbances. Figure 7.11 shows two control systems whose open-loop transfer functions are identical and both systems

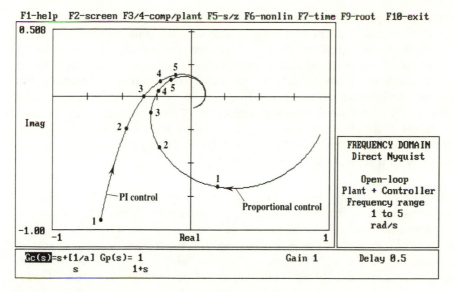

Figure 7.10   The effect of reset action in the direct Nyquist plane

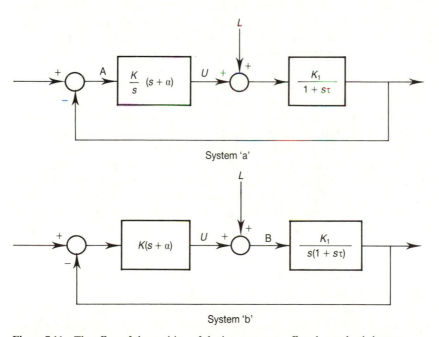

Figure 7.11   The effect of the position of the integrator on offset due to load changes

are Type-1. System 'a' has a PI controller and system 'b' is typical of a servo system with lead compensation. The fact that the form of the lead compensator is different from the type dealt with in Chapter 4 does not invalidate the argument. In fact this type of 'lead compensator' will be dealt with shortly under the topic of proportional plus derivative control.

Both systems in Fig. 7.11 are subject to load disturbances. As far as set-point changes are concerned the behaviour of the two loops is identical, but consider their respective performance as regulators. Suppose there is a change in load. The presence of an integrator in the loop will cause the signal immediately 'upstream' of the integrator to be zero, i.e. points A and B respectively for the two systems. Thus in system 'a' the error will be zero; which is what one expects. In system 'b', however, the signal at point B will be zero in the steady-state, thus the control effort must be equal and opposite to the load change in steady-state ($U = -L$). This means that there must be a finite error to produce that control effort and so there will be a finite offset notwithstanding the presence of the integrator in the forward loop.

Reset windup There is one last aspect about integral action that must be considered and that is, what happens when an increase in control effort does not reduce the deviation? Suppose a PI controller is in use in a particular loop and something happens whereby the controller cannot eliminate the deviation. There are many possible examples where this situation could arise, for example a control valve could stick, or the steam supply pressure to a heater could be too low and so the steam valve opens fully. Under these circumstances the controller output will grow due to the integral action, but without effect. When the cause of the trouble is eliminated, the deviation will drop rapidly, but because of the build up of reset action the controller output cannot change until there is a deviation present of the opposite sense and of sufficient duration to cancel the historical build up of control effort. This build up is termed *reset windup*. The result is a large overshoot and a considerable delay before the system is under control once more.

The adverse effects of reset windup can be reduced considerably by stopping the integrating action as soon as the control effort reaches the normal range limits. This is a very simple scheme to limit windup; other more sophisticated methods are available which sense changes in the actuating element. Controllers which employ techniques for limiting reset windup are said to incorporate *anti-reset windup*.

To illustrate the effects of windup graphically an extreme example is depicted in Fig. 7.12. The situation illustrated is of a total loss of control agent (e.g. loss of steam in a temperature control system) and then subsequently the control agent is regained. At point A the control agent is lost and the measured value (temperature) starts to drop. The controller opens the control valve in a vain attempt to admit more steam. After a short time the controller output reaches 100% but the temperature continues to drop. At point B, the control agent is restored. After the transport delay has elapsed the temperature starts to rise. With no anti-reset windup, a huge temperature overshoot occurs and the temperature does not reach the set-point until much later.

The reset build up during the initial positive deviation phase is proportional to the hatched area shown in Fig. 7.12. The integral contribution built up during this phase can only be cancelled by an equal and opposite deviation area. Hence the sustained period of negative deviation which lasts for a long time. With anti-reset windup incorporated in the controller, the reset action is suspended as soon as the controller output saturates and so the integral build up is far less. The result is that the controller comes out of saturation much sooner, there is virtually no overshoot and control is restored much more quickly.

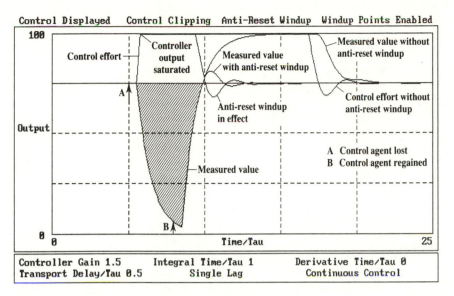

Control Displayed   Control Clipping   Anti-Reset Windup   Windup Points Enabled

Figure 7.12 The effect of reset windup and anti-reset windup

Proportional plus integral plus derivative control Integral action can be considered the steady cart-horse looking after long persistent errors. Proportional action responds immediately to changes in deviation but it is insensitive to how rapidly the deviation is changing. *Derivative action* adds in an extra term which is proportional to the rate of change of deviation. The variable control effort produced by a proportional plus integral plus derivative (PID) controller is given by

$$u_c = K\left[e + \frac{1}{T_r} \int e\, dt + T_d \frac{de}{dt} \right] = u_p + u_r + u_d \qquad (7.17)$$

where u_d is the derivative contribution. A factor is used to weight the amount of derivative action relative to the proportional action in a similar manner to reset time with integral action. The weighting factor, T_d, is called the *derivative time*. Note that derivative time appears in the numerator whereas reset time is in the denominator. Hence a derivative time of zero removes the derivative contribution, whereas an infinite reset time is required to eliminate the reset action.

Again it is instructive to examine the open-loop behaviour of the controller. To keep things simple, it is assumed that the integral contribution is negligible, i.e. a proportional plus derivative (PD) controller will be considered. In order to study its behaviour a triangular deviation transient is considered as shown in Fig. 7.13. The reason for choosing a triangular transient rather than a square pulse is that, in theory, differentiating a step change in deviation will produce an infinite change in control effort.

Referring to Fig. 7.13, the proportional contribution, u_p, is the same shape as the deviation but scaled by the proportional gain. The derivative contribution, u_d, is a positive constant whilst the deviation is changing positively, but when the rate of change of deviation is negative, the derivative contribution changes sign. The net effect is that control effort is augmented during the first half of the deviation transient and diminished

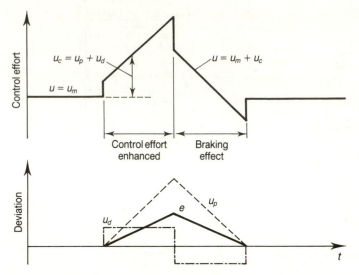

Figure 7.13 Response of a PD controller to a triangular transient deviation

during the second half of the deviation pulse which produces a very lively control effort. The analogy of the behaviour of a PD controller is of the car driver who wants to get from one set of lights to the next in a hurry. During the first part of the journey he applies the accelerator pedal hard and just before he gets to the lights he slams on the brakes.

In closed-loop the effect of derivative action is to increase the speed of response and, provided the derivative time is correctly chosen, it improves the damping. To understand how derivative action works in closed-loop, examine the response of the control system to a set-point change as shown in Fig. 7.14. In this example a system has been simulated with two lags. The one trace shows the response of the system under proportional control and the other with proportional plus derivative control. Note that the offset is unaffected by the introduction of derivative action since the steady-state derivative contribution is zero.

The response with derivative action $(T_d = 0.3\tau)$ is better damped, has a faster response and so settles more quickly. The reasons for the improved behaviour can be understood better by considering sections of the response in more detail. From B to C the deviation is positive and getting smaller, i.e. its rate of change is **negative**. In other words the measured value is moving in the right direction and the derivative action acts as a brake to reduce the overall control effort and so reduce the size of the overshoot. From C to D the rate of change of deviation is **positive**, i.e. it grows positively. During this phase the derivative action adds to the proportional control effort so helping to reverse the situation again and reduce the error. At point E the cycle of events repeats once more.

Dynamic effects of derivative action The transfer function of a PID controller can be found by applying Laplace transforms to Eq. (7.17). In order to study the effects of the derivative contribution it will again be assumed that the reset effect is negligible, i.e.

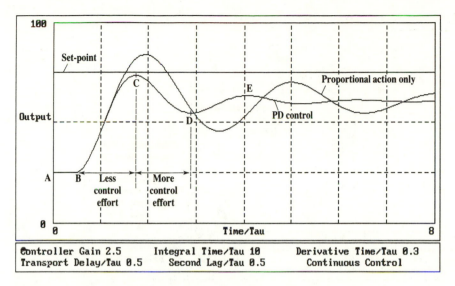

Figure 7.14 Proportional plus derivative control

$$G_c(s) = \frac{U_c(s)}{E(s)} = 1 + sT_d \qquad (7.18)$$

The first thing that is clear from Eq. (7.18) is that pure derivative control is not physically realizable (the transfer function is not proper). Thus in practice one or more lags must be present, but the poles associated with these lags will be situated well away from the numerator so that the above transfer function can be used for the purpose of evaluating and understanding derivative action.

The frequency response function of a PD controller is:

$$G_c(j\omega) = 1 + j\omega T_d$$

and hence the corresponding gain and phase functions are

$$|G_c(j\omega)|^2 = 1 + (\omega T_d)^2 \qquad \angle\, G_c(j\omega) = \tan^{-1}(\omega T_d)$$

Thus a PD controller acts as an extreme case of a phase lead compensator where the denominator pole is located at infinity. The gain of a PD controller is unity at low frequencies and the phase shift is zero. At high frequencies the gain becomes very big (tending to infinity) and the phase lead increases to 90°. Thus in the Nyquist plane the effect is to elongate vectors and move them **anti-clockwise**.

This means that the gain-crossover and phase-crossover frequencies are increased. Figure 7.15 shows the Nyquist plot of the system used to produce the traces in Fig. 7.14. With proportional control the phase crossover frequency is about 2.3 rad/s and the gain margin is about 1.5. When derivative action is incorporated correctly (trace 'b'), the gain margin is increased to nearly 2 and the phase-crossover frequency has increased by about 40%. If too much derivative action is employed stability margins will become worse because vectors are elongated more than the natural rate of

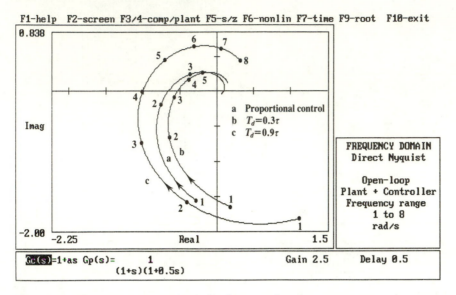

Figure 7.15 Effect of derivative action in the frequency domain

attenuation of the system at higher frequencies. Eventually with too much derivative action, the system can become unstable (see trace 'c').

Note: When using CODAS to simulate the time response of PD controllers, a problem arises because the transfer function of the controller is not proper. One solution is to define a lead compensator of the form

$$G_c(s) = \frac{1 + sT_d}{1 + s\alpha T_d}$$

where α is about 0.1. The result will not be significantly different from an idealized PD controller and indeed is rather nearer the truth for a practical controller. Another possibility if there is no need to view the control effort explicitly and the plant is at least second order is to include the $(1 + sT_d)$ term with the plant.

Noise The subject of noise has been mentioned before especially with reference to excessive bandwith when using phase lead compensation. A parallel situation arises in process controllers which employ derivative action. High frequency components in the input signal are amplified by the derivative term and appear exaggerated in the controller output signal. This noisy output signal will cause violent movements of the actuator which is undesirable from the point of view of wear and it may cause disturbances in the power supply. On the whole the plant output will not be affected too much because the dynamics of the plant act like a filter and high frequencies will be attenuated.

Figure 7.16 shows two traces which illustrate the effects of plant noise. The upper trace shows the regulator performance of a PD controller when the disturbance is just noise, and the lower trace is the identical situation (lower operating point) but with proportional control only.

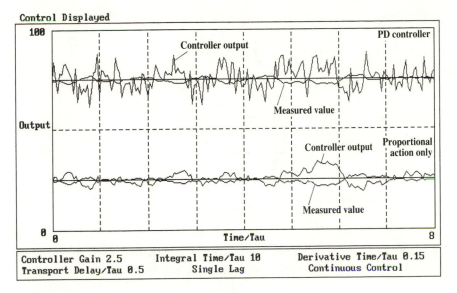

Figure 7.16 The effect of derivative action in the presence of noise

Derivative action should be avoided where there is a significant amount of plant noise or where there are noisy measurements. An example of a measurement which is inherently noisy is flow measurement using orifice plates where turbulence in the flow gives rise to high frequency components in the differential pressure. Filtering can be used to reduce the level of noise but care must be taken not to introduce significant new lags into the process. Furthermore, excessive filtering may render the derivative action ineffective by introducing a pole which essentially cancels out the derivative zero.

Set-point kicks The derivative contribution has been defined as a term proportional to the rate of change of **deviation**. Now the deviation is the difference between the set-point, r, and the measured value, c, i.e.

$$e = r - c$$

Thus

$$\frac{de}{dt} = \frac{dr}{dt} - \frac{dc}{dt}$$

Changes in the measured value are slow because of the dynamics of the plant and so, in the absence of noise, the rate of change of measured value will therefore also be small. It is possible, however, to make sudden changes to the set-point, r, which upon differentiation cause very large changes to the control effort. These changes are referred to as set-point kicks. One way of avoiding a set-point kick is to filter the set-point signal so allowing only gradual changes to the set-point. However, this affects the other actions of the controller where there is no problem. Another way of overcoming set-point differentiation is to apply the derivative action only to the measured value. This solution

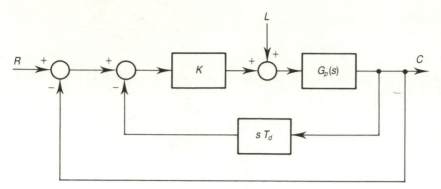

Figure 7.17 PD control with no set-point kick

is the one adopted by many commercial controllers and is in fact used in the PCS package. The derivative term in Eq. (7.17) becomes modified to

$$u_d = -KT_d \frac{dc}{dt} \qquad (7.19)$$

This type of derivative action is equivalent to velocity feedback that was discussed in Chapter 4. The block diagram of a PD controller of this type is shown in Fig. 7.17 which is very similar to the block diagram of a servo system with velocity feedback.

7.7 TUNING THREE-TERM CONTROLLERS

In order to get the best performance from a three-term controller the amount of each action has to be selected carefully. If a perfect model of the plant were available, then the selection process could be done through simulation or other analytical techniques. With process plant the model is only vague and liable to change, and so the selection procedure has to be done *in situ* and this is termed control-loop tuning. Books have been written on this subject alone, and articles appear regularly in process control journals claiming ever better techniques.

   A method widely used in industry is based on an article published by Ziegler and Nichols dating from 1942 which established a set of empirical rules for tuning controllers. This tuning procedure is called the 'continuous cycling method' and uses results from a **closed-loop** test to tune the controller. There is a second method derived from the same source called the 'reaction curve method' which uses an **open-loop** test. The recommended settings for each method are summarized in Table 7.1.

   The objective of both methods is to produce settings which result in transient responses with a decay ratio of 1/4. The quarter decay ratio was considered 'optimum' for regulators correcting for **load** changes. It is not clear from the original paper what target systems were used to obtain the settings, but judging from the curves and other hints one system at least had two lags and a small transport delay.

Table 7.1 Ziegler–Nichols recommended settings for process controllers

(a) Continuous cycling method

Type of controller	Controller gain	Controller PB	T_r	T_d
P	$K_u/2$	$2PB_u$	—	—
PI	$K_u/2.2$	$2.2PB_u$	$\dfrac{T_u}{1.2}$	—
PID	$K_u/1.7$	$1.7PB_u$	$\dfrac{T_u}{2}$	$\dfrac{T_u}{8}$

PB_u = Ultimate proportional band.
K_u = Ultimate proportional gain.
T_u = Ultimate period.

(b) Reaction curve method

Type of controller	Controller gain	Controller PB	T_r	T_d
P	$\dfrac{\Delta u}{ND}$	$\dfrac{ND}{\Delta u}$	—	—
PI	$0.9\,\dfrac{\Delta u}{ND}$	$\dfrac{ND}{0.9\Delta u}$	$\dfrac{D}{0.3}$	—
PID	$1.2\,\dfrac{\Delta u}{ND}$	$\dfrac{ND}{1.2\Delta u}$	$\dfrac{D}{0.5}$	$\dfrac{D}{2}$

N = Maximum slope of reaction curve as a fraction of measurement range.
D = Effective delay.
Δu = Fractional change at process input.

Continuous Cycling Method

The method is based on determining by experiment the phase-crossover frequency and the gain of the system for marginal stability. The procedure is as follows:

1. With the controller in manual settle the plant near its normal operating condition.
2. Remove all actions except proportional (i.e. $T_d = 0$ and T_r = maximum).
3. Select a wide proportional band.
4. Switch to auto and introduce a slight set-point change (5–10%).
5. Observe the response.
6. Switch back to manual, restore the plant to its normal operating point and the set-point back to its original setting.
7. Reduce the proportional band and repeat 4, 5 and 6 until the plant exhibits sustained oscillations which neither grow nor decay.

The value of proportional band which caused the sustained oscillations is termed the *ultimate proportional band*, PB_u, and the corresponding period of oscillation is termed the *ultimate period*, T_u. From a knowledge of these two parameters the controller can be tuned by referring to Table 7.1(a).

Reaction Curve Method

The method is based on modelling the open-loop system as a pure delay, D, and a first-order lag, τ. The procedure is as follows:

1. With the controller in manual settle the plant near its normal operating condition.
2. Apply a small step change to the controller output and record the response.

The open-loop response curve is called the *reaction curve* because it shows how the plant reacts to a step change in control effort. The reaction curve is usually an S-shaped curve of the form shown in Fig. 7.18. The first step is to find the maximum slope of the reaction curve, N, and to draw a tangent at that point. The next step is to determine the 'effective delay', D, by finding the time between applying the step change in the control effort and where the line of maximum slope crosses the initial operating point of the plant as shown in Fig. 7.18. The proportional band setting (Table 7.1(b)) depends on 'ND' rather than 'N' or 'D' alone, and this factor can be determined by extrapolating the line of maximum slope back to the time when the step was applied and measuring the difference between the point of intersection on the vertical scale and the initial value of the plant output as shown in Fig. 7.18.

For the reaction curve in Fig. 7.18, the effective delay, D, is about 0.4τ. The step of control effort that was applied was 50% of the span, hence Δu is 0.5 or 50%. Extrapolating back the line of maximum slope one finds that 'ND' is 0.07 or 7%. Hence the $ND/\Delta u$ is 0.14 or 14%.

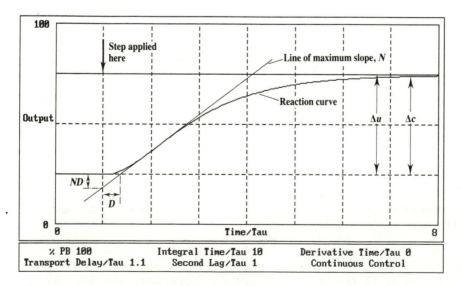

Figure 7.18 Reaction curve and associated parameters

Comparison of the Two Tuning Methods

In order to compare the two tuning methods three typical plants will be considered, namely:

Plant A:
$$G_p(s) = \frac{e^{-0.1\tau s}}{(1+s\tau)^2}$$

Plant B:
$$G_p(s) = \frac{e^{-\tau s}}{(1+s\tau)^2}$$

Plant C:
$$G_p(s) = \frac{e^{-2\tau s}}{(1+s\tau)}$$

The key parameters for the three plants and the settings of proportional plus integral (PI) are summarized in Table 7.2.

The table shows that the two methods lead to quite different and inconsistent values for the settings. For plants A and B the reset times are similar but the proportional bands are quite different; wider in the one case and narrower in the other. For plant C both the gain setting as well as the reset times are vastly different.

Figures 7.19(a), (b) and (c) show the responses of the three systems to load changes (not all the same magnitude) with a proportional plus integral controller tuned using both methods. Case A with the small transport delay shows that neither method leads to a particularly good settling time but the reaction curve response is slightly better damped. The other two cases are interesting in that the continuous cycling method gives a much faster settling time. In case B, however, the reaction curve method produces a much more oscillatory response whereas for case C the response is more heavily damped.

Table 7.2

Plant	%PB_u	$T_u\tau$	%$ND/\Delta u$	D/τ
		Tuning parameters		
A	4.8	1.4	14.0	0.4
B	36.9	4.8	47.0	1.3
C	65.8	5.5	192.0	2.0

	Continuous cycling		Reaction curve	
	Controller settings (PI)			
Plant	%PB	T_r/τ	%PB	T_r/τ
A	10.6	1.2	15.6	1.3
B	81.2	4.0	52.2	4.3
C	144.8	4.6	213.3	6.7

(a)

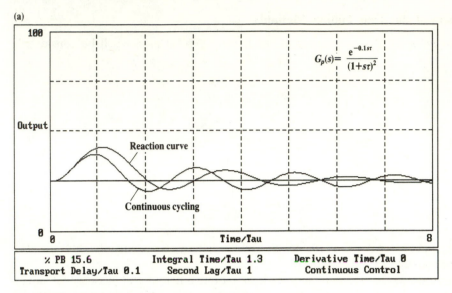

Figure 7.19(a)   Regulation performance of system A

(b)

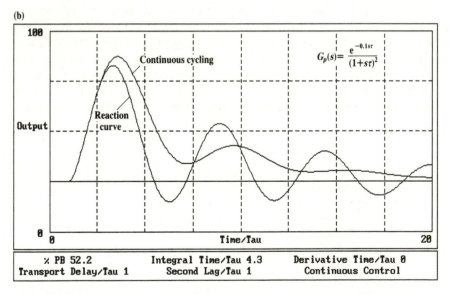

Figure 7.19(b)   Regulation performance of system B

Conclusions Regarding the Two Tuning Methods

The three cases studied do not provide sufficient data alone to draw general conclusions. However it is evident from the responses that where there is significant transport delay, the results using reaction curve are poor. Other tests on a wider range of plants support

(c)

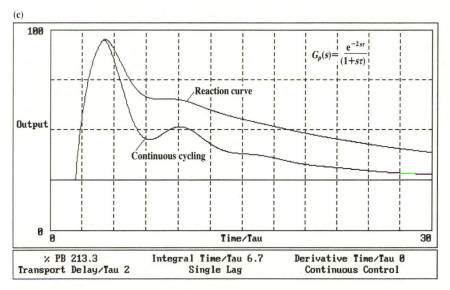

$$G_p(s)= \frac{e^{-2st}}{(1+s\tau)}$$

| % PB 213.3 | Integral Time/Tau 6.7 | Derivative Time/Tau 0 |
| Transport Delay/Tau 2 | Single Lag | Continuous Control |

Figure 7.19(c)   Regulation performance of system C

this tentative conclusion, and that continuous cycling provides more consistent and acceptable results.

From a theoretical point of view the continuous cycling method is sounder because at least with proportional control the tuning approximates to a 1/4 decay ratio which was the design objective. In the reaction curve method there is a greater degree of empiricism and there is no simple theoretical relationship between the tuning factor *ND* and established stability criteria.

The ultimate period that is used in the continuous cycling method is equal to the phase-crossover period which has a well defined relationship with frequency domain design methods. The reaction curve method bases the tuning of the integral and derivative times solely on the effective dead time of the system, which is less easy to justify theoretically for a general case.

The continuous cycling method is time-consuming and disruptive. It cannot work with systems which are unconditionally stable such as pure first- and second-order systems. In practice this latter aspect is not a significant factor as the majority of process control systems have significant transport delays in addition to any other dynamics. However this fact helps to explain why the continuous cycling method was not very good for case A where there was hardly any transport delay.

The reaction curve method is quick to perform, but finding the point of maximum slope is very subjective and obtaining accurate or consistent values of the effective transport delay, even in computer simulations, is difficult. In real plants with noisy measurements it is even more hit and miss. Evidence shows that on the whole it produces poorer results than the continuous cycling method except with systems where the transport delay is small. But if the transport delay is small, controller tuning is less critical anyway because higher controller gains can be employed with safety.

The general advice is therefore to use continuous cycling if possible, but use an open-loop method (reaction curve) if the duration of disruption is to be minimized.

Example 7.2 The temperature of a plant is controlled by a steam heater. The amount of steam admitted into the heater is controlled by means of a valve with a 4–20 mA input range. The signal to the control valve is obtained from a three-term controller. An open-loop test of the system showed that when a step change of current of 5 mA was applied to the control valve the temperature of the plant increased by 60°C after the initial transient had died away. Examination of the response indicated that the plant dynamics could be modelled approximately as a transport delay of 2 minutes and two cascaded first-order lags of time constants 2 and 4 minutes respectively. Given that the plant is controlled by an electronic three-term controller with a measurement range of 0°C to 500°C determine the settings of the controller using both the Ziegler–Nichols reaction curve and the continuous cycling method.

SOLUTION The transfer function of the plant is

$$G_p(s) = \frac{K_1 \, e^{-2s}}{(1+2s)(1+4s)}$$

where K_1 is the steady-state gain of the plant and 's' has the dimensions of minutes^{-1}. From the step-response test, K_1 is 12 mA/°C. This problem can be tackled either using CODAS or PCS. If PCS were to be used the time constants, transport delay and plant gain would have to be normalized first, results obtained for the normalized system, and then translated to engineering units subsequently. With CODAS the advantage is that one can work directly in engineering units but one cannot directly include reset and derivative time. In this example CODAS will be used to show how to get over the restrictions.

Figure 7.20 shows the open-loop response of the system to a unit step change

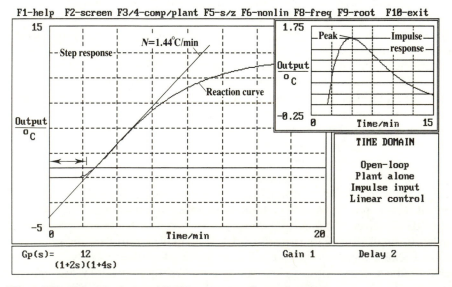

Figure 7.20 Reaction curve and impulse response of example system

Table 7.3

Controller		K_u 0.32	T_u 11.1			$ND/\Delta u$ 4.13	D 2.7	
		Continuous cycling				Reaction curve		
	Gain, K	T_r/min	T_d/min		Gain, K	T_r/min	T_d/min	
P	0.160	—	—		0.242	—	—	
PI	0.145	9.25	—		0.218	9.00	—	
PID	0.188	5.55	1.39		0.291	5.40	1.35	

in control effort, i.e. 1 mA. The maximum slope, N, and the effective delay, D, can be measured. These turn out to be about 1.44°C/min and 2.75 min respectively. It is interesting to obtain the impulse response of the system (shown in the small window). The impulse response is the derivative of the step response, hence the point at which it peaks tells us what the maximum slope of the reaction curve is, and when it occurs. This confirms that N is 1.5°C/min and that the maximum slope occurs 4.8 min after the step is applied. The slight discrepancy between the two figures for N is simply due to the difficulty of drawing the line of maximum slope. The figures that will be used in the reaction curve calculations will be 1.5°C/min for N and 2.75 min for D. Hence $ND/\Delta u$ is 4.125°C/mA.

The ultimate gain and ultimate period can be found by trial and error or from a frequency response plot. They are 0.32 mA/°C and 11.1 minutes respectively. Table 7.3 shows the whole range of settings for both the continuous cycling and the reaction curve data for this plant. Don't forget that in this example we are working with values of gain and so refer to the gain column in Table 7.1 to see how the gain figures are obtained from the ultimate gain, K_u, and $ND/\Delta u$.

The proportional band can be determined from the recommended gain settings by using the relationship of Eqs (7.5) or (7.6). In the example the measurement span is 500°C and the output span is 16 mA. Hence for this controller

$$\%PB = \frac{1}{K} \times \frac{16}{500} \times 100 = \frac{3.2}{K}$$

In order to examine the responses, a controller has to be defined of the following (proper) form

$$G_c(s) = \frac{K}{(1 + s\tau)}\left(1 + \frac{1}{sT_r} + sT_d\right) = K\frac{(1 + sT_r + T_rT_ds^2)}{sT_r(1 + s\tau)}$$

where τ is a small value relative to the derivative time. Typical values of τ are about $0.1T_d$ to $0.2T_d$. Figure 7.21 illustrates the responses for the two different controller settings. In this case τ was chosen as 0.1 min for convenience. It is very apparent that in this example the reaction curve method has led to a very underdamped response whereas the response with the continuous cycling settings is much more acceptable.

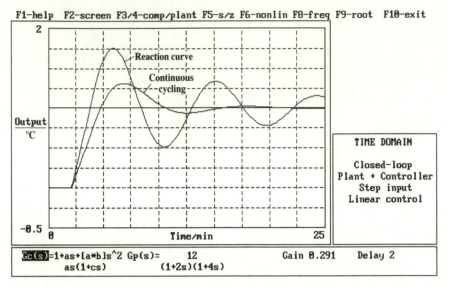

Figure 7.21 Responses of example system with the controller tuned using two different methods

PROBLEMS

7.1 The open-loop transfer function of a plant relating its outlet pressure, p (bar), to the controller output, u (volts) is:

$$\frac{P}{U}(s) = \frac{12\,e^{-3s}}{(1+6s)(1+15s)}$$

The controller has an output range of 1 to 5 V and the span of the pressure transmitter is 50 bar.

Confirm that the Ziegler–Nichols reaction curve tuning procedure results in a PI controller with a proportional band setting of around 22% and a reset time of about 20 s. Obtain settings using the continuous cycling method as well and confirm that it results in a similar value for the reset time but that the proportional band is significantly wider.

Compare the effective damping ratios and settling times of the closed-loop system for both controller settings.

7.2 The outlet temperature of a gas-fired furnace is controlled by a pneumatic three-term temperature controller which modulates the fuel flow by means of a pneumatic diaphragm control valve. The output range of the controller is 0.2 to 1.0 barg and the controller temperature range is 0 to 1000°C. A suddenly applied step change of 0.4 barg to the control valve produced the results shown in Fig. P7.2.

Draw a block diagram of the system. Obtain the reaction rate tuning parameters (i.e. ND and D). Determine controller settings assuming that all three terms are employed.

The actual dynamics of the process consist essentially of a pure time delay of 5 seconds and two identical cascaded lags of 4 seconds. Model the overall response of the closed-loop system and comment on the stability of the loop.

7.3 The dynamics of a stirred reactor can be adequately described as two identical cascaded lags of time constant 10 s each and a 5 s transport delay. The plant gain is K_1. The temperature of the plant is controlled by a PID controller of the form

$$G_c(s) = K\left(1 + \frac{1}{sT_r} + \frac{sT_d}{(1+0.1sT_d)}\right)$$

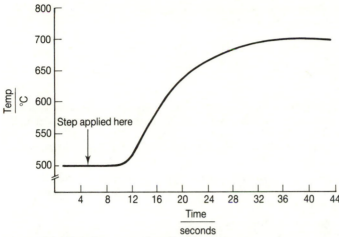

Figure P7.2 Reaction curve of gas-fired furnace

The proportional gain, K, is adjusted so that the steady-state loop gain, KK_1, is exactly 2 and then is maintained at this value.

(a) With proportional action only is the closed-loop system stable, and if so what is the gain margin?

(b) Reset action alone is added with a reset time, T_r, of 35 seconds. What is the effect on the gain margin? What is the peak overshoot and 5% settling time of the closed-loop system?

(c) Finally a little derivative action is added ($T_d = 2.5$ seconds). What effect does this have on the performance measures obtained in part (b)?

If you had been starting from scratch using the Ziegler–Nichols continuous cycling method what settings would have been used for two-term and three-term control and how would the results have compared with those found in parts (b) and (c) above.

7.4 A control system consists of a proportional controller, a first-order lag and a pure transport delay. When the system is marginally stable the period of oscillation, T_u, is exactly three times the transport delay. Prove that the ratio of transport delay to the time constant of the first-order lag is $2\pi/(3\sqrt{3})$. Hence show that with unity steady-state loop gain, the gain margin is 2.

The proportional controller is replaced by a PI controller. The reset time is fixed at $0.8T_u$. What steady-state loop gain will maintain the gain margin of 2?

7.5 Figure P7.5 shows a simple temperature control system. The liquid flows into and out of a stirred vessel

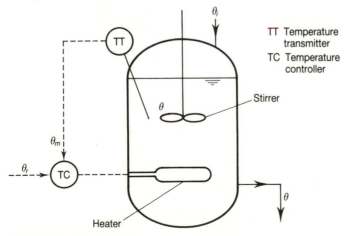

Figure P7.5 Stirred reactor

at a steady rate (m) of 2.0 kg/s at an inlet temperature θ_i. The specific heat (c) of the liquid is 2.5 kg(kJ K). The vessel holds 100 kg of liquid (M) at a uniform temperature θ. The output of the heater is Q watts. Write down a heat balance for the system.

The temperature transmitter has a 25 s time constant, i.e.

$$\frac{\theta_m}{\theta}(s) = \frac{1}{1 + 25s}$$

A proportional controller is employed ($Q = K_c(\theta_r - \theta_m)$) to keep the temperature in the vessel constant. Draw a block diagram for the overall system. Show that a proportional gain, K_c, of 6.5 watts/K will result in an overall damping ratio of about 0.7.

What is the change in outlet temperature if the inlet temperature changes by 10°C? Is the control system performance satisfactory as a regulator?

7.6 Figure P7.6 shows the block diagram of a cascade control system designed to counteract disturbances D.

(a) Initially consider the situation with no cascade loop, i.e. link AB open. What steady-state loop gain ($K_1 K_2$) will give the overall system a gain margin of 3?

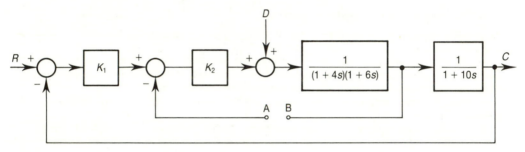

Figure P7.6 Cascade control system

With this value of gain determine the steady-state change in output due to a unit step change in the disturbance.

(b) Link AB is closed to implement the cascade loop and the slave loop gain K_2 is fixed at 3. Show that the previous gain margin is maintained with a master controller gain, K_1, of about 2.4. Show the steady-state change in output is now only about 0.089 due to a unit step disturbance.

Hint: this problem can be done using CODAS-II which allows you to look at the dynamic aspects of the cascade loop. Be careful, however, in part (a) to put the steady-state gain in the compensator, G_c, and then select the non-unity feedback option.

Part (b) can also be done using CODAS-II, but first rearrange the block diagram by moving the disturbance to the left of the inner summing junction and then represent the cascade loop and the 10 second lag as a single transfer function, i.e.

$$\frac{3}{(4 + 10s + 24s^2)(1 + 10s)}$$

If you do have access to CODAS-II, how long does it take in each case for the control system to settle down after the disturbance has occurred?

EIGHT

DISCRETE TIME SYSTEMS

8.1 INTRODUCTION

This chapter is concerned with the application of digital computers to the control of dynamic systems and the monitoring of signals. The material covered here deals with open-loop systems and will lay the foundation for the next chapter which deals with the design of computer-controlled feedback systems.

The use of digital computers to control and monitor processes has grown considerably over the past 25 years so that now we see digital computers and microprocessors involved with most industrial (and domestic) machines and processes. The reasons for the wide use of digital control and monitoring include improved performance, versatility and reliability. In many cases installation and operation costs are also reduced. Over the last decade many new applications have been made possible only because of digital computer technology. Digital computers can perform complex calculations and make decisions which would be impossible to implement using other means. Examples are found in robotics, signal analysis, process optimization and the developing area of adaptive control.

Perhaps the most conspicuous developments have occurred at the man–machine interface. Computers can process and filter information gathered so that a clear picture of the state of a plant or process is presented to operators. Graphical displays showing plant mimic diagrams and time histories enable the information to be presented in a helpful form. Up-to-date management information such as throughput and efficiency figures can also be made available. Intelligent monitoring for fault conditions or potentially dangerous situations can be implemented giving more reliable warning systems.

The recent development of low cost microprocessor systems means that computer control can now be cost effective in the simplest of applications. Perhaps more importantly, the availability of powerful software development tools such as modern programming languages and debugging aids are reducing development costs for computer-based control and monitoring systems.

Classification of Signals

The signals which have so far been discussed have been *continuous time signals*. This means that they are defined over a continuous range of time and can change value at any time or change smoothly with time. *Discrete time signals* on the other hand, are defined only at certain instants of time and can change value only at those instants. Quite often discrete time signals come about as a consequence of *sampling* continuous time signals. The sample instants are usually equally spaced and are separated by the *sample period*, T.

An *analogue signal* is one which has a continuous range of values. Theoretically, an analogue signal can change value by infinitesimally small amounts, i.e. its amplitude is infinitely resolvable. Conversely, a *digital signal* is quantized, i.e. it has only a finite set of values and can only change in clearly defined steps. The most commonly encountered digital signal is a *binary signal* which has only two possible values or states (on/off, high/low, etc.). However not all digital signals are binary.

Analogue signals carry information by virtue of their amplitude and are therefore prone to errors caused by drift, pickup and noise. Any change in the amplitude of an analogue signal due to these effects is indistinguishable from changes of the transmitted value. On the other hand the precise amplitude of a digital signal is relatively unimportant. For example, the value represented by a binary voltage signal depends only on whether the signal voltage is above or below a certain threshold. Noise and drift therefore do not affect digital signals to the same extent as analogue signals.

The classification of signals into analogue or digital and continuous time or discrete time are independent and should not be confused. Figure 8.1 shows that it is quite possible to have digital continuous time signals as well as analogue discrete time signals.

As an example, consider a simple thermostat monitoring temperature; the contacts of a thermostat open or close to represent temperatures above or below the reference setting. The monitored temperature is an analogue continuous time signal since it will change continuously and smoothly with time. The output signal from the thermostat depends upon the state of the electrical contacts (i.e. whether open or closed) and is thus a digital signal, in this case binary. The actual value of the contact resistance is relatively unimportant, what matters is whether the resistance is large (open contacts) or small (closed contacts). The contacts can change state at any time and thus the thermostat output is also a continuous time signal.

Another example is the automatic counting of components on a conveyor. The count will be limited to whole numbers (integers) and thus the resolution is clearly defined as one count. The components can arrive at random and thus the counter can change state at any time, therefore the counter output is a digital continuous time signal. The counter output although digital may not be binary: mechanical counters often count in decimal where each digit has ten possible states.

When a continuous time analogue signal is sampled the resulting sample values form a discrete time signal. There are many different continuous time signals which can produce the same sample values (Fig. 8.2). Thus without making certain assumptions about the original signal it is impossible to reconstruct the original continuous time signal from the sampled version. This is a very important principle which we will return to in Chapter 9.

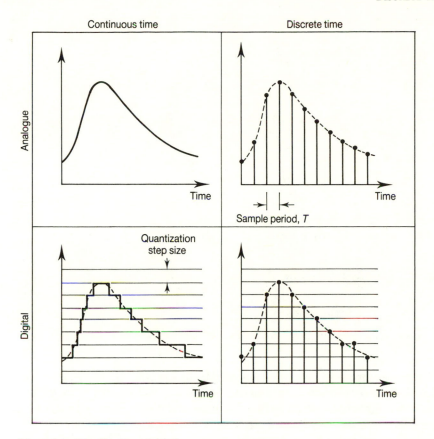

Figure 8.1 Classification of signals

Quite often sampling is carried out by a digital system (e.g. a computer or microprocessor system) and so the sampled signal will also be *quantized* in the sampling process. Quantization occurs when a value is rounded up or down to the next discrete level. Digitization of analogue signals thus causes *quantization errors*, since changes in the analogue signal of less than the digital resolution go undetected.

Exercise Classify the following signals into analogue or digital and continuous time or discrete time:

(a) the voltage from a thermocouple;
(b) the readout from a digital watch;
(c) the daily rainfall measured at a weather station;
(d) the peak voltage of an amplitude-modulated sinusoidal signal;
(e) rotational speed calculated as the number of revolutions occurring in a fixed time interval.

Classes of System

A *continuous time system* is one which exclusively involves continuous time signals. As

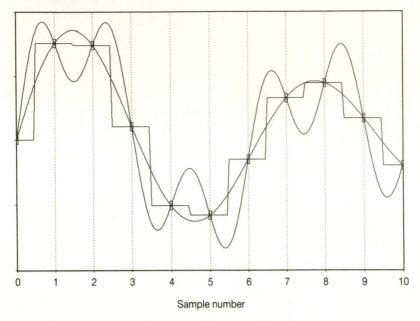

Figure 8.2 Many different continuous time signals can produce the same sample values

we have seen, continuous time systems can be modelled by differential equations. *Discrete time systems* exclusively involve discrete time signals and as we shall see can be modelled by *difference equations*. Systems in which both continuous and discrete time signals exist are called *sampled-data systems* since they invariably involve some form of sampling.

Computer control and monitoring applications where the signals from a plant or process are sampled by a digital computer are sampled-data systems. The process signals are continuous time but the computer is dealing with discrete time signals. In addition the process signals will generally be analogue whereas the computer works with digital signals and so quantization effects will also be involved. However, sampled-data systems do not necessarily involve digital signals.

8.2 REAL-TIME SYSTEMS

Time is an important variable in any industrial computer application. The computer must often react to external events as they occur or at least within a time critical interval. The computer, as it gathers data from sensors, may have to relate each sample to the time at which it was taken. Such applications require *real-time systems*.

Real-time systems can be defined as systems in which either

(a) the order of program execution depends upon events which are external to the computer or the passage of time; or
(b) the data manipulated by the computer includes time or time interval.

Timers are an important part of any industrial computer system. Most systems include several timers which can be used for accurate timing, synchronizing sampling, etc. A timer can be a simple oscillator generating pulses at regular intervals or it can be a sophisticated device with time of day and date information and/or programmable capability for generating accurate delays. It is particularly important to realize that these timers operate independently of any program and keep track of time while the program is operating.

In the main, computers use *interrupts* to provide real-time operation. An interrupt can be generated by an external event such as the closure of a switch or a pulse from a timer. The occurrence of an interrupt causes the normal program operation to be suspended and the state of the computer (i.e. the next program instruction and the contents of important registers) to be saved in computer memory. The computer then executes an *interrupt service routine* which is a short section of program designed to determine the source of, and deal with, the event which caused the interrupt. At the end of the interrupt service routine the 'saved' computer state is restored and execution of the original program continues as if nothing had happened.

Interrupt service routines normally execute very rapidly (typically taking less than 10 ms) and so to the observer it can appear that the computer is doing two tasks at the same time. Interrupt service routines can themselves be interrupted and so interrupts can be several layers deep.

8.3 ANALOGUE SIGNAL CONVERSION

A digital to analogue converter (DAC) transforms a binary number inside the computer into an external proportional signal (a voltage or current). The full-scale output is usually determined by an external reference voltage. The *resolution* is defined as the smallest change in output. For an N-bit converter the resolution is

$$\text{Resolution} = \frac{100}{2^N}\,\% \tag{8.1}$$

For example a 12-bit DAC has a resolution of $100/4096 = 0.024\%$.

The DAC as well as converting from digital signal to analogue signal performs the task of constructing a continuous time signal from a discrete time signal. Since the DAC holds the previous value until the next conversion, the continuous time output signal appears as a series of steps or rectangular pulses which change only at the instant the computer outputs a new value (Fig. 8.3). Such rectangular interpolation between samples can be considered as an attempt at reconstruction of a continuous time signal from its sample values. Devices which use this method of reconstruction are termed *zero-order holds*.

Analogue to digital conversion is rather more complex than digital to analogue conversion and is considerably slower. Typically, the conversion time can be from a few microseconds to several hundred milliseconds depending on the type of converter and its resolution. The resolution of an analogue to digital converter (ADC) is defined as the smallest detectable change in input. This resolution is again dictated by the number of bits and so Eq. (8.1) can be applied to ADCs also.

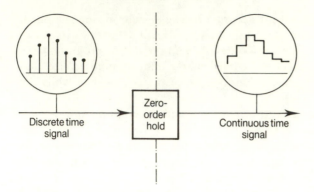

Figure 8.3   Signal reconstruction using a zero-order hold

8.4 DIFFERENCE EQUATIONS

Linear Models of Discrete Systems

Chapter 2 introduced the idea of modelling continuous time systems by means of differential equations and eventually transfer functions. In discrete time systems the signals exist only at the sample instants and therefore the differentials of signals are indeterminate and differential equations cannot be used. *Difference equations* are used to model discrete time systems.

Consider the flow chart for a computer program shown in Fig. 8.4. The program takes samples, x, from an analogue to digital converter, calculates the running total, y, and sends the result to a digital to analogue converter. The samples are taken at regular intervals of time, T.

The symbol '$\leftarrow$' represents the assignment of a value to a variable, i.e. $y \leftarrow y + x$ means the sum of y and x replaces y in the computer memory. If we denote the *i*th

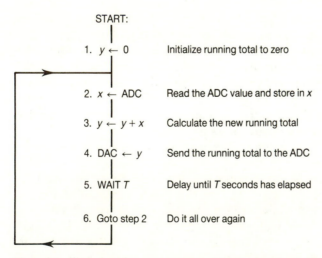

Figure 8.4   Flow chart for a program to calculate a running total

sample value as $x(i)$ then the operation of the above program can be expressed by the equation

$$y(i) = y(i-1) + x(i) \qquad (8.2)$$

This equation says that the current value of y, $y(i)$, is obtained by adding the previous value, $y(i-1)$ to the current value of x, $x(i)$. The sample number, i, is an integer quantity and symbolizes discrete time. An increment of one in discrete time corresponds to a change of T in real-time.

Equation (8.2) is a difference equation which describes the operation of the computer program. The value sampled by the ADC can be considered as an input to the equation and the value written to the DAC as an output. This difference equation is linear in that all terms have a proportional contribution. Of course the system is only linear if y is within the working range of the DAC. When the DAC receives a value outside its working range the output signal will stop at the limit of the converter.

Exercise Write down the difference equation which describes the process of finding the running average of the last three samples taken.

The program shown in Fig. 8.4 illustrates some of the points concerned with real-time operation. For consistent operation the samples should be timed to occur at regular intervals. The inexperienced programmer may be tempted to initiate a delay of T seconds in the program at step 5. This method can produce erratic results since the computation time and ADC conversion time must be added to this programmed delay. Unfortunately, computation time can vary depending upon the complexity of the calculations; in addition interrupts may occur which will add an indeterminate time to the overall duration.

One solution to the above timing problem is to use a timer to generate synchronization pulses at regular intervals, T. The sampling program simply waits at the delay statement until the next timer pulse is detected. Because the timer operates independently of the program, the time between samples will be correct.

Another solution is to implement the section of the program within the loop as an interrupt service routine. The timer in this case needs to generate an interrupt at regular intervals so that the interrupt service routine is repeatedly executed at the correct times. This solution has the advantage that the computer can be doing other tasks such as updating the screen display or printing out some results between interrupts.

Approximation of Differentiation and Integration

The astute reader may have recognized that the program of Fig. 8.4 operates in a similar manner to an analogue integrator, i.e. the output, y, increases at a rate proportional to x. Since the output is incremented at each sample, as the sample time T gets smaller the rate of change of the output gets larger. By multiplying the input by T the output rate can be made independent of the sample time and a discrete time approximation to integration is obtained.

$$y(i) = y(i-1) + Tx(i) \qquad (8.3)$$

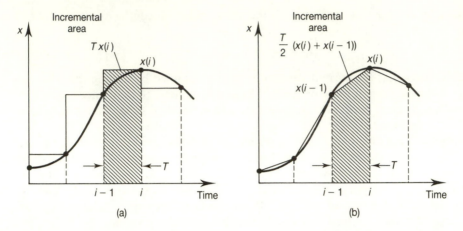

Figure 8.5 Discrete time approximations of integration. (a) Rectangular rule. (b) Trapezoidal rule

Integration of a function is the same as finding the area under the curve. This difference equation approximates the area for each sample as a rectangle of width T and height $x(i)$. The method is therefore termed the *rectangular approximation to integration*. Figure 8.5(a) illustrates the idea. As the sample time diminishes the approximation becomes more accurate since the difference between the rectangle area and the true area becomes less.

A more accurate estimation of area can be derived by approximating the area under the curve as a trapezoidal shape (Fig. 8.5(b)). The area of the trapezoid is given by

$$\frac{T}{2}(x(i) + x(i-1))$$

resulting in the *trapezoidal approximation to integration*

$$y(i) = y(i-1) + \frac{T}{2}(x(i) + x(i-1)) \tag{8.4}$$

Another example of the manipulation of discrete signals is an approximation of differentiation. Differentiation can be approximated by calculating the slope of the line joining two adjacent samples (Fig. 8.6), i.e.

$$\frac{dx}{dt}(t) \approx y(i) = \frac{x(i) - x(i-1)}{T} \tag{8.5}$$

Rearranging Eq. (8.5) to give $x(i)$ in terms of $y(i)$ results in

$$x(i) = x(i-1) + Ty(i)$$

which is the same as the rectangular approximation to integration (Eq. (8.3)). This differencing equation is thus the exact converse of the rectangular approximation to integration.

Using a similar argument the approximation of differentiation which corresponds

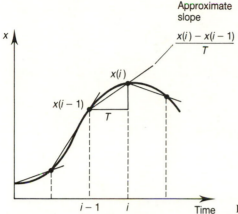

Figure 8.6 Differencing approximation of differentiation

to the trapezoidal rule can be obtained from Eq. (8.4) as

$$\frac{dx}{dt}(t) \approx y(i) = \frac{2(x(i) - x(i-1))}{T} - y(i-1) \tag{8.6}$$

Exercise A unit ramp signal when sampled once per second produces the discrete time series

$$\{x(i)\} = 0, 1, 2, 3, 4, \ldots \text{ etc.}$$

Apply the discrete approximations of differentiation based upon the rectangular and trapezoidal rules to the series. Compare the discrete results with those obtained by differentiating the continuous time ramp. Repeat for the two approximations to integration. What do you conclude about the suitability of each rule for pure differentiation and integration?

Example 8.1 Discrete approximation of a phase lead compensator Using the rectangular rule derive the difference equation for a discrete time approximation to the phase lead compensator

$$\frac{U}{E}(s) = \frac{1 + a\tau s}{1 + \tau s} \tag{8.7}$$

SOLUTION The differential equation relating u to e is obtained from the transfer function as

$$u(t) + \tau \frac{du}{dt}(t) = e(t) + a\tau \frac{de}{dt}(t)$$

Replacing each continuous time signal with its discrete time equivalent (i.e. $u(t)$ becomes $u(i)$) and also using the rectangular approximation to differentiation for each of the derivative terms gives:

$$u(i) + \tau \frac{(u(i) - u(i-1))}{T} = e(i) + a\tau \frac{(e(i) - e(i-1))}{T}$$

Rearranging to give the current output, $u(i)$, we have

$$u(i)\left(1 + \frac{\tau}{T}\right) = \frac{\tau}{T} u(i-1) + e(i)\left(1 + a\frac{\tau}{T}\right) - a\frac{\tau}{T} e(i-1)$$

or

$$(T + \tau)u(i) = \tau u(i-1) + (T + a\tau)e(i) - a\tau e(i-1) \tag{8.8}$$

8.5 DISCRETE TIME TRANSFORMATION

The Delay Operator, q

The advantages of using an operator to simplify differential equations for continuous systems expounded in Chapter 2 apply equally to discrete systems. So what transformation will be applicable to difference equations? Difference equations invariably involve terms which are delayed by one or more samples, so an operator which represents a delay of one sample period is appropriate. We will use the symbol 'q' for this, i.e.:

$$qx(i) = x(i-1) \tag{8.9}$$

Using the delay operator allows us to obtain an input/output relationship as a function of q. For example, applied to Eq. (8.8), the discrete phase lead compensator can be represented by

$$(T + \tau)u = \tau qu + (T + a\tau)e - a\tau qe$$

which can now be rearranged as

$$(T + \tau - \tau q)u = (T + a\tau - a\tau q)e$$

or

$$\frac{u}{e}(q) = \frac{T + a\tau - a\tau q}{T + \tau - \tau q} \tag{8.10}$$

The purpose of this equation is similar to that of the transfer operator developed in Chapter 2. To emphasize the discrete time nature of the system we will call it a *pulse transfer operator*. A pulse transfer operator can be considered to be a dynamic sensitivity for a linear discrete time system. In Chapter 2 we replaced the D operator with the more widely used Laplace operator via the Laplace transform. Again we find that control engineers tend to use the *z-transform* to deal with discrete time systems rather than the delay operator.

The z-transform

The z-transform is a formal transformation for discrete time signals which transforms functions of discrete time, i, into functions of a new variable, z.

Given a discrete time signal

$$x(i) = x(0), x(1), x(2), \ldots \text{ etc.}$$

the z-transform of $x(i)$ is written as the infinite power series in terms of the complex variable 'z':

$$Z\{x(i)\} = x(0) + x(1)z^{-1} + x(2)z^{-2} \ldots$$

or
$$Z\{x(i)\} = X(z) = \sum_{i=0}^{\infty} x(i)z^{-i} \tag{8.11}$$

Notice the convention of using lower case symbols to represent functions of discrete time and upper case for their z-transforms, or functions of z. The infinite series given by this summation will only converge if the terms $x(i)z^{-i}$ eventually become smaller as i becomes large. The series in Eq. (8.11) will converge only if the magnitude of the complex variable 'z' is sufficiently large, i.e.

$$|z| > R$$

This restriction of z to a given region of convergence is of concern only when deriving z-transforms or their inverses and does not in practice cause any restrictions. The table in Appendix B lists a few of the more commonly used z-transforms.

Example 8.2 z-transform of a discrete time unit step The discrete unit step is shown in Fig. 8.7 and can be written as

$$x(i) = \begin{cases} 0 \text{ for } i < 0 \\ 1 \text{ for } i \geqslant 0 \end{cases}$$

The z-transform is therefore given by

$$X(z) = \sum_{i=0}^{\infty} x(i)z^{-i} = 1 + z^{-} + z^{-2} + z^{-3} \ldots$$

In order to evaluate this series in a closed form, the binomial expansion for $(1 - x)^{-1}$ can be used:

$$\frac{1}{1 - x} = 1 + x + x^2 + x^3 \ldots \text{ where } |x| < 1$$

so that
$$X(z) = \frac{1}{1 - z^{-1}} = \frac{z}{z - 1} \tag{8.12}$$

The region of convergence is thus $|z| > 1$ for this transform.

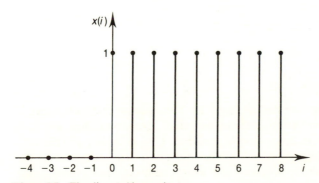

Figure 8.7 The discrete/time unit step

Discrete Time Transfer Functions

We have seen several examples of discrete systems which can be described by difference equations involving discrete time signals which are delayed by one or more samples. In order to develop the z-transform method to deal with such systems we need to consider the z-transform of a signal which is delayed by 'n' samples

$$Z\{x(i-n)\} = \sum_{i=0}^{\infty} x(i-n)z^{-i}$$

Substituting j for $i-n$

$$Z\{x(i-n)\} = \sum_{j=-n}^{\infty} x(j)z^{-(j+n)}$$

$$= z^{-n} \sum_{j=-n}^{\infty} x(j)z^{-j}$$

and if $x(i)$ is zero for $i < 0$ then

$$Z\{x(i-n)\} = z^{-n}X(z) \tag{8.13}$$

Delaying a signal by 'n' samples corresponds to multiplication of its z-transform by z^{-n} providing the signal is zero for negative time. There is thus an equivalence between z^{-1} and the delay operator, q.

$$qx(i) \Rightarrow z^{-1}X(z) \tag{8.14}$$

The *pulse transfer function* of a discrete time system is defined as the z-transform of the output divided by the z-transform of the input.

$$G(z) \equiv \frac{Z\{y(i)\}}{Z\{x(i)\}} = \frac{Y}{X}(z) \tag{8.15}$$

We can use the idea of z^{-1} as a delay operator to obtain the pulse transfer function of a discrete time system. For example the phase lead compensator will have a pulse transfer function identical to the pulse transfer operator (Eq. (8.10)) with z^{-1} replacing q, i.e.

$$\frac{U}{E}(z) = \frac{T + a\tau - a\tau z^{-1}}{T + \tau - \tau z^{-1}} = \frac{(T + a\tau)z - a\tau}{(T + \tau)z - \tau} \tag{8.16}$$

The z-Transform Method

The response of a discrete system can be obtained by z-transform methods using a similar approach to that used for continuous systems with the Laplace transform method. The technique is useful if an analytical solution is required. Unfortunately the method is mathematically laborious for higher-order systems and gives little insight into the dynamics of discrete time systems, however we will examine one example for completeness.

Example 8.3 Step response by the z-transform method As an example of the z-transform method we will find the open-loop unit step response of the first-order discrete system given by the difference equation

$$y(i) = ay(i-1) + bx(i-1) \tag{8.17}$$

where a and b are constants.

First the pulse transfer function of the discrete time system must be obtained. Taking z-transforms of each term we have

$$Y(z) = az^{-1}Y(z) + bz^{-1}X(z)$$

From which

$$\frac{Y}{X}(z) = \frac{bz^{-1}}{1 - az^{-1}} = \frac{b}{z-a} \tag{8.18}$$

The z-transform of a unit step input is given by

$$X(z) = \frac{z}{z-1}$$

The z-transform of the system output is thus

$$Y(z) = \frac{Y}{X}(z) \cdot X(z) = \frac{bz}{(z-1)(z-a)} \tag{8.19}$$

The discrete time response can be obtained from the inverse z-transform of $Y(z)$. From Appendix B we can see that

$$Z\{1 - \beta^i\} = \frac{(1-\beta)z}{(z-1)(z-\beta)}$$

By comparison with Eq. (8.19) we have

$$\beta = a$$

so that

$$y(i) = \frac{b}{1-a}(1 - a^i) \tag{8.20}$$

Exercise Determine the unit step response of the discrete time system of Eq. (8.17) with $a = 0.7$ and $b = 0.15$ over the first five samples ($i = 0, 1 \ldots 5$):
(a) Analytically, using Eq. (8.20).
(b) By simulation, using Eq. (8.17). (The input $x(i)$ will be 1.0 for all i values and the initial value of $y(0)$ can be taken as zero. Subsequent values for $y(i)$ can be found iteratively by substitution in the difference equation.)
(c) By simulation, using CODAS-II and Eq. (8.18). (The pulse transfer function can be entered into CODAS-II in the same way as a continuous time transfer function. A sample time must be set to obtain the step response. The actual value is immaterial for this example but it is suggested that 1 second is used.)

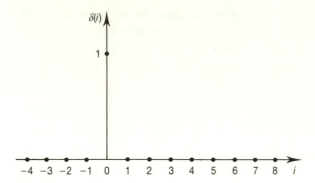

Figure 8.8 The discrete/time unit impulse

The Discrete Time Impulse

The *discrete time impulse* or *Kronecker delta function* is defined as

$$\delta(i) = \begin{Bmatrix} 1; & i = 0 \\ 0; & \text{elsewhere} \end{Bmatrix} \tag{8.21}$$

Figure 8.8 shows that the discrete time impulse is very different to the continuous time impulse introduced in Chapter 3 since that had an infinite amplitude. However the discrete time impulse plays a similar role in discrete time systems to that played by the continuous time impulse in continuous systems.

The z-transform of a discrete time impulse is

$$Z\{\delta(i)\} = \sum_{i=0}^{\infty} \delta(i) z^{-i}$$

$$= \delta(0) + \delta(1) z^{-1} + \delta(2) z^{-2} \ldots \text{etc.}$$

$$= 1$$

Compare this with the Laplace transform of continuous time impulse, $\delta(t)$, which is also unity. The impulse response of a discrete time transfer function, $G(z)$, is given by

$$y(i) = Z^{-1}\{Y(z)\} = Z^{-1}\{G(z) \cdot X(z)\}$$

$$= Z^{-1}\{G(z)\} = g(i)$$

That is, the impulse response of a discrete time system is the inverse transform of its pulse transfer function.

8.6 THE z-PLANE

The ideas developed in Chapter 3 about the importance of the roots of a system's characteristic equation in determining the transient response apply equally to discrete time systems. The roots of the characteristic equation are the poles of the system transfer function, however for discrete time systems the transfer function is a function of 'z'

rather than 's' and thus the poles and zeros of the pulse transfer function must be plotted on the *z-plane*. Before examining the *z*-plane in detail it is worth spending a little time examining the role of the characteristic equation on the transient dynamics of a discrete time system.

The Discrete Time Characteristic Equation

Consider the general discrete time transfer function

$$\frac{Y}{X}(z) = G(z) = \frac{N(z)}{D(z)} \tag{8.22}$$

$$\frac{Y}{X}(z) = \frac{b_0 + b_1 z + \ldots b_m z^m}{a_0 + a_1 z + \ldots a_n z^n} \tag{8.23}$$

The roots of the characteristic equation

$$D(z) = 0$$

exclusively determine the form of the transient response of the discrete time system.

Each root of the discrete time system characteristic equation at $z = \lambda$, gives rise to a transient response term of the form

$$\tilde{y}(i) = C\lambda^i \tag{8.24}$$

where i is discrete time and C is an arbitrary constant.

To see that this is so, consider Eq. (8.22) with the input set to zero so that the system response contains only the transient component

$$\tilde{Y}(z)D(z) = 0$$

$$\tilde{Y}(z)(a_0 + a_1 z + \ldots a_n z^n) = 0$$

or $\qquad a_0 \tilde{Y}(z) + a_1 z \tilde{Y}(z) + \ldots a_n z^n \tilde{Y}(z) = 0$

Dividing both sides by z^n we have

$$a_0 z^{-n} \tilde{Y}(z) + a_1 z^{-(n-1)} \tilde{Y}(z) + \ldots a_n \tilde{Y}(z) = 0$$

Considering z^{-1} to be equivalent to the unit delay operator we have

$$a_0 \tilde{y}(i-n) + a_1 \tilde{y}(i-n-1) + \ldots a_n \tilde{y}(i) = 0 \tag{8.25}$$

Now from Eq. (8.24)

$$\tilde{y}(i-1) = C\lambda^{(i-1)} = \frac{1}{\lambda} C\lambda^i = \frac{1}{\lambda} \tilde{y}(i)$$

Similarly $\qquad \tilde{y}(i-2) = \frac{1}{\lambda^2} \tilde{y}(i)$

so that Eq. (8.25) becomes

$$\tilde{y}(i)\left(\frac{a_0}{\lambda^n} + \frac{a_1}{\lambda^{(n-1)}} + \ldots a_n\right) = 0$$

Finally multiplying throughout by λ^n and ignoring the trivial solution that $\tilde{y}(i) = 0$, we arrive at

$$a_0 + a_1\lambda + \ldots a_n\lambda^n = 0$$

which is indeed the system characteristic equation with λ replacing z. This confirms that the transient response of a discrete time system consists of terms of the form given by Eq. (8.24) where the λ values are the roots of the characteristic equation.

> **Example 8.4 Transient response of a first-order system** The first-order system of Example 8.3 had a pulse transfer function
>
> $$\frac{Y}{X}(z) = \frac{b}{z - a}$$
>
> The system characteristic equation has a single root at $\lambda = a$ and so the transient will be of the form
>
> $$\tilde{y}(i) = Ca^i$$
>
> Examination of the solution for the step response of this system using z-transforms (Eq. (8.20)) confirms that the transient is of this form.

Relationship Between Pole Location and Transient Response

A simple pole on the real axis at $z = a$ produces a transient term of the form

$$\tilde{y}(i) = Ca^i$$

$$\{a^i\} = 1, a, a^2, a^3, \ldots \text{ etc.} \tag{8.26}$$

Figure 8.9 shows the transient response for various values of a. For small positive values of a the transient response is similar to that of a stable continuous time first-order system, i.e. it decays away monotonically with time. With $a > 1$ the terms in Eq. (8.26) become successively larger with increasing 'i' and so the transient increases exponentially without limit. A pole on this section of the real axis corresponds to an unstable system. When the pole lies on the negative real axis the numbers in the sequence in Eq. (8.26) alternate between positive and negative values. The transient is therefore oscillatory and looks very different to that of a continuous first-order system. As the pole becomes more negative the oscillations take longer to die away and for $a < -1$ they no longer die away but increase in amplitude with time, again corresponding to an unstable system.

Consider next a pair of complex conjugate poles located at a radius, r, and angle, θ, i.e.

$$z = r\,e^{\pm j\theta} \tag{8.27}$$

The corresponding transient will be of the form

$$\tilde{y}(i) = C_1(r\,e^{j\theta})^i + C_2(r\,e^{-j\theta})^i$$

$$= r^i[C_1\,e^{j\theta i} + C_2\,e^{-j\theta i}]$$

The exponential terms in the square brackets give rise to a sinusoid of frequency

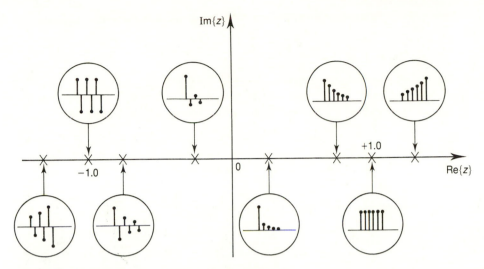

Figure 8.9 Transient response corresponding to a real pole

θ rad per sample. For any real system, the constants C_1 and C_2 are complex conjugates and produce a constant phase shift ϕ (the details of the proof follow similar lines to that developed for continuous systems in Section 3.7), and so

$$\tilde{y}(i) = Ar^i \cos(i\theta + \phi) \qquad (8.28)$$

The transient response is therefore a sinusoid of frequency θ rad per sample which decays, or increases, at a rate governed by the pole radius.

Since the time between samples is T seconds the frequency of the transient is

$$\omega = \frac{\theta}{T} \qquad \text{rad/s} \qquad (8.29)$$

Figure 8.10 shows how the z-plane pole location relates to the corresponding transient term.

Note that the transient resulting from a simple real pole on the negative real axis can be considered as a special case of Eq. (8.28). A pole at $z = -a$ has a radius $r = a$ and an angle $\theta = \pi$. The transient for this case therefore oscillates at a frequency of π rad per sample, which is one cycle every two samples, hence agreeing with Eq. (8.26).

Exercise Obtain a pulse transfer function for the difference equation

$$y(i) = 0.64[x(i-2) - y(i-2)]$$

where y is an output and x is an input. Plot the system pole zero pattern on a z-plane diagram and hence estimate the main characteristics of the system impulse response. Check the open-loop impulse response using CODAS-II.

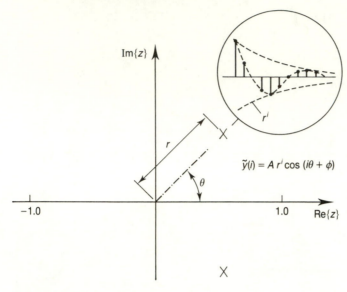

Figure 8.10   Relationship between complex pole location and transient response

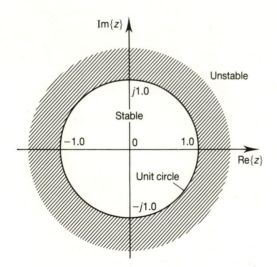

Figure 8.11   Stability region on the z-plane

Stability on the z-plane

The envelope for the oscillatory transient produced by a complex pole pair (Eq. (8.28)) decays in exactly the same way as for a real pole at radius $r = a$ (Eq. (8.26)). A pole pair at a radius of less than 1.0 will produce a decaying (stable) transient whereas a radius of greater than 1.0 produces an increasing (unstable) response. The region of stability on the z-plane is thus inside a circle of unit diameter (Fig. 8.11).

For a discrete time system to be stable all the system poles must lie inside the unit circle, i.e.

$$|r| < 1.0 \tag{8.30}$$

A pole which appears on the unit circle corresponds to a marginally stable system. The system will oscillate continuously with a frequency of θ radian per sample, where θ is the angle of the pole around the unit circle.

Relationship Between s-plane and z-plane

A z-plane pole at $z = \lambda$ produces a transient of the form

$$\tilde{y}(i) = C\lambda^i$$

Section 3.5 showed that a continuous time system with an s-plane pole at $s = \alpha$ exhibits a transient

$$\tilde{y}(t) = C\,\mathrm{e}^{\alpha t}$$

The continuous and discrete time systems will have the same transient at the sample instants if

$$\tilde{y}(t)|_{t=iT} = \tilde{y}(i)$$

Thus for identical values at the sample instants

$$\lambda^i = \mathrm{e}^{\alpha i T}$$

or

$$\lambda = \mathrm{e}^{\alpha T}$$

In other words a continuous time system with a pole at 's' will possess the same underlying dynamic characteristics as a discrete time system with a pole at

$$z = \mathrm{e}^{sT} \tag{8.31}$$

Now this equation must be used with care since its development was based on the equivalence between a continuous time signal and the discrete time signal at the sample instants. We have seen however that many different continuous time signals can produce the same discrete time signal when sampled.

Interestingly, inverting both sides of Eq. (8.31) yields

$$z^{-1} = \mathrm{e}^{-sT} \tag{8.32}$$

which confirms that z^{-1} corresponds to a pure delay of 1 sample, T seconds, i.e. it is equivalent to the unit delay operator, q.

Expressing the Laplace operator in terms of real and imaginary parts

$$s = \sigma + j\omega$$

the equivalent z-plane pole location is

$$z = \mathrm{e}^{(\sigma + j\omega)T} = \mathrm{e}^{\sigma T} \cdot \mathrm{e}^{j\omega T}$$

or

$$z = r\,\mathrm{e}^{j\theta} \tag{8.33(a)}$$

where
$$r = e^{\sigma T}$$
(8.33(b))

and
$$\theta = \omega T$$
(8.33(c))

The real part of the s-plane maps to the radius on the z-plane and the imaginary part of the s-plane maps to the angle on the z-plane.

Chapter 3 showed that for a continuous time system to be stable all the roots of the characteristic equation must lie in the negative half of the s-plane. Equation (8.33(b)) allows us to map this stability region into the z-plane. If the real part of the s-plane pole is negative then the z-plane pole must be at a radius of less than unity. This confirms the result expressed by Eq. (8.30).

Poles at the z-plane Origin

A pole at $z = 0$ corresponds to an s-plane pole at $s = -\infty$, i.e. an infinitely fast transient term. However this gives a false impression of the effect of such a pole. Such a pole derives from the term z^{-1} in the transfer function which represents a delay of one sample. Poles at the z-plane origin thus introduce pure delay with no dynamic effect.

Free Integrators and Poles at z = 1

A free integrator is a common building block for dynamic systems and in feedback control systems can be used to remove offset in Type-0 systems. A free integrator has the transfer function

$$\frac{Y}{X}(s) = \frac{1}{s}$$

which has a pole at the origin of the s-plane. The equivalent discrete time pole corresponds to $z = e^0 = 1$.

The discrete time approximation of integration based on the rectangular rule which was developed in Section 8.4 gave rise to the difference equation

$$y(i) = y(i-1) + Tx(i)$$

This corresponds to the pulse transfer function

$$\frac{Y}{X}(z) = \frac{T}{1 - z^{-1}} = \frac{Tz}{z - 1}$$

Sure enough, a pole at $z = 1$. The reader may like to confirm that the trapezoidal approximation to integration (Eq. (8.4)) also results in a pole at $z = 1$.

Design Regions on the z-plane

Damping ratio In Chapter 3 we saw that contours of constant damping ratio are radial lines on the left hand side of the s-plane. The slope of these lines, m, is related to the damping ratio by

$$m = -\frac{\omega}{\sigma} = \frac{\sqrt{1 - \zeta^2}}{\zeta}$$

where
$$s = \sigma + j\omega$$

Equation (8.33(c)) relates ω to the angle of a z-plane pole, θ

$$\omega = \frac{\theta}{T}$$

and Eq. (8.33(b)) relates σ to the z-plane pole radius

$$\sigma = \frac{\ln(r)}{T}$$

Therefore
$$m = -\frac{\theta}{T}\frac{T}{\ln(r)}$$

or
$$\ln(r) = -\frac{\theta}{m} = \frac{-\theta\zeta}{\sqrt{1-\zeta^2}} \tag{8.34}$$

Therefore lines of constant damping ratio are logarithmic spirals in the z-plane. These spiral contours are called Jury contours (Fig. 8.12).

Jury contours are useful in that they give an idea of the decay rate of the transient, however we must be careful since damping ratio is a property derived from a continuous time second-order system. If the peaks of the continuous time response do not correspond to the sample instants then we will not see those peaks in the discrete time response (Fig. 8.13). The discrete time transient response will generally only show the same peaks when the sample time is small compared to the damped natural frequency, i.e. for small θ. The theory developed from continuous time logarithmic decrement should not generally be used to predict overshoot or decay ratio for a discrete time system.

Damped natural frequency The damped natural frequency of a pair of complex s-plane poles is given by their imaginary part. Equation (8.33(b)) shows that the imaginary part of s maps to the angle, θ, on the z-plane and therefore

$$\omega_d = \frac{\theta}{T} \qquad \text{rad/s} \tag{8.35}$$

or
$$\omega_d = \theta \qquad \text{rad/sample} \tag{8.36}$$

Lines of constant damped natural frequency are therefore radial lines on the z-plane emanating from the origin (Fig. 8.14).

Settling time The 5% settling time of a continuous second-order system is related to the real part of the s-plane poles, σ, and can be approximated by

$$t_s \approx -\frac{3}{\sigma}$$

Equation (8.33(b)) maps σ into a radius on the z-plane, and so

$$\sigma = \frac{\ln(r)}{T}$$

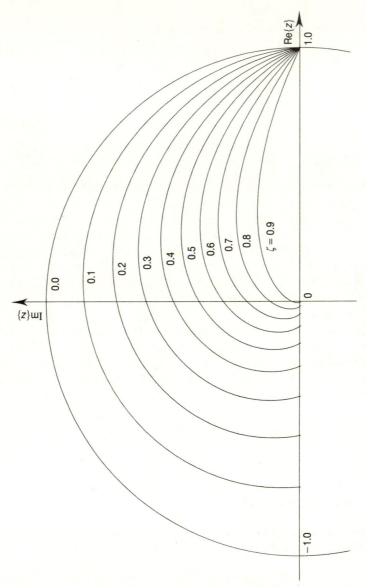

Figure 8.12 Jury contours—lines of constant damping ratio on the z-plane

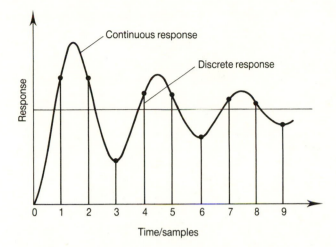

Figure 8.13 Damping ratio can give misleading results for the decay ratio of discrete time systems

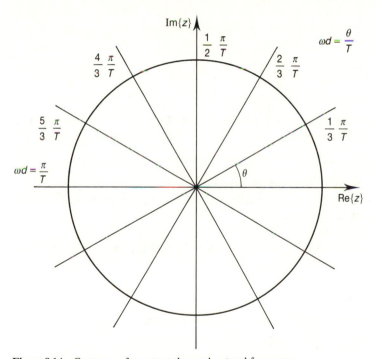

Figure 8.14 Contours of constant damped natural frequency

giving the approximate 5% settling time for a discrete time system as

$$t_s \approx \frac{-3T}{\ln(r)}$$

(8.37)

Thus contours of constant settling time are circles on the z-plane (Fig. 8.15). The above design contours are shown together on a z-plane diagram in Fig. 8.16.

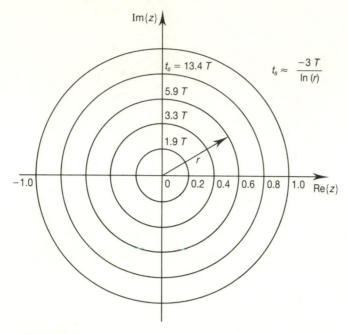

Figure 8.15   Contours of constant settling time

Example 8.5 Second-order filter design on the z-plane A discrete time filter with a sample frequency of 10 Hz can be described by the second-order difference equation

$$y(i) = a_1 y(i-1) + a_2 y(i-2) + bx(i-1)$$

(a) Obtain the pulse transfer function of the filter.
(b) Determine the pole locations which would give the filter a step response with
  i) A 5% settling time of 0.5 second;
  ii) A damping ratio of 0.5.
(c) Determine the coefficients a_1 and a_2 and use CODAS-II to verify the filter performance with the coefficient b set to unity.

SOLUTION

(a) Taking z-transforms of the filter difference equation gives

$$Y(z) = a_1 z^{-1} Y(z) + a_2 z^{-2} Y(z) + bz^{-1} X(z)$$

or
$$\frac{Y}{X}(z) = \frac{bz}{z^2 - a_1 z - a_2}$$

(b) The settling time requirement determines the pole radius. Equation (8.37) with $T = 0.1$ second gives

$$t_s = \frac{-0.3}{\ln(r)} = 0.5$$

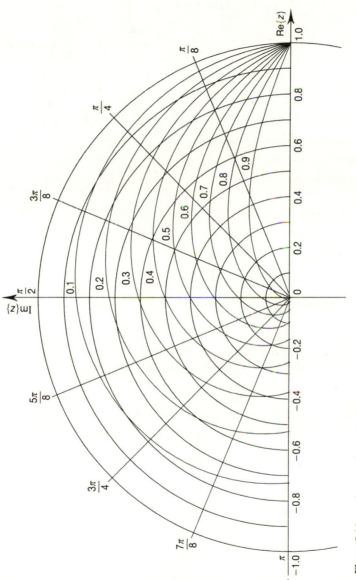

Figure 8.16 z-plane design contours

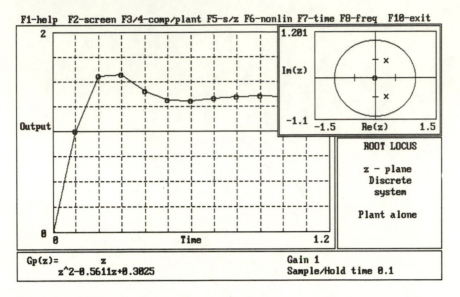

Figure 8.17   Step response of a discrete time filter

or
$$\ln(r) = -3/5$$

The required pole radius is therefore
$$r = e^{-3/5} = 0.55$$

The damping ratio requirement can be used to determine the angle of the pole, θ. Substituting $\zeta = 0.5$ and $r = 0.55$ into Eq. (8.34) gives

$$\ln(0.55) = \frac{-0.5\theta}{\sqrt{1 - 0.5^2}}$$

or
$$\theta = 1.035 \text{ radian} \qquad (59.3°)$$

(c) The filter characteristic equation can now be written as

$$(z - r\, e^{j\theta})(z - r\, e^{-j\theta}) = 0$$

or
$$z^2 - r(e^{j\theta} + e^{-j\theta})z + r^2 = 0$$

$$z^2 - 2r\cos(\theta)z + r^2 = 0$$

$$z^2 - 0.5611z + 0.3025 = 0$$

The filter transfer function with $b = 1.0$ is thus

$$\frac{Y}{X}(z) = \frac{z}{z^2 - 0.5611z + 0.3025}$$

Figure 8.17 shows the open-loop step response of the filter.

8.7 SYSTEM CLASSIFICATION

Much useful information about system response can be obtained without the need for simulation or analysis by simple examination of its pulse transfer function. Most of the classifications developed in Chapter 3 for continuous time systems can be extended to discrete time systems.

Steady-state and Instantaneous Gain

The simplest way to deal with the problem of finding steady-state and instantaneous system gains is to use the relationship between z and s developed above (Eq. (8.31)).

Recall that the steady-state gain of a continuous system is obtained by setting s to zero in the transfer function. Now as s tends to zero, z tends to unity, since

$$\lim_{s \to o} \{e^{sT}\} = 1$$

The steady-state gain of a discrete time system is therefore given by setting z to unity in the pulse transfer function.

$$\text{Steady-state gain} = \lim_{z \to 1} G(z) = G(1) \tag{8.38}$$

Similarly the instantaneous gain of a discrete time system is obtained when s tends to infinity, making z also tend to infinity

i.e.

$$\lim_{s \to \infty} \{e^{sT}\} = \infty$$

$$\text{Instantaneous gain} = \lim_{z \to \infty} G(z) = G(\infty) \tag{8.39}$$

Exercise Determine the steady-state and instantaneous gains of the filter derived in Example 8.5 and confirm your answers from Fig. 8.17.

Properness of Discrete Systems

Not all discrete time transfer functions are physically realizable. Systems which operate in **real-time** must be *causal*; that is the output can depend only upon current or past values of inputs or outputs.

Consider the general pulse transfer function with polynomial numerator and denominator of order m and n respectively:

$$\frac{Y}{X}(z) = \frac{b_0 + b_1 z \ldots b_m z^m}{a_0 + a_1 z \ldots a_n z^n} \tag{8.40}$$

assuming for the moment that the denominator is of higher order than the numerator

$(n > m)$. Dividing by z^n and cross multiplying produces

$$Y(z)\{a_0 z^{-n} + a_1 z^{-(n-1)} \ldots + a_n\} = X(z)\{b_0 z^{-n} + b_1 z^{-(n-1)} \ldots + b_m z^{-(n-m)}\}$$

which corresponds to the difference equation:

$$a_0 y(i-n) + a_1 y(i-n+1) \ldots + a_n y(i) = b_0 x(i-n) + b_1 x(i-n+1) \ldots + b_m x(i-n+m)$$

or

$$y(i) = -\frac{a_0}{a_n} y(i-n) - \frac{a_1}{a_n} y(i-n+1) \ldots + \frac{b_0}{a_n} x(i-n) + \frac{b_1}{a_n} x(i-n+1) \ldots$$

$$+ \frac{b_m}{a_n} x(i-n+m) \tag{8.41}$$

This equation expresses the current output, $y(i)$, in terms of past output values and, providing $n > m$, past input values. Such a system is causal and can be implemented in real-time. We would have severe problems implementing the system if the numerator were of higher order than the denominator (i.e. $m > n$) since calculation of the current output would require knowledge of future inputs! The definitions for system properness developed for continuous time systems is therefore also useful for discrete time systems. When the pulse transfer function has more poles than zeros it is strictly proper and is fully realizable in real-time. A pulse transfer function with more zeros than poles is not proper and cannot be implemented in real-time.

If the time taken to calculate the discrete system output is short compared to the sample time then the constraint of strict properness can be relaxed slightly. We can allow the current output to depend upon the **current** input. Examination of Eq. (8.41) shows that this corresponds to a system with numerator and denominator of the same order ($m = n$). That is, a proper system is realizable if the calculation time is ignored. This will be the case with many computer control and filtering applications where the sample period is relatively slow and the small delay between sampling the current input and producing the current output is negligible.

It is possible to have discrete time systems which are not proper acting on data which has been previously sampled and stored, i.e. not in real-time. This is because all past and future inputs would be available from the data store. Such schemes are no good for computer control systems since these must operate in real-time on up-to-date sampled data.

Transport Delay

Examination of Eq. (8.41) indicates that the delay between an input sample and its effect on the output depends upon the relative orders of the numerator and denominator. Every z-plane pole introduces a delay of one sample, and conversely every zero an advance of one sample. Thus a system with an excess of poles over zeros ($n - m$) will exhibit a transport delay of ($n - m$) samples. If there are the same number of zeros as poles ($m = n$) the system is proper but not strictly proper and there will be no delay in the response, i.e. the instantaneous response will be non-zero.

Exercise For the following pulse transfer functions determine the properness of the systems and the existence of any transport delay. For the proper systems obtain a difference equation for calculating the current output, $y(i)$.

(a) $\dfrac{Y}{X}(z) = \dfrac{(z+1)(z+1)}{(z-1)(z-1)}$ (b) $\dfrac{Y}{X}(z) = z - 2$

(c) $\dfrac{Y}{X}(z) = \dfrac{0.5z^{-1}}{1 - 0.5z^{-1}}$ (d) $\dfrac{Y}{X}(z) = \dfrac{1}{z^{-1} + z^{-2}}$

System Type Number

Following the definition of system type number for continuous time systems, the type number of a discrete time system is equal to the number of free integrators in its transfer function. Now as we have seen, a free integrator corresponds to a z-plane pole at $z = 1$. The type number of a discrete system can therefore be defined as the number of z-plane poles located at $z = 1$.

8.8 SAMPLED-DATA SYSTEMS

Figure 8.18 shows a computer interfaced to a continuous time process for the purpose of control. The computer must sample the process output via the ADC, calculate the

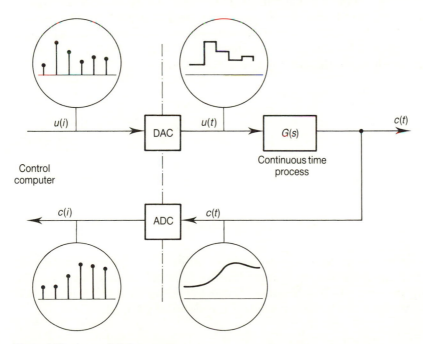

Figure 8.18   A sampled-data system

required control effort and apply this to the process via the DAC. As far as the computer is concerned, it is sending out one discrete time signal, $u(i)$, and receiving back another, $c(i)$. The process 'looks' to the computer as though it too is a discrete time system even though the discrete response is derived from a continuous process.

We have seen that discrete time systems can behave at the sample instants in a similar manner to continuous time systems. It is not surprising that it is generally possible to devise a difference equation to generate the same response at the sample instants as that from the sampled plant. However, we must be careful since the DAC will normally incorporate some form of signal reconstruction, such as a zero-order hold, which will appear as part of the process.

Pulse Transfer Function for Sampled-data Systems

The system in Fig. 8.18 shows a continuous time process, $G(s)$, with input supplied from a DAC and output sampled by an ADC. The DAC input, $u(i)$, which is the discrete time computer output, is reconstructed by a zero-order hold into the step-like continuous signal, $u(t)$, which is applied to the process. The continuous time process response, $c(t)$, is sampled by the ADC to produce the discrete time signal $c(i)$.

A simple way to obtain the pulse transfer function of the process is to consider the sampled response of the system to a discrete time impulse applied at $u(i)$. Since the z-transform of the discrete time impulse is unity the z-transform of the response, $c(i)$, will give the pulse transfer function directly.

A unit discrete time impulse applied to the DAC input will produce a rectangular pulse at the DAC output due to the inherent zero-order hold (Fig. 8.19(a)). The Laplace transform of such a rectangular pulse can be obtained by considering the pulse as a unit step at $t = 0$ followed by a negative unit step at $t = T$ (Fig. 8.19(b)).

The Laplace transform of the DAC output pulse is thus

$$\mathscr{L}\{u(t)\} = \frac{1}{s} - \frac{1}{s}e^{-sT}$$

$$= \frac{1}{s}(1 - e^{-sT}) \tag{8.42}$$

This pulse is applied to the process and so the Laplace transform of the continuous time response will be

$$\mathscr{L}\{c(t)\} = \frac{1}{s}(1 - e^{-sT})G(s)$$

The ADC samples the continuous time response to produce the discrete time signal, $c(i)$, which, neglecting quantization, is identical to $c(t)$ at the sample instants. The z-transform of the response which is the overall system pulse transfer function is obtained by taking the z-transform of the inverse Laplace transform of $C(s)$, i.e.

$$C(z) = Z\{c(i)\} = Z\left\{\mathscr{L}^{-1}\left[\frac{1}{s}(1 - e^{-sT})G(s)\right]\right\}$$

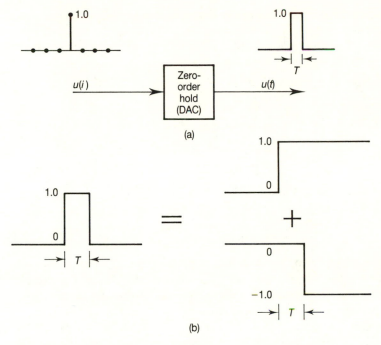

Figure 8.19 Response of a zero-order hold to a discrete time impulse

As the input to the system was a unit impulse, the pulse transfer function, $G(z)$, is equal to $C(z)$. For simplicity the equation is usually written as

$$G(z) = Z\left\{\frac{1}{s}(1 - e^{-sT})G(s)\right\}$$

Finally, since e^{-sT} is just a delay of one sample its z-transform is simply z^{-1} and the term $(1 - e^{-sT})$ can be transformed separately, giving

$$G(z) = (1 - z^{-1})Z\left\{\frac{G(s)}{s}\right\}$$

or
$$G(z) = \frac{(z - 1)}{z}Z\left\{\frac{G(s)}{s}\right\}$$
(8.43)

The z-transformation is easily obtained using the look-up table in Appendix B.

Example 8.6 First-order sampled-data system A first-order system with time constant 10 seconds is controlled by an industrial computer using a sample time of 2 seconds. Determine the pulse transfer function of the plant as seen by the computer and compare the open-loop step response of the continuous time and discrete time transfer functions. Also examine the open-loop ramp response of the two systems. Why are there differences in the two ramp responses but not in the step responses?

SOLUTION The discrete time transfer function is obtained using Eq. (8.43) as

$$G(z) = \frac{(z-1)}{z} Z\left\{\frac{1}{s(1+\tau s)}\right\}$$ (8.44)

Using the look-up table in Appendix B gives

$$G(z) = \frac{(z-1)}{z} \frac{(1-\beta)z}{(z-1)(z-\beta)}$$

$$= \frac{(1-\beta)z}{z-\beta}$$

Now $\beta = e^{-aT} = e^{-T/\tau} = e^{-0.2} = 0.8187$, thus

$$G(z) = \frac{0.1813}{z - 0.8187}$$

Figure 8.20(a) shows the open-loop step response of the original continuous time system and the equivalent pulse transfer function. The two are identical at the sample instants. Figure 8.20(b) shows the open-loop ramp responses of the two systems and here there is a clear difference at the sample instants.

The two step responses are identical because the discrete system was obtained by placing a zero-order hold in front of the plant, which is equivalent to making the control effort constant between samples. A step input is constant between samples and so the responses of the continuous and discrete systems are identical. This is clearly not the case with a ramp input.

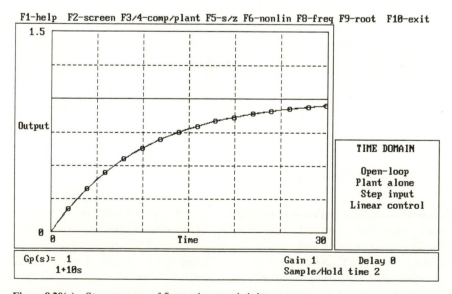

Figure 8.20(a)   Step response of first-order sampled-data system

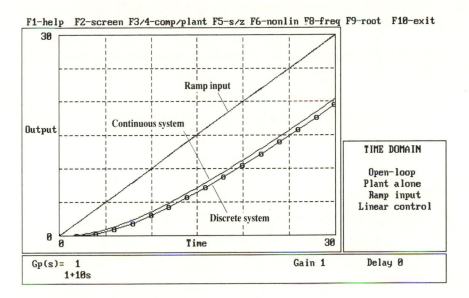

F1-help F2-screen F3/4-comp/plant F5-s/z F6-nonlin F8-freq F9-root F10-exit

Figure 8.20(b)   Ramp response of first-order sampled-data system

Exercise Repeat Example 8.6 with sample times of 0.1 and 20 seconds. What differences are now apparent in the open-loop step and ramp responses and why?

Transport Delay Systems and the Modified z-transform

The term z^{-k} in a pulse transfer function corresponds to a delay of k samples. Therefore any continuous time system with a transport delay of kT gives rise to a term z^{-k} in the pulse transfer function. In reality however, transport delays are rarely an exact multiple of the sample interval and cannot be dealt with in this simple way. *Modified z-transforms* allow us to obtain the z-transform of signals and systems which incorporate delays which are a fraction of the sample interval. The modified z-transform of a fractionally delayed sampled signal, $x(iT - \delta)$, is written as a function of the variable 'm' as well as z:

$$Z_m\{x(iT - \delta)\} = X(z, m)$$

where

$$m = 1 - \delta/T; \qquad 0 \geqslant \delta > T$$

$$Z_m\{x(iT - \delta)\} = \sum_{i=0}^{\infty} x(iT - \delta)z^{-i}$$

or

$$X(z, m) = \sum_{i=0}^{\infty} x(T(i - 1 + m))z^{-i}; \qquad 0 > m \geqslant 1$$

A detailed coverage of modified z-transforms is outside the scope of this book, however it is worth looking at a simple example and spending a little time examining the general effects of fractional delay on a simple sampled-data process.

Example 8.7 Modified z-transform of a delayed function Consider the simple exponential function

$$x(t) = e^{-at}; \qquad t \geqslant 0$$

When the signal is delayed by a small amount, δ, and sampled every T seconds we have

$$x(iT - \delta) = e^{-a(iT - \delta)}$$

The modified z-transform of this signal is given by

$$Z_m\{x(iT - \delta)\} = x(-\delta) + x(T - \delta)z^{-1} + x(2T - \delta)z^{-2} + \cdots \text{ etc.}$$

$$= 0 + e^{-a(T-\delta)}z^{-1} + e^{-a(2T-\delta)}z^{-2} \cdots$$

Substituting $\qquad m = \dfrac{T - \delta}{T}$

we have $\qquad X(z, m) = e^{-amT}z^{-1} + e^{-amT}e^{-aT}z^{-2} + e^{-amT}e^{-2aT}z^{-3} \cdots$

$$= e^{-amT}z^{-1}(1 + e^{-aT}z^{-1} + e^{-2aT}z^{-2} \cdots)$$

$$= \frac{e^{-amT}z^{-1}}{1 - e^{-aT}z^{-1}}$$

thus $\qquad X(z, m) = \dfrac{e^{-amT}}{z - e^{-aT}} = \dfrac{e^{-a(T-\delta)}}{z - e^{-aT}}$

When the delay is zero, $\delta = 0$, and the modified z-transform reduces to the normal z-transform

$$X(z) = \frac{e^{-aT}}{z - e^{-aT}}$$

The modified z-transform technique can be applied to sampled-data systems with transport delay which is not divisible by the sample time. Consider the general system

$$\frac{Y}{X}(s) = G(s) e^{-sT_D} \tag{8.45}$$

where $G(s)$ is a delay-free transfer function. The transport delay is split into two parts, an integer delay which is divisible by the sample time and a remaining fractional part.

i.e. $\qquad\qquad\qquad\qquad T_D = kT + \delta \tag{8.46}$

Using Eq. (8.43) the pulse transfer function is given by

$$G(z) = \frac{(z - 1)}{z} Z \left\{ \frac{G(s) e^{-ksT} e^{-s\delta}}{s} \right\}$$

or
$$G(z) = \frac{(z-1)}{z} z^{-k} Z_m \left\{ \frac{G(s) e^{-s\delta}}{s} \right\} \tag{8.47}$$

Thus the integer part of the transport delay introduces 'k' extra poles at the origin of the z-plane. The fractional part, δ, introduces discrete time zeros into the pulse transfer function but does not affect the system poles. The calculation of $G(z)$ from Eq. (8.47) is rather involved for all but the simplest of systems. Appendix C gives a few simple systems and their equivalent modified z-transform transfer functions.

The z-transform option in CODAS-II (F5) calculates the sampled-data plant pulse transfer function using Eq. (8.47) with a modified z-transform and thus deals correctly with transport delay.

Effect of Delay on a First-order Plant

In order to obtain a feel for the behaviour of a sampled-data system with fractional delay we will examine a simple first-order system with transport delay.

$$G(s) = \frac{1}{1 + \tau s} e^{-sT_D}$$

expressing the transport delay as an integer part and fractional part

$$T_D = kT + \delta$$

The pulse transfer function can be found from Eq. (8.47)

$$G(z) = \frac{(z-1)}{z} z^{-k} Z_m \left\{ \frac{e^{-s\delta}}{s(1 + \tau s)} \right\}$$

Appendix C gives the answer for $k = 0$ and so with non-zero k we have

$$G(z) = z^{-k} \frac{(1 - \Gamma)z + (\Gamma - \beta)}{z(z - \beta)}$$

or
$$G(z) = \frac{(1 - \Gamma)z + (\Gamma - \beta)}{z^{(k+1)}(z - \beta)}$$

where
$$\beta = e^{-T/\tau}$$

and
$$\Gamma = e^{-(T - \delta)/\tau}$$

When the transport delay is zero, $k = 0$ and $\delta = 0$ therefore the pulse transfer function reduces to

$$G(z) = \frac{(1 - \beta)z}{z(z - \beta)} = \frac{(1 - \beta)}{(z - \beta)}$$

with a solitary pole at $z = \beta$ due to the system time constant.

Upon introducing a small delay of less than one sample period, k remains zero

Table 8.1 Zero location for a first-order sampled-data system with fractional transport delay

δ/T	$\tau = T$	$\tau = 10T$
0.0	0.0	0.0
0.2	-0.148	-0.238
0.4	-0.401	-0.634
0.6	-0.917	-1.427
0.8	-2.487	-3.806
1.0	$-\infty$	$-\infty$

but δ is equal to the delay giving

$$G(z) = \frac{(1 - \Gamma)z + (\Gamma - \beta)}{z(z - \beta)}$$

The discrete transfer function now has one additional pole at the z-plane origin and an additional zero. As the fractional delay is increased from 0 to T the value of Γ grows from being equal to β to unity. Thus the relative weighting of the coefficients in the pulse transfer function numerator alters drastically with the delay. The transfer function zero is located at

$$z = \frac{(\Gamma - \beta)}{(\Gamma - 1)}$$

Table 8.1 shows how the zero migrates from the origin along the negative real axis as the fractional delay is increased. As the delay is increased past the sample time a second pole appears at the origin and the zero starts once again to move from 0 to $-\infty$. Figure 8.21 shows the open-loop step response of the pulse transfer function with different values of transport delay.

The behaviour of this simple example is typical of many sampled-data systems which include transport delay. Zeros which rapidly migrate to the left with small increases in delay are common. A short sampling interval in comparison to the transport delay will lead to multiple poles at the z-plane origin and as we will see in the next chapter this can cause problems with the design of digital feedback controllers.

Computer Simulation of Continuous Time Systems

It is worth spending a little time on the problem of using a digital computer to simulate continuous time dynamic systems. There are several well known techniques, however, they all approach the problem by dividing continuous time into small discrete steps. The problem is then the one of finding the response at the next step giving the current response.

One set of algorithms directly uses the system differential equation with estimates of the appropriate derivatives to determine the next step. The accuracy of these methods

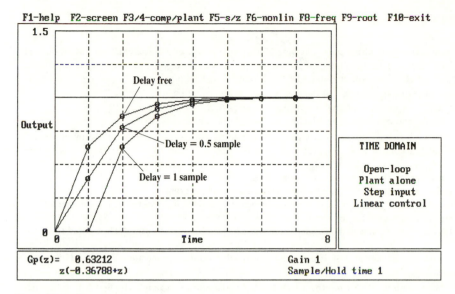

Figure 8.21 Open-loop step response of sampled-data first-order system with transport delay

depends very much on the step size in relation to the rate of change of the derivatives. One such widely used technique is called the Runge–Kutta method which can give accurate results but requires considerable calculation at each step.

An alternative method is to consider the system as a sampled-data system with the sample time equal to the step size. The equivalent pulse transfer function can then be used to simulate the system. This technique demands considerable calculation before the simulation can be started but has the advantage that the simulation calculations are simple and therefore fast. The method however, places a zero-order hold in front of the plant which, for changing inputs, can cause errors. The accuracy again depends upon the step size in relation to the speed or the system dynamics. This is the method used in CODAS to obtain the response of continuous time systems. CODAS does use a compensation technique to reduce the errors to an acceptable value, however, in closed-loop with a large step size the errors can be significant.

At very small step lengths another problem occurs concerned with the way in which the computer stores numbers. Very small step lengths give rise to very small incremental values in the simulation. Roundoff errors in the computer can cause inaccuracies at each step which, because of the iterative nature of the solution, accumulate to produce large errors.

Exercise Use CODAS to simulate the closed-loop step response of the system

$$G(s) = \frac{1}{s(1 + 2s)(1 + 0.2s)}$$

with the default x-axis. Successively reduce the number of points (to increase the step size) and observe any difference in the simulated response. Repeat, but this time successively increase the number of points above the default value.

PROBLEMS

8.1 A temperature transmitter with a range covering 50 to 150°C produces a linearly related output current in the range 4 to 20 mA. The transmitter output current is passed through a precision 250 Ω resistor to provide a voltage signal for use with a 10-bit ADC with range 0 to 5 volts.

   (a) What temperature resolution will the system give?
   (b) What ADC output (as a decimal integer) corresponds to a temperature of 100°C?
   (c) What would you conclude if the ADC produced a decimal output of 100?

8.2 A 12-bit ADC with range 0 to 10 volts is used to provide the measured value for a computer control system. The calculated control effort is output via a 10-bit ADC of range 0 to 10 volts. The computer implements simple proportional control with a gain of 32 volt/volt. The reference value for the controller is stored in the computer memory as a 12-bit unsigned integer.

   (a) Determine the multiplicative gain as used in the computer program.
   (b) By considering all possible values of measured value and reference determine the number of bits required to store the discrete time error signal.
   (c) What is the smallest change in measured value which will produce an output? How could this limit be reduced?
   (d) What is the largest change in measured value which will produce a proportional output? How could this limit be increased?

8.3 A popular puzzle with children concerns a frog trying to jump across a pond. His first leap takes him exactly half-way across. Each subsequent leap is only half the size of the previous one. Write down a difference equation for the size of the leap, x. By considering x as the incremental distance travelled by the frog, i.e. $x = \Delta y$, show that the frog's total progress can be represented by the difference equation

$$y(i) = 1.5y(i-1) - 0.5y(i-2)$$

Substitute the initial conditions $y(0) = 0$ and $y(1) = 0.5$ and confirm that the difference equation then gives the expected result. Explain why the step response of the system

$$G(z) = \frac{0.5}{z - 0.5}$$

gives the same response as the difference equation.

8.4 When home computers first became available, a favourite game was the 'moon lander'. The program simulated the Apollo Lunar Excursion Module approaching the surface of the moon; thrusters could be fired (by pressing a key) to slow the vehicle in its descent. The aim of the game was to land the module on the moon with a final velocity as low as possible without running out of fuel.

   Assuming the lander velocity and thrust are always vertical, obtain a continuous time differential equation relating the height of the lander above the moon's surface, h, to the thrust, F. Show that by using the rectangular rule, the simulation game can be modelled by the difference equation

$$h(i) = T^2\left(\frac{F}{m} - g\right) + 2h(i-1) - h(i-2)$$

where m is the module mass and g is the acceleration due to the moon's gravity.

8.5 Obtain the pulse transfer function Y/X for each of the following difference equations. Comment upon the properness, and stability of each example.

   (a) $2y(i) = 3(y-1) + x(i-2)$
   (b) $y(i+1) = -0.5[y(i) + x(i)]$
   (c) $y(i) = y(i-2) + x(i)$
   (d) $y(i) = 0.7y(i-1) + 0.3x(i) - x(i-1)$
   (e) $y(i) = \sqrt{2}[y(i-1) + x(i-1)] + x(i) + x(i-2) - y(i-2)$

8.6 Determine the difference equation for the current output, $y(i)$, for each of the following pulse transfer functions. Comment upon the possibility of their realization in real time.

(a) $\dfrac{Y}{X}(z) = \dfrac{z}{z-2}$

(b) $\dfrac{Y}{X}(z) = \dfrac{0.5}{z^{-1} - 0.5z^{-2}}$

(c) $\dfrac{Y}{X}(z) = z + 1$

(d) $\dfrac{Y}{X}(z) = \dfrac{z+1}{z^2}$

8.7 A discrete time system has a characteristic equation given by

$$z^2 - a_1 z + a_2 = 0$$

Show that when the poles are complex they are at a radius of $\sqrt{a_2}$ and the system has a damped natural frequency of

$$\omega_d = \frac{1}{T}\cos^{-1}\left\{\frac{a_1}{2\sqrt{a_2}}\right\} \qquad \text{rad/s}$$

8.8 By considering all possible values of 'a' show that the following system can never be absolutely stable in open-loop.

$$\frac{Y}{X}(z) = \frac{1}{z^2 + az + 1}$$

8.9 Figure P8.9 shows a computer flow chart. Write down the difference equation which describes the relationship between the output, y, and the input, x.

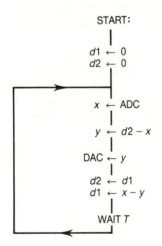

START:

$d1 \leftarrow 0$
$d2 \leftarrow 0$

$x \leftarrow \text{ADC}$

$y \leftarrow d2 - x$

$\text{DAC} \leftarrow y$

$d2 \leftarrow d1$
$d1 \leftarrow x - y$

WAIT T

Figure P8.9   Computer flow chart

8.10 With the aid of Fig. 8.16, sketch the region of the z-plane in which the poles of a second-order system must lie in order to meet the following specifications:

(a) The damping ratio must be at least 0.5.
(b) The damped natural frequency must be at least $\pi/4$ rad/sample.

Determine the pulse transfer function of a second-order system which meets the above specifications and has the largest posible settling time. The system should have a steady-state gain of unity and should exhibit a delay of one sample period. Test your design using CODAS-II.

COMPUTER CONTROL SYSTEMS

9.1 INTRODUCTION

Chapter 8 covered the basic tools needed for modelling discrete time and sampled-data systems. Here we will apply these techniques to the analysis and design of feedback control systems. The use of digital computers for the control and monitoring of processes was justified in the previous chapter on the general grounds of cost, flexibility and improved user interface. In this chapter we will see that, with the correct design approach, the use of quite simple discrete time control laws can lead to system performance which would be difficult to achieve with analogue implementations.

Digital computers are sometimes used in a supervisory role where the feedback control of individual loops is carried out by subordinate controllers. The supervisory computer sends set-point information to slave controllers and receives back information on performance, product quality, process throughput, etc. The term *set-point control* is often used to describe this type of computer control. Such schemes were common in the process industries in the early days of computerization when computers were less reliable. The advantage of set-point control is that the controllers continue to operate even if the computer fails.

Direct digital control (DDC) describes the use of digital computers to calculate the actual control effort that is applied to the process or plant. Increasingly DDC is replacing analogue technology either as microprocessor-based single loop controllers or as large multi-loop control computers. Direct digital control is the main concern of this chapter. We will examine the design and performance of DDC systems as well as details of practical implementation. The frequency response of discrete time systems will be covered and the associated topic of digital filtering.

Approximation of Analogue Controllers or Direct Discrete Design?

There are two basic philosophies which are used to arrive at direct digital control laws. One method is to take a successful analogue design and approximate it with a suitable discrete time control law. This method has one main advantage, namely that of familiarity. A vocabulary has developed to describe controller characteristics and behaviour which is based upon analogue concepts. Terms such as integral action, derivative time, phase lead compensation, etc., are all based upon analogue technology. Traditional operators and control engineers are happy dealing with adjustments involving 'reset time' or 'proportional band'.

A different philosophy for the design of digital controllers is to deal with the control problem directly in discrete time. Here the plant is considered as an equivalent discrete time system and the design is carried out without reference to the underlying analogue plant. This approach has the advantage that the limitations of analogue implementation do not have to be observed, the design is not limited to constructs such as that of the phase lead compensator or velocity feedback, etc.

Perhaps the major advantage of direct discrete design methods concerns transport delays. In the continuous or Laplace domain transport delay is an odd man out, an irrational term in transfer functions. In the discrete time domain transport delay is natural, discrete time systems are based upon the ideas of pure delays of one or more samples. This means that controller designs can easily take account of transport delay and give performance comparable to delay free systems.

We must be careful in promoting the direct discrete design method since it is the performance of the controlled plant in continuous time which is important. We will see that it is quite possible to achieve excellent control performance in discrete time, i.e. at the sample instants, while the controlled variable oscillates wildly between samples.

In order to appreciate some of the characteristics and possibilities of direct digital control we will examine a simple first-order system with discrete time control.

Discrete Time Proportional Control

Figure 9.1 shows a plant under proportional control from a digital computer. Figure 9.2 shows a flowchart for the section of computer program concerned with the calculation of proportional control. At each sample instant the computer samples the plant output

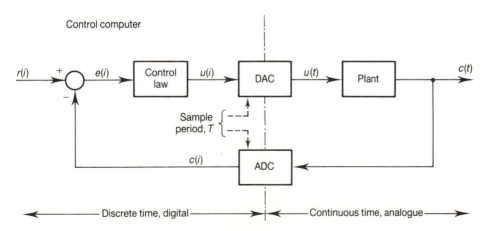

Figure 9.1 Feedback control of a sampled-data system

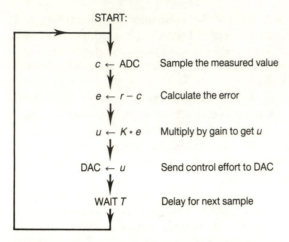

Figure 9.2 Flow chart for a proportional controller

via the analogue to digital converter (ADC) to produce the sampled measured value, c. The computer then subtracts this from the reference value, r, to form the error, e. This error is then multiplied by the controller gain, K, to produce the control effort, u, which is then sent to the digital to analogue converter (DAC) to drive the plant.

The values of reference and gain are assumed here to be available in the computer memory, having been set up by some other part of the program. The timing of the sampling is of course critical and so the delay must be synchronized to a real-time clock as discussed in Section 8.2 of the previous chapter. It can be assumed that the calculation time between the sampling of c and the output of u is very small compared to the sample interval, T, and so the control effort appears almost at the same time as the sample is taken.

The variables c, r, e and u all relate to the current sample, i, and so the control calculation can be described by the discrete time equations

$$e(i) = r(i) - c(i) \tag{9.1}$$

$$u(i) = Ke(i) \tag{9.2}$$

Consider the case where the plant is a simple first-order system

$$\frac{C}{U}(s) = \frac{1}{1 + \tau s} \tag{9.3}$$

Previously we showed that the pulse transfer function for a first-order system together with a zero-order hold (DAC) has a pulse transfer function given by

$$\frac{C}{U}(z) = \frac{(1 - \beta)}{z - \beta}$$

where

$$\beta = e^{-T/\tau}$$

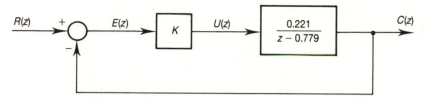

Figure 9.3   Block diagram of first-order sampled-data system with proportional control

For a sample time of 0.25 second and a time constant of 1 second

$$\beta = 0.779$$

so that
$$\frac{C}{U}(z) = \frac{0.221}{z - 0.779} \tag{9.4}$$

We must be careful here since this transfer function neglects any gains or sensitivities associated with the DAC and ADC. For simplicity we can assume that the ADC sensitivity is the reciprocal of the DAC sensitivity (not unreasonable if the converters have the same resolution and voltage range).

Figure 9.3 shows the block diagram of the discrete time system obtained from Eqs (9.1), (9.2) and (9.4). The closed-loop pulse transfer function is obtained by block diagram reduction as

$$\frac{C}{R}(z) = \frac{0.221K}{z - 0.779 + 0.221K} \tag{9.5}$$

The overall system has a single pole at

$$z = 0.779 - 0.221K$$

The root locus method can be used directly with the open-loop transfer function to examine the closed-loop z-plane pole migration as the gain is varied. Figure 9.4 shows the root locus together with the closed-loop step response for various gain values. For small gain values the closed-loop pole is on the positive real axis (points A and B). The response here shows a reducing offset and reducing response time as the pole moves to the left. This is exactly the same as expected for a first-order continuous time system with proportional control.

The pole reaches the origin (point C) when the gain has a value 3.52. Here the system behaves like a pure delay of one sample, i.e.

$$\frac{C}{R}(z) = \frac{0.779}{z} = 0.779z^{-1}$$

This value of gain produces just the correct amount of control effort to make the response reach the steady-state value of 0.779 in one sample period. This is a special type of response which cannot normally be achieved with continuous control. So-called *deadbeat control* occurs when the closed-loop poles are all at the z-plane origin. The transient response with general deadbeat control does not display asymptotic behaviour but instead settles exactly in a finite number of samples.

As the gain is increased above the deadbeat value the pole moves onto the negative

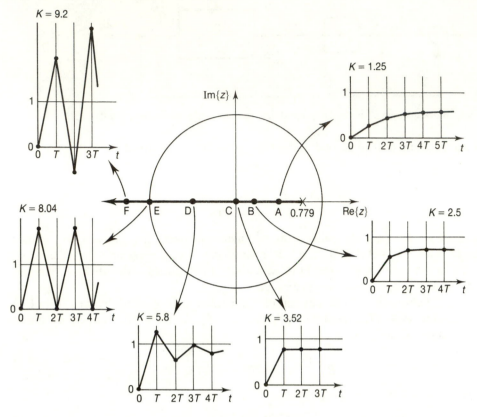

Figure 9.4 Root locus diagram for discrete time proportional control of a first-order system

real axis (point D). Here the system exhibits an oscillatory response at half the sample frequency. As the gain increases the oscillations become more violent until at a gain of 8.04 the pole reaches the unit circle at $z = -1$ (point E). Here the system is marginally stable. Finally when the gain is greater than the critical value, the pole is outside the unit circle producing an unstable response (point F). It is interesting to note that the response of a pure first-order system under continuous time proportional control never shows an oscillatory response or signs of instability even with very high gain.

The control system designer has the choice of placing the closed-loop pole anywhere on the real axis by choosing the appropriate gain. Deadbeat control looks very attractive at first sight but, particularly at higher sample rates, small errors in the plant model can lead to a large degradation in performance, also the control effort can be excessive. A more gentle action with the pole somewhere on the positive real axis may be preferable.

Discrete Time Proportional Plus Integral Control

A proportional controller does not involve any discrete time dynamics. Here we will examine the use of a discrete time approximation of a proportional plus integral controller.

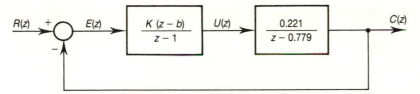

Figure 9.5 Block diagram for discrete proportional plus integral control of a first-order system

A discrete time version of a proportional plus integral controller can be described by the difference equation

$$u(i) = u(i-1) + K[e(i) + be(i-1)] \tag{9.6}$$

The parameter K is the gain and b is related to the controller reset time.

The transfer function can be obtained directly as

$$\frac{U}{E}(z) = \frac{K(z-b)}{z-1} \tag{9.7}$$

Figure 9.5 shows the overall system block diagram. The overall open-loop transfer function is now

$$\frac{C}{E}(z) = \frac{0.221K(z-b)}{(z-1)(z-0.779)} \tag{9.8}$$

The open-loop system now has an extra pole at $z = 1$ corresponding to the integral action of the controller, and in addition a zero has been introduced at $z = b$. This new zero can be placed anywhere along the real axis by simply choosing the value of 'b' used in the control program.

One interesting choice of b places the zero at $z = 0.779$ to cancel the effect of the plant pole. The open-loop transfer function is then

$$\frac{C}{E}(z) = \frac{0.221K}{z-1} \tag{9.9}$$

Now just as for proportional control, the gain can be chosen to place the closed-loop pole anywhere on the real axis, including at the origin (for deadbeat control). The advantage of this controller over simple proportional control is that there is no offset since the system is now Type-1.

Exercise For the discrete time proportional plus integral controller (Eq. (9.7) with $b = 0.779$) determine the controller gain, K, which gives the first-order system (Eq. (9.3)) a closed-loop step response which has

(a) a deadbeat characteristic;
(b) a time constant of 0.25 second.

Hint: pole at $z = e^{-T/\tau}$.

Simulate the continuous time plant and your discrete time controllers on CODAS-II and verify the step responses. Examine the peak control effort for each controller. What is the effect of an increase in system time constant of 50% on the performance of both controllers?

9.2 FREQUENCY RESPONSE OF DISCRETE SYSTEMS

The advantages of frequency response methods for the analysis and design of continuous systems have been discussed in length in Chapters 4 and 5. Here we will examine the frequency response of discrete time systems.

The frequency response of a system is its steady-state response to a sinusoidal input,

$$x(t) = \cos(\omega t)$$

This sinusoid is a continuous function of time. To determine the frequency response of discrete systems we require a discrete time input. This can be obtained by sampling the sinusoid every T seconds, i.e.

$$x(i) = x(t)|_{t=iT} = \cos(\omega i T) \tag{9.10}$$

The frequency response of a discrete time system is expressed in terms of the gain and phase of the system steady-state response to this sampled input.

Frequency Response from Pulse Transfer Function

The frequency response of a continuous time system can be obtained by evaluating the transfer function at $s = j\omega$. Since, as we have seen, the s-plane maps to the z-plane via the relationship

$$z = e^{sT}$$

then it is apparent that the frequency response of a discrete time system is obtained by evaluating the pulse transfer function at $z = e^{j\omega T}$, i.e.

$$\text{Frequency response} = G(z)|_{z=e^{j\omega T}} = G(e^{j\omega T}) \tag{9.11}$$

Because $G(e^{j\omega T})$ is rather a mouthful we will simply write the frequency response as $G(\omega)$.

Now $z = e^{j\omega T}$ is a unit length vector at an angle, $\theta = \omega T$, drawn on the z-plane. As ω increases from zero frequency the vector sweeps around the unit circle (Fig. 9.6). This implies that the frequency response of discrete time systems repeats as θ becomes greater than 2π (i.e. at frequencies greater than $2\pi/T$).

Example 9.1 Frequency response of a first-order system A standard first-order system with time constant $\tau = 1.0$ second when sampled at a frequency, $f_s = 5$ Hz, can be modelled as the pulse transfer function

$$G(z) = \frac{0.1813}{z - 0.8187}$$

Substituting $z = e^{j\omega T}$ into the pulse transfer function gives the frequency response

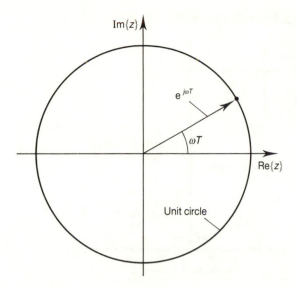

Figure 9.6 Evaluation of the frequency response of a discrete system

of the system

$$G(\omega) = \frac{0.1813}{e^{j\omega T} - 0.8187}$$

or

$$G(\omega) = \frac{0.1813}{\cos(\omega T) + j\sin(\omega T) - 0.8187}$$

the gain of the system is therefore given by

$$|G(\omega)| = \frac{0.1813}{\sqrt{(\cos(\omega T) - 0.8187)^2 + (\sin(\omega T))^2}}$$

and the phase by

$$\angle G(\omega) = -\tan^{-1}\left\{\frac{\sin(\omega T)}{\cos(\omega T) - 0.8187}\right\}$$

Figure 9.7 shows the gain for the system in Example 9.1 plotted on a Bode diagram. The low frequency behaviour is very similar to that for the continuous time system, the gain dropping away from 0 dB as the frequency increases. As the frequency rises further a significant difference emerges between the discrete time and continuous time system responses. At frequencies above half the sample frequency the gain in fact increases and climbs to unity at the sample frequency. As hinted above, the response then repeats exactly at multiples of the sample frequency. The picture is distorted on the Bode diagram because of the logarithmic frequency scale. Figure 9.8 shows the same frequency response with a linear frequency scale which indicates that there is in fact perfect symmetry about half the sample frequency. The system phase response also shows the same symmetry.

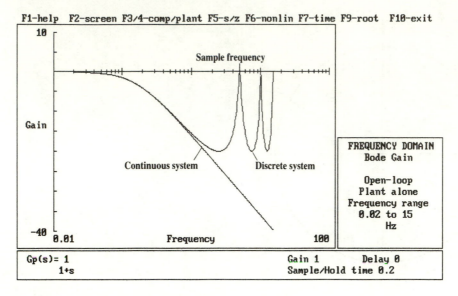

Figure 9.7   Bode diagram of a discrete first-order system

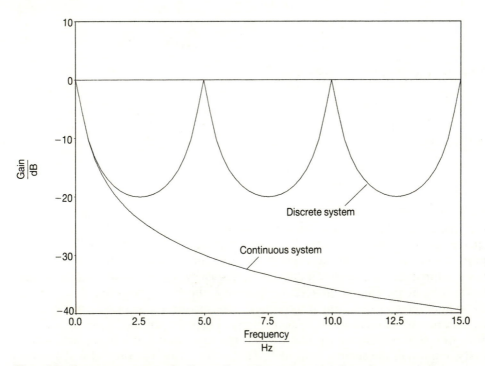

Figure 9.8   Frequency response of a discrete first-order system with a linear frequency scale

Aliasing and the Nyquist Frequency

We have seen that a discrete time system (or sampled-data system) has a frequency response which repeats at multiples of the sample frequency and shows symmetry about the half sample frequency point. To obtain a fuller understanding of this behaviour it is worth looking at the sampled sinusoidal input.

Figure 9.9 shows several sampled sinusoids of successively higher frequencies. When the sinusoid is at a low frequency in comparison with the sample frequency there is little question that the sample values give a good representation of the continuous sinusoid. However at half the sample frequency the sample values could easily belong to a triangular wave or square wave at the same frequency. At frequencies above half the sample frequency we are in real trouble. Figure 9.9(c) shows a signal at 0.9 of the sample frequency; the sample values look like they originate from a signal at a much lower frequency. When the cosine frequency is equal to the sample frequency, the wave passes through a complete cycle between samples and the sample values are identical to those from a constant input (i.e. zero frequency). Finally, Fig. 9.9(e) shows a frequency of 1.1 times the sample rate; the sample values are the same as those obtained at 0.1 times the sample frequency (Fig. 9.9(a)).

The sampled sinusoid, at frequencies higher than half the sample rate, appears identical to a sinusoid at lower frequency. This lower frequency equivalent is called an *alias* and the frequency distortion caused by sampling is called *aliasing*. The frequency of half the sample frequency is called the *Nyquist frequency*. Aliasing will occur with all signal components which are above the Nyquist frequency.

The reason that the frequency response of a discrete time system shows periodicity should now be clear. At frequencies higher than the Nyquist frequency the system is responding to the low frequency alias of the excitation signal.

Exercise Using CODAS-II simulate the open-loop time response of the sampled-data first-order system examined in Example 9.1. Set up a user defined input function as a sinusoid of frequency 'a' Hz, "cos(2*pi*a*t)". Examine the amplitude of the response at frequencies of 1, 2.5, 4.9, 5 and 10 Hz and compare your results with those obtained from the frequency response function.

Shannon's Sampling Theorem

Since sinusoidal signals at frequencies above the Nyquist frequency become aliased into a lower frequency band, we can conclude that to avoid aliasing a signal should be sampled at more than twice the signal frequency. Important work by Shannon showed that it is possible to recover any continuous time signal from its sample values provided the sampling rate is at least twice the highest frequency present in the signal. This principle is known as *Shannon's sampling theorem*. Because of the difficulty in establishing the highest frequency present in a general signal and also because the ideal reconstruction proposed by Shannon is computationally intensive, sampling rates of more than this theoretical limit are generally used. For practical signals we can say the sample rate should be at least 5 to 20 times the highest frequency in order to avoid aliasing.

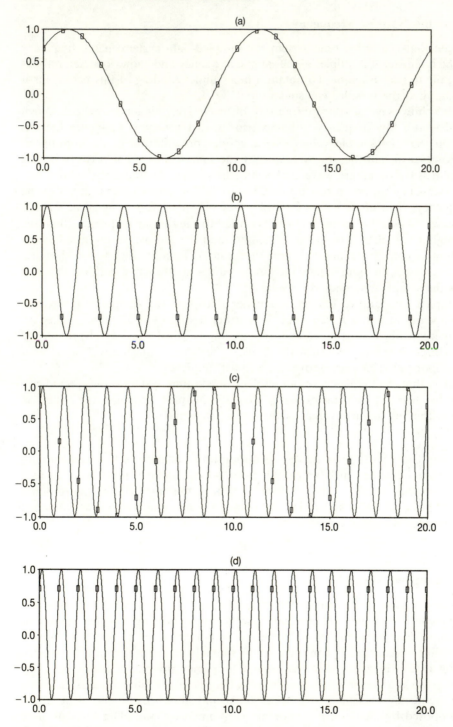

Figure 9.9 Sampled sinusoids showing aliasing (a) $f = 0.1fs$ (b) $f = 0.5fs$ (c) $f = 0.9fs$ (d) $f = fs$ (e) $f = 1.1fs$

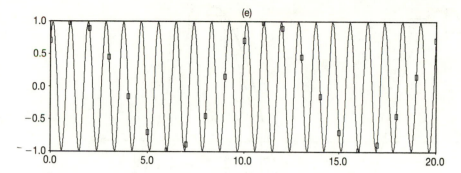

9.3 DIGITAL FILTERS

A filter is normally considered to be a system designed to pass certain frequencies and block or reject others. The compensators and controllers used in feedback control systems can be considered to be filters in a general sense. They are required to have certain gain **and** phase characteristics over a specified frequency range. When the filter function is implemented digitally as a discrete time system it is called a digital filter. Many of the techniques developed for the design of digital filters are thus applicable to controller or compensator design.

Here we consider the problem of finding a discrete time system with approximately the same characteristics as a given continuous time system. The problem can be considered as one of converting from the s-domain to the z-domain and the methods for achieving this are called *mapping techniques.*

One possible approach is to consider the continuous filter as though it were a sampled-data system. To obtain the discrete time transfer function we would place a zero-order hold in front of the filter and take z-transforms. Unfortunately, this approach does not generally give good results in the frequency domain. The reason for this is that the zero-order hold does its job too well, i.e. it holds the sampled value of the sinusoidal input constant over a complete sample period and produces an additional phase lag (Fig. 9.10).

At low frequencies compared to the sample rate the zero-order hold produces an effect similar to a transport delay of half the sample period, $T/2$. That is it introduces a phase lag of $\omega T/2$. The discrete filter obtained using a zero-order hold therefore shows an extra phase lag of $\omega T/2$ over and above that of the continuous filter and thus is not a good approximation in the frequency domain.

Exercise Why is it that the discrete system obtained using a zero-order hold gives exactly the same step response at the sample instants as the underlying continuous time system, yet the frequency response is so different?

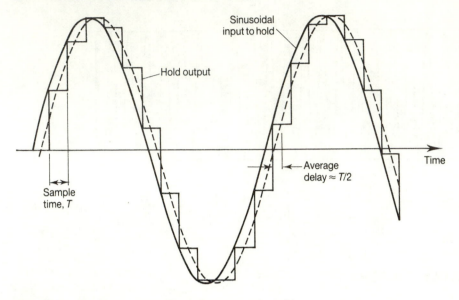

Figure 9.10 Effective delay introduced by a zero-order hold

Approximation of Continuous Systems

Chapter 8 introduced the idea of approximating integration or differentiation by using the rectangular and trapezoidal rules. These rules can be used to find a $G(z)$ which will have a response approximating to that of a given $G(s)$.

Consider the rectangular rule applied to the approximation of differentiation

$$y(t) = \frac{dx}{dt}(t) \qquad \rightarrow \qquad \frac{Y}{X}(s) = s \qquad\qquad (9.12(a))$$

$$y(i) \approx \frac{x(i) - x(i-1)}{T} \quad \rightarrow \quad \frac{Y}{X}(z) \approx \frac{1 - z^{-1}}{T} \qquad (9.12(b))$$

Comparison of Eqs (9.12(a)) and (b) leads to the idea of approximating 's' with a function of z.

$$s \approx \frac{1 - z^{-1}}{T}$$

or $$s \approx \frac{z - 1}{Tz} \qquad\qquad (9.13)$$

This approximation can be used to derive a pulse transfer function with similar characteristics to a continuous system. The discrete approximation using the rectangular rule is obtained by replacing 's' in the continuous transfer function by $(z - 1)/Tz$, i.e.

$$\tilde{G}_r(z) = G(s)|_{s = \frac{z-1}{Tz}} \qquad\qquad (9.14)$$

Example 9.2 Approximation using the rectangular rule As an example we will use the rectangular rule to obtain a discrete approximation of the first-order system

$$G(s) = \frac{1}{1 + \tau s}$$

with a sample period $T = \tau$.

Substituting 's' in the transfer function with the rectangular rule (Eq. (9.14)) gives the discrete time approximation

$$\tilde{G}_r(z) = \frac{1}{1 + \tau \left[\dfrac{z-1}{Tz} \right]} = \frac{z}{2z - 1}$$

Figure 9.11 shows the frequency response of this approximation together with that of the sampled-data system obtained using a zero-order hold. Both these discrete systems show considerable differences from the continuous system especially at higher frequencies. Interestingly, at low frequencies the rectangular rule gives a phase characteristic which is considerably closer to the continuous system than that from the zero-order hold system. On the other hand, the gain characteristic obtained from the rectangular rule is considerably worse than that from the zero-order hold.

Tustin's Rule and the Bilinear Transform

The trapezoidal rule gives a more accurate approximation of integration than the rectangular rule and might therefore be expected to lead to a better approximation of a continuous system. Using a similar derivation to that for the rectangular rule, the trapezoidal rule leads to the approximation

$$s \approx \frac{2(z - 1)}{T(z + 1)} \tag{9.15}$$

This approximation is often referred to as *Tustin's rule* or, because of its mathematical form, the *bilinear transform*. The discrete approximation of a continuous system using Tustin's rule is

$$\tilde{G}_t(z) = G(s)|_{s = \frac{2(z-1)}{T(z+1)}} \tag{9.16}$$

Tustin's rule has the property of mapping the imaginary axis of the s-plane exactly on to the unit circle of the z-plane. Now the frequency response of a continuous system is obtained by evaluating the transfer function as 's' runs up the imaginary axis and similarly for a discrete system as 'z' runs around the unit circle. One might therefore think that the frequency response of the two systems is identical. Unfortunately this is not the case since the frequencies are not mapped correctly.

Consider the frequency response of the discrete approximation, i.e. $\tilde{G}_t(z)$ evaluated at $z = e^{j\omega T}$

$$\tilde{G}_t(\omega) = G(s)|_{s = \frac{2(e^{j\omega T} - 1)}{T(e^{j\omega T} + 1)}} \tag{9.17}$$

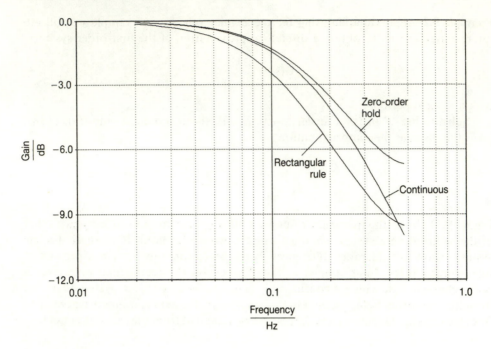

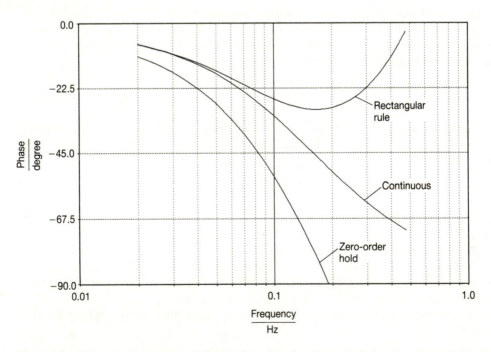

Figure 9.11 Gain and phase curves of discrete time approximations of a first-order system. (i) Continuous system. (ii) Zero-order hold. (iii) Rectangular approximation

The expression in the substitution can be manipulated as follows

$$s = \frac{2(e^{j\omega T} - 1)}{T(e^{j\omega T} + 1)} = \frac{2(e^{j\omega T/2} - e^{-j\omega T/2})}{T(e^{j\omega T/2} + e^{-j\omega T/2})} = \frac{2(j \sin(\omega T/2))}{T(\cos(\omega T/2))}$$

or

$$s = j\frac{2}{T}\tan\left(\frac{\omega T}{2}\right) = j\Omega \qquad (9.18)$$

i.e.

$$\tilde{G}_t(\omega) = G(s)|_{s=j\Omega} \qquad (9.19)$$

Now the continuous time system has a frequency response

$$G(j\omega) = G(s)|_{s=j\omega} \qquad (9.20)$$

Comparing Eqs (9.19) and (9.20) we can see that the gain and phase of the continuous system will be identical to that of the system obtained via Tustin's approximation but at a different frequency

$$\Omega = \frac{2}{T}\tan\left(\frac{\omega T}{2}\right) \qquad (9.21)$$

This equation gives the frequency distortion or *warping* caused by mapping the imaginary axis of the s-plane on to the unit circle of the z-plane via Tustin's rule. At low frequencies when ωT is small, the frequency distortion is small, but as ω approaches the Nyquist frequency (i.e. as $\omega T/2$ approaches $\pi/2$) Ω approaches infinity and the distortion becomes very large. As ω increases above the Nyquist frequency Ω becomes negative. The response to a frequency of $-\Omega$ is the same at the response to a frequency of Ω except that the sign of the phase is reversed. Remember though that the response at frequencies above the Nyquist frequency represents the response to an alias of the input frequency.

Because the frequency response of Tustin's approximation is the same as the continuous time system (apart from the frequency warping) the **shape** of the Nyquist plots for both systems will be identical for frequencies below the Nyquist frequency. In fact Tustin's rule will produce loci of identical shape on any plot which does not explicitly show frequency (Nyquist, Nichols or inverse Nyquist diagrams).

Frequency Prewarping

Equation (9.21) can be used to allow for the frequency distortion prior to applying Tustin's rule. By choosing some critical frequency, ω_c, and shifting this by the appropriate amount then after applying Tustin's rule the critical frequency will be correct. The frequency prewarping is accomplished by scaling the coefficients associated with 's' in a similar manner to that for time scaling (Chapter 3).

The method is as follows:

1. Decide on the critical frequency, ω_c, and write the continuous transfer function in a form where every 's' appears divided by ω_c.

2. Replace each occurrence of ω_c with $\Omega_c = \frac{2}{T}\tan\left(\frac{\omega_c T}{2}\right)$.

3. Replace every 's' with $\dfrac{2(z-1)}{T(z+1)}$.

The choice of critical frequency, ω_c, depends upon the application. For a filter, ω_c could be its bandwidth or centre frequency. For a phase lead compensator we might choose the frequency of maximum phase lead. It must be remembered that the discrete approximation will have the correct frequency response only at ω_c. All other frequencies will still be distorted.

Example 9.3 Tustin's rule with and without prewarping We will apply Tustin's rule to the first-order system used in Example 9.2.

(a) Firstly with no prewarping: substituting Tustin's rule directly to the first-order transfer function we have

$$\tilde{G}_t(z) = \frac{1}{1 + \tau \left[\dfrac{2(z-1)}{T(z+1)} \right]} = \frac{T(z+1)}{T(z+1) + 2\tau(z-1)}$$

$$= \frac{z+1}{3z-1}$$

(b) Secondly using prewarping: we will choose the critical frequency, ω_c, as the system corner frequency, $1/\tau$. The continuous system can be written in terms of ω_c as

$$G(s) = \frac{1}{1 + \left[\dfrac{s}{\omega_c} \right]}$$

Next the prewarping formula gives

$$\Omega_c = \frac{2}{T} \tan\left(\frac{\omega_c T}{2} \right) = \frac{2}{T} \tan\left(\frac{T}{2\tau} \right) = \frac{2}{T} 0.546$$

and so the transfer function with prewarped critical frequency is

$$G_w(s) = \frac{1}{1 + \left[\dfrac{T}{2} \dfrac{s}{0.546} \right]}$$

Finally applying Tustin's rule gives

$$\tilde{G}_w(z) = \frac{1}{1 + \left[\dfrac{1(z-1)}{0.546(z+1)} \right]} = \frac{0.546(z+1)}{0.546(z+1) + (z-1)}$$

$$= \frac{0.546(z+1)}{1.546z - 0.453}$$

Figure 9.12 shows the frequency response of both these systems. Notice that both responses are considerably more accurate than those obtained from either the

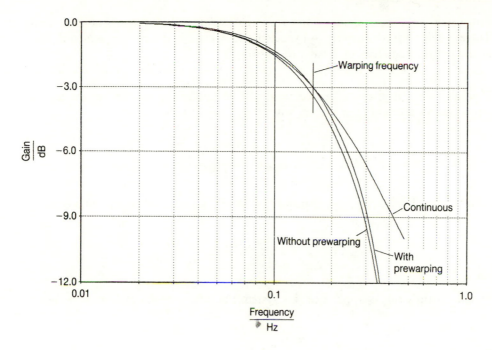

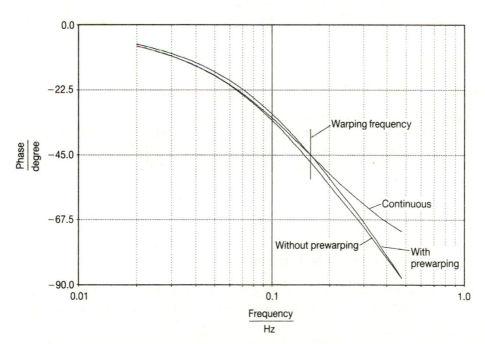

Figure 9.12 Gain and phase curves of discrete time approximations of a first-order system using Tustin's rule. (i) Continuous system. (ii) Without prewarping. (iii) With prewarping

zero-order hold equivalent or the rectangular rule. The system obtained using Tustin's rule with prewarping, as expected, shows exactly the correct gain and phase at the corner frequency of 1 rad/s (0.159 Hz).

Tustin's rule can give unexpected results when applied to systems which are not proper. Consider a continuous system with one more zero than pole

$$G(s) = \frac{b_0 + b_1 s + \dots b_n s^n + b_{(n+1)} s^{(n+1)}}{a_0 + a_1 s + \dots a_n s^n}$$

This can be considered as a proper transfer function in cascade with a pure differentiator, i.e.

$$G(s) = \frac{b_0 + b_1 s + \dots b_n s^n + b_{(n+1)} s^{(n+1)}}{a_0 s + a_1 s^2 + \dots a_n s^{(n+1)}} \cdot s$$

When Tustin's rule is applied to this system the differentiator produces the term

$$s \rightarrow \frac{2(z-1)}{T(z+1)}$$

and so the discrete system will have a pole at $z = -1$, which is right on the stability margin. Tustin's rule should therefore not be used to map continuous transfer functions which are not proper.

Summary

A reasonably accurate discrete time approximation of continuous frequency response characteristics can be obtained using Tustin's rule. The method of prewarping produces exactly the correct gain and phase at one specific frequency. When this critical frequency is much lower than the Nyquist frequency, the correction introduced by prewarping is small and can be omitted. The foregoing discussion was based on frequency domain considerations only and ignored the resulting time response. Tustin's rule, even with prewarping, will not produce a discrete system with the same step response as a continuous system. If an identical step response is the aim, then the method of using a zero-order hold and z-transformation will give the correct results, but remember that the response will not be correct for input functions other than a step.

Exercise Examine the open-loop Nyquist plots and step responses of the discrete approximations to the first-order system of Example 9.2 which were obtained using:

(a) The rectangular rule.
(b) Zero-order hold and z-transform.
(c) Tustin's rule.

9.4 DIGITAL IMPLEMENTATION OF ANALOGUE CONTROLLERS

The majority of digital controllers used in industry are based upon analogue prototypes. For many control problems standard three-term controllers give quite adequate performance. Most industrial control software has the facility to implement three-term control, phase lag or phase lead, etc. Usually the parameters of such controllers can be adjusted or tuned on-line to achieve the best performance and so the discussion of which particular mapping technique to use is rather academic.

It must not be forgotten that the controlled process will behave differently under direct digital control than with analogue control because the process is effectively in open-loop between samples. Furthermore, the zero-order hold inherent in most digital to analogue converters introduces an extra delay of approximately half the sample interval into the process dynamics. For these reasons, any digital design carried out in the frequency domain must be based on the frequency response of the sampled-data system (i.e. with the zero-order hold).

A digital controller will never give exactly the same performance as any analogue original, however, provided the sample rate is reasonably fast compared to the process dynamics the digital approximation should give similar results.

Tustin's Rule Applied to Frequency Domain Controller Design

Tustin's rule, when combined with prewarping, is able to give a controller with exactly the same gain and phase as an analogue model at one specific frequency. This method is therefore well suited to implement many of the frequency domain design methods covered in Chapter 5. The critical frequency chosen for prewarping will depend upon the design objective. For example, the closed-loop resonant frequency would be suitable when designing for closed-loop peak magnification, or the gain-crossover frequency when phase margin is specified.

The general frequency domain design procedure is to use the frequency response of the sampled-data system (i.e. with zero-order hold) and design a continuous time compensator to achieve the desired frequency domain performance objective. The compensator is then mapped into the z-domain using Tustin's rule with prewarping of the design frequency.

Example 9.4 Discrete phase lead compensator design A process has the open-loop transfer function

$$G_p(s) = \frac{1}{s(1 + 3s)}$$

Design a digital phase lead compensator with a sample interval of 1 second which gives the system a phase margin of 50° without reducing the velocity error constant.

SOLUTION The phase lead compensator can be designed using the method developed in Chapter 5 based upon the characteristic curves of a continuous phase

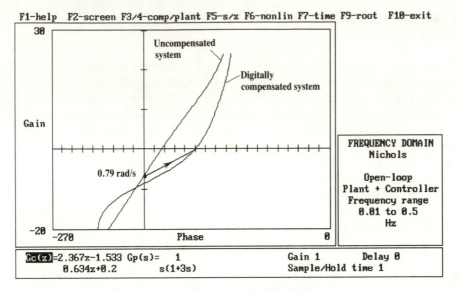

Figure 9.13   Discrete phase lead compensator design example

lead compensator. The compensator can then be converted into a discrete time system using Tustin's rule. To maintain the phase margin at 50°, prewarping of the gain crossover frequency must be applied.

Figure 9.13 shows the open-loop frequency response of the sampled-data system. A suitable continuous phase lead compensator is

$$G_c(s) = \frac{1 + b\sqrt{a}(s/\omega_s)}{1 + (b/\sqrt{a})(s/\omega_s)}$$

with $a = 9$, $b = 0.65$ and $\omega_s = 0.79$ rad/s, i.e.

$$G_c(s) = \frac{1 + 1.95(s/0.79)}{1 + 0.217(s/0.79)}$$

The critical frequency for prewarping is the design frequency, 0.79 rad/s. Applying the prewarping formula

$$\Omega_c = \frac{2}{T}\tan\left(\frac{\omega_c T}{2}\right) = \frac{2}{T}\tan\left(\frac{0.79}{2}\right) = \frac{2}{T}0.417$$

As the sample interval is 1 second, the prewarped compensator is

$$G_{cw}(s) = \frac{1 + 1.95(s/0.417)}{1 + 0.217(s/0.417)}$$

Finally applying Tustin's rule

$$\tilde{G}_{cw}(z) = \frac{1 + 1.95\left[\dfrac{1(z-1)}{0.417(z+1)}\right]}{1 + 0.217\left[\dfrac{1(z-1)}{0.417(z+1)}\right]}$$

$$\tilde{G}_{cw}(z) = \frac{2.367z - 1.533}{0.634z + 0.200}$$

The compensated frequency response (Fig. 9.13) shows that the phase margin criterion of 50° is indeed met.

The Digital Three-term Controller

The three-term or PID controller is the work horse of process control and for many processes gives good performance with reasonable robustness. The digital implementation of three-term control is therefore of major interest.

The general equation for the output of a PID control law was developed in Chapter 7 as

$$u_c = K\left[e + \frac{1}{T_r}\int e\,dt + T_d\frac{de}{dt}\right]$$

where
$$e = r - c$$

This corresponds to the transfer function

$$G_c(s) = K\left[1 + \frac{1}{sT_r} + sT_d\right] \tag{9.22}$$

Applying Tustin's rule to each of the terms in Eq. (9.22) gives

$$\tilde{G}_{ct}(z) = K\left[1 + \frac{T(z+1)}{2(z-1)T_r} + \frac{2(z-1)T_d}{T(z+1)}\right]$$

Now, the derivative term has a z-plane zero at $z = -1$ corresponding to a marginally stable oscillatory mode. This problem can be overcome in several ways. We can introduce a lag to make the derivative term proper, indeed this is the case in practical analogue controllers. Alternatively the rectangular rule can be used to approximate the derivative term. There is little to choose between these approaches but the latter is more usual and leads to simpler equations.

Using Tustin's rule for the integral action and the rectangular rule for derivative action gives rise to the digital controller

$$\tilde{G}_c(z) = K\left[1 + \frac{T(z+1)}{2(z-1)T_r} + \frac{(z-1)T_d}{Tz}\right] \tag{9.23}$$

which reduces to

$$\tilde{G}_c(z) = \frac{K(b_0 + b_1 z + b_2 z^2)}{z(z - 1)} \qquad (9.24(a))$$

where

$$b_0 = \frac{T_d}{T} \qquad (9.24(b))$$

$$b_1 = \left(\frac{T}{2T_r} - \frac{2T_d}{T} - 1 \right) \qquad (9.24(c))$$

and

$$b_2 = \left(\frac{T}{2T_r} + \frac{T_d}{T} + 1 \right) \qquad (9.24(d))$$

This digital PID controller has a pole at $z = 1$ corresponding to the integral action. The transfer function is proper and realizable providing the computation time is short compared to the sample interval.

Various techniques can be used to derive digital three-term controllers and each will give slightly different controllers and/or coefficients.

Exercise Show that the pulse transfer function for a digital PID controller based wholly upon the rectangular rule is also of the form

$$\tilde{G}_{cr}(z) = \frac{K(b_0 + b_1 z + b_2 z^2)}{z(z - 1)}$$

and hence find the coefficients b_0, b_1 and b_2 in terms of the reset time, derivative time and sample interval.

Practical Implementation of Discrete Controllers

Here we consider the problem of converting the discrete controller transfer function into a working computer program. The discussion will initially be limited to three-term controllers but many of the ideas developed are applicable to general discrete time controllers and compensators. The controller software should normally have facilities for auto/manual operation and should be able to cope adequately with practicalities such as control saturation and reset windup.

The discrete three-term controller considered above can be implemented as a difference equation in many different ways. Two alternatives will now be examined.

Positional algorithm The discrete three-term controller in its basic form is expressed in Eq. (9.23). The total control effort can be seen to be made up from the three separate actions

$$u_c(i) = u_p(i) + u_r(i) + u_d(i) \qquad (9.25)$$

Taking each term in turn:

(a) The discrete proportional action is

$$u_p(i) = Ke(i) \qquad (9.26)$$

(b) The integral action or reset term, u_r, is obtained from

$$\frac{U_r}{E}(z) = \frac{KT(z+1)}{2T_r(z-1)} = \frac{KT(1+z^{-1})}{2T_r(1-z^{-1})}$$

Multiplying both sides by $(1 - z^{-1})$ and taking z^{-1} as a unit sample delay we have

$$u_r(i) - u_r(i-1) = \frac{KT}{2T_r}[e(i) + e(i-1)]$$

or
$$u_r(i) = u_r(i-1) + \frac{KT}{2T_r}[e(i) + e(i-1)] \qquad (9.27)$$

(c) The derivative action is similarly obtained from

$$\frac{U_d}{E}(z) = KT_d\frac{(z-1)}{Tz}$$

as
$$u_d(i) = \frac{KT_d}{T}[e(i) - e(i-1)] \qquad (9.28)$$

The algorithm described by Eqs (9.25) to (9.28) is called a *positional algorithm* because the actual value of the control effort (i.e. its position) is calculated at each sample instant.

Incremental algorithm An alternative implementation of the same controller can be obtained directly from the reduced transfer function, Eq. (9.24(a))

$$\frac{U_c}{E}(z) = \frac{K(b_0 + b_1z + b_2z^2)}{z(z-1)}$$

which directly gives the difference equation

$$u(i) - u(i-1) = K[b_2e(i) + b_1e(i-1) + b_0e(i-2)] \qquad (9.29)$$

Defining the incremental control effort as

$$\Delta u(i) = u(i) - u(i-1) \qquad (9.30)$$

The control calculation can be written as

$$\Delta u(i) = K[b_2e(i) + b_1e(i-1) + b_0e(i-2)]$$

and
$$u(i) = u(i-1) + \Delta u(i)$$

This implementation of the controller is termed an *incremental algorithm* or *velocity algorithm* since it is the increment in control effort which is calculated at each sample. Integral action is inherent in the incremental algorithm.

The positional algorithm has the advantage over the incremental form that the controller parameters (reset time and derivative time) are used directly in the calculation. However it is a simple matter to calculate the new coefficients (b_0, b_1 and b_2) in the velocity algorithm when the controller settings are changed. There is of course no need to perform these calculations every sample. Because this positional algorithm calculates each action separately it is relatively easy to place special conditions or restrictions on each term. For example, to prevent set-point kick, the derivative term could be calculated from the measured value rather than from the error signal. Similarly the integral action can be inhibited if desired (not so easy in the velocity algorithm).

Dynamically, the incremental and positional algorithms are identical. However there are some inherent advantages to the incremental algorithm when auto/manual transfer and reset windup are considered.

Figure 9.14 shows the flow chart for a program to implement the digital three-term

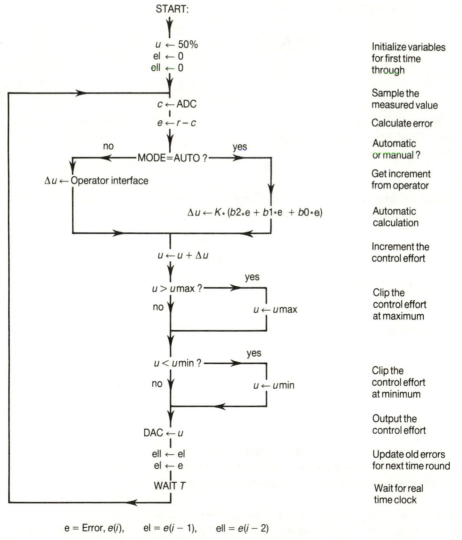

START:

$u \leftarrow 50\%$	Initialize variables
$el \leftarrow 0$	for first time
$ell \leftarrow 0$	through
$c \leftarrow$ ADC	Sample the measured value
$e \leftarrow r - c$	Calculate error
MODE = AUTO ?	Automatic or manual ?
$\Delta u \leftarrow$ Operator interface	Get increment from operator
$\Delta u \leftarrow K*(b2*e + b1*e + b0*e)$	Automatic calculation
$u \leftarrow u + \Delta u$	Increment the control effort
$u >$ umax ? / $u \leftarrow$ umax	Clip the control effort at maximum
$u <$ umin ? / $u \leftarrow$ umin	Clip the control effort at minimum
DAC $\leftarrow u$	Output the control effort
$ell \leftarrow el$ / $el \leftarrow e$	Update old errors for next time round
WAIT T	Wait for real time clock

e = Error, $e(i)$, el = $e(i - 1)$, ell = $e(i - 2)$

Figure 9.14   Flow chart for a digital three-term controller using an incremental algorithm

controller in incremental form. The flow chart incorporates an auto/manual switch such that when the controller is in manual mode the control calculation is bypassed and the control effort can be incremented manually. Whether in manual or automatic mode, the incremental control effort is always added to the previous controller output. One advantage of the incremental algorithm is now evident, namely *automatic bumpless transfer*. On switching from manual to auto or vice versa the previous value of the output is always used as an initial value for the control effort in the new mode. Thus there will be no sharp discontinuity in control effort upon switching mode.

In order to avoid the case where the controller output is outside the range of the DAC the raw control effort is limited to the maximum and minimum output values. This control effort clipping reflects the fact that the final control elements (actuator or control valve, etc.) cannot apply more than 100% or less than 0% control effort. Chapter 7 discussed the problem of reset windup caused when the integral action of a three-term controller continues to build up even though the control effort is limited. The inherent integral action performed by the incremental algorithm is however inhibited whenever the control effort clipping is in effect. Thus anti-reset windup is implicitly included in the incremental algorithm when control effort clipping is incorporated.

Positional algorithms do not intrinsically incorporate the features of auto/manual bumpless transfer and anti-reset windup, but these can of course be added to the control program. Bumpless transfer from manual to automatic can be achieved by adding the automatic control effort, u_c, to the manual value, u_m. However in addition the integral term, u_r, must be initialized to zero at switch over. Bumpless transfer from auto to manual is achieved by making u_m equal to the total controller output at the moment of switch over. Anti-reset windup can be achieved in the positional algorithm by inhibiting the integral term summation (Eq. (9.27)) whenever the control effort reaches the clipping limit.

9.5 DIRECT DISCRETE CONTROLLER DESIGN

We will now turn our attention to the direct design of digital controllers without reference to any underlying analogue design. Digital compensators are not limited by the same considerations of implementation that apply to analogue compensators. Any real-time compensator can only react to disturbances or changes in reference value after they are detected; this means that the digital compensator must be proper. Apart from the restrictions of properness and of course stability we are relatively free to choose the controller to give whatever compensation is required. Because digital controllers are not restricted by realization using physical elements, the control law can include just the correct compensation necessary to achieve the desired closed-loop characteristics.

Direct design methods are based on the simple principle of finding a compensator which gives the system a specified closed-loop transfer function. Consider the discrete time feedback control system comprising a controller and a plant as shown in Fig. 9.15. The closed-loop transfer function is

$$\frac{C}{R}(z) = \frac{G_c G_p}{1 + G_c G_p} = F(z) \tag{9.31}$$

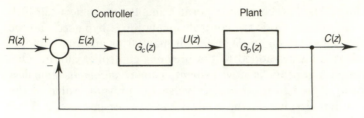

Figure 9.15 Feedback configuration for the direct digital design method

This relationship can be rearranged to give an expression for the controller in terms of the closed-loop transfer function, $F(z)$, and the plant, $G_p(z)$

$$G_c(z) = \frac{1}{G_p} \frac{F(z)}{1 - F(z)} \tag{9.32}$$

This direct controller design equation gives a controller which firstly cancels the plant dynamics and then includes additional terms necessary to give the desired closed-loop transfer function.

The main problems are concerned with finding a suitable target closed-loop transfer function, $F(z)$, to aim for. The design emphasis is placed firmly on the closed-loop system. The chosen closed-loop transfer function should give the required performance and yet must be practical and realizable. Performance is usually expressed in terms of steady-state accuracy, settling time, percentage overshoot, time constant, etc. Additional specifications such as velocity error constant or maximum control effort may be required. We will now tackle the problem of finding a realistic closed-loop transfer function which achieves the design requirements and constraints.

The direct design method treats the controlled plant purely as a discrete time system. The plant output is thus only considered at the sample instants and the response between samples is ignored at the design stage. It is therefore very important to simulate the final system design as a sampled-data system so that intersample behaviour can be checked.

The Causality Constraint

A plant which includes transport delay will not respond to an input immediately. The effect of any control effort will only be seen at the plant output after the transport delay has elapsed. Therefore no matter what controller we employ, the closed-loop system must also exhibit this same transport delay with respect to the reference. If we ask for a closed-loop system with less transport delay than the plant then the controller would have to anticipate any changes in reference value before they occurred. Such a prophetic compensator would be non-causal and would be impossible to implement. The closed-loop transfer function, $F(z)$, in Eq. (9.31) must therefore have a transport delay at least as large as that of the plant.

In discrete transfer functions, transport delay appears as an excess number of poles over zeros. Therefore the constraint of controller causality can be stated as

Constraint $F(z)$, must have at least the same number of excess poles over zeros $(P - Z)$ as the plant, $G_p(z)$.

Performance Requirements

Closed-loop dynamics and pole location The system dynamic characteristics are determined by the closed-loop poles. Quite often a first-order characteristic is adequate, with a time constant chosen to achieve the desired speed of response. The closed-loop time constant must be chosen realistically in relation to the plant dynamics otherwise the resulting control effort will be excessive.

When the plant incorporates transport delay we have seen that the closed-loop system must exhibit at least the same delay. Thus additional poles may need to be added to $F(z)$ to achieve the correct pole-zero excess. Since poles at the origin contribute pure delay without other dynamics any additional poles should be added at $z = 0$.

For systems with accurately known dynamics or where the dynamics are fast compared to the sample rate, deadbeat control can be a realistic aim. With deadbeat control the system settles exactly in a number of samples equal to the system order. For example a second-order discrete time system will have a deadbeat response which settles in two samples. For deadbeat response all the closed-loop poles must be located at the origin of the z-plane.

It should be remembered that the direct design method explicitly uses the plant transfer function in the design equation. The controller is thus directly matched to the plant transfer function and any inaccuracy in the plant model can lead to poor control performance. Generally, the more we try to speed up the response of the closed-loop system the greater the sensitivity to model inaccuracy. The system performance must therefore be traded against robustness. Systems which have significant transport delay are particularly sensitive to modelling errors.

Steady-state accuracy One of the basic requirements of many control systems is that of accurate steady-state reference following. This means that the closed-loop transfer function should have a steady-state gain of unity.

For zero steady-state error in closed-loop

$$F(z)|_{z=1} = F(1) = 1$$

The controller direct design equation (9.32) contains the term $(1 - F(z))$ in the denominator. The requirement of zero steady-state error ensures that

$$1 - F(1) = 0 \qquad (9.33)$$

In other words the controller is given a pole at $z = 1$ (a free integrator). If the plant $G_p(z)$ already has a pole at $z = 1$ then the two factors will cancel in the design equation leaving the controller without integral action. The direct design method thus forces the controller to incorporate an integrator only when it is required.

Example 9.5 Direct design method A process can be modelled as having first order dynamics with time constant 45 seconds plus a transport delay of 10 seconds. Design a discrete time controller with a sample interval of 10 seconds which gives the closed-loop system first-order dynamics with a time constant of less than 15 seconds and accurate steady-state reference following.

SOLUTION The sampled-data process with zero-order hold has a pulse transfer function

$$G_p(z) = \frac{z-1}{z} Z\left\{\frac{e^{-10s}}{s(1+45s)}\right\}$$

$$G_p(z) = \frac{0.2}{z(z-0.8)}$$

The closed-loop system must have a time constant of less than 15 seconds therefore $F(z)$ must have a pole at

$$z < e^{-T/\tau}$$

For $\tau = 15$ seconds and $T = 10$ seconds $e^{-T/\tau} = 0.513$, therefore a pole at $z = 0.5$ will satisfy the design specification. The plant has a pole/zero excess of 2 and thus $F(z)$ needs at least one extra pole at $z = 0$.

Putting

$$F(z) = \frac{b_0}{z(z-0.5)}$$

the coefficient b_0 can be chosen to satisfy the steady-state requirement

$$F(z)|_{z=1} = \frac{b_0}{1-0.5} = 1.0$$

therefore
$$b_0 = 0.5$$

The direct design formula then gives the controller as

$$G_c(z) = \frac{1}{G_p} \frac{F(z)}{1-F(z)}$$

Now
$$F(z) = \frac{0.5}{z(z-0.5)}$$

and
$$1 - F(z) = \frac{z(z-0.5) - 0.5}{z(z-0.5)} = \frac{z^2 - 0.5z - 0.5}{z(z-0.5)}$$

so that
$$G_c(z) = \frac{z(z-0.8)}{0.2} \frac{0.5}{z^2 - 0.5z - 0.5}$$

$$= \frac{2.5z(z-0.8)}{(z-1)(z+0.5)}$$

As expected, the controller incorporates integral action to ensure steady-state reference following. Figure 9.16 shows the step response of the resulting sampled-data system showing a first-order characteristic with transport delay.

Velocity error requirements In some cases there will be a requirement for steady-state accuracy with a ramp input. For Type-1 systems (i.e. with a free integrator in the forward path) steady-state accuracy is usually expressed in terms of a velocity error constant, K_v.

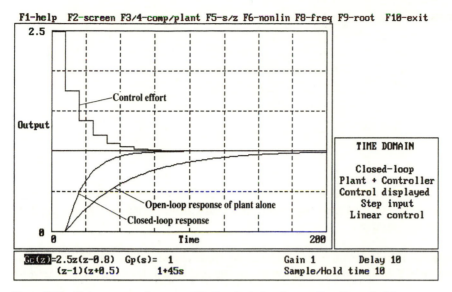

F1-help   F2-screen F3/4-comp/plant F5-s/z F6-nonlin F8-freq F9-root   F10-exit

Figure 9.16   Step response of system with directly designed controller, Example 9.5

A unity feedback system when subjected to a ramp input of slope 'A' units per second will display a steady-state error of

$$\bar{e} = \frac{A}{K_v}$$

Adopting an approach similar to that taken for continuous systems (Section 3.12), the velocity error constant for the configuration of Fig. 9.15 can be shown to be

$$K_v = \lim_{z \to 1} \frac{(z-1)}{Tz} G_c G_p \tag{9.34}$$

However, Eq. (9.32) allows $G_p G_c$ to be expressed in terms of the closed-loop transfer function

$$G_c G_p = \frac{F(z)}{1 - F(z)}$$

Therefore

$$K_v = \lim_{z \to 1} \frac{(z-1)}{Tz} \frac{F(z)}{(1 - F(z))}$$

Now for a Type-1 system $F(1) = 1$ and so

$$TK_v = \lim_{z \to 1} \frac{(z-1)}{(1 - F(z))}$$

This fraction is indeterminate in the limit but L'Hospital's rule may be used to evaluate the expression

$$TK_v = \lim_{z \to 1} \frac{\dfrac{d}{dz}(z-1)}{\dfrac{d}{dz}(1-F(z))}$$

so that

$$\frac{1}{TK_v} = -\frac{dF(z)}{dz}\Big|_{z=1} \tag{9.35}$$

The velocity error constant has units of second^{-1} but TK_v is dimensionless. $1/TK_v$ can be visualized as the steady-state error when the system is subjected to a ramp input of slope one unit per sample.

Example 9.6 Direct design with velocity error requirement Repeat the direct design of Example 9.5 but with the additional requirement of a velocity error constant of 0.1 second^{-1}.

SOLUTION In order to satisfy both steady-state requirements (i.e. $F(1) = 1$ and $K_v = 0.1$ s^{-1}) an extra degree of freedom is required. This can be achieved by adding an extra numerator coefficient. Or course an extra pole must also be added at the origin to maintain the pole–zero excess at 2.

Putting
$$F(z) = \frac{b_0 + b_1 z}{z^2(z-0.5)}$$

the steady-state reference following requires that

$$F(1) = 1 = \frac{b_0 + b_1}{1 - 0.5}$$

so that
$$b_0 + b_1 = 0.5 \tag{9.36}$$

The velocity error requirement with $T = 10$ and $K_v = 0.1$ means that

$$\frac{1}{TK_v} = -\frac{dF(z)}{dz}\Big|_{z=1} = 1$$

$$\frac{dF}{dz} = \frac{z^2(z-0.5)b_1 - (b_0 + b_1 z)(3z^2 - z)}{(z^2(z-0.5))^2}$$

which when evaluated at $z = 1$ gives

$$\frac{0.5b_1 - 2(b_0 + b_1)}{0.5^2} = -1 \tag{9.37}$$

Substituting Eq. (9.36) into Eq. (9.37) we have

$$0.5b_1 - 1 = -0.25$$

or $$b_1 = 1.5$$

and therefore $$b_0 = -1.0$$

The closed-loop transfer function is therefore

$$F(z) = \frac{1.5z - 1}{z^2(z - 0.5)}$$

and

$$1 - F(z) = \frac{z^2(z - 0.5) - (1.5z - 1)}{z^2(z - 0.5)} = \frac{z^3 - 0.5z^2 - 1.5z + 1}{z(z - 0.5)}$$

so that

$$G_c(z) = \frac{z(z - 0.8)}{0.2} \frac{1.5z - 1}{z^3 - 0.5z^2 - 1.5z + 1}$$

$$= \frac{5z(z - 0.8)(1.5z - 1)}{(z - 1)(z^2 + 0.5z - 1)}$$

Figures 9.17(a) and (b) show the step and ramp responses of the resulting sampled-data system. The zero which has been added to the closed-loop transfer function has reduced the velocity lag to the required value. The effect of the zero on the step response is quite noticeable. The zero causes such a large kick when the step is applied that the initial response overshoots the steady-state value considerably. The ramp response has been improved to the detriment of the step response.

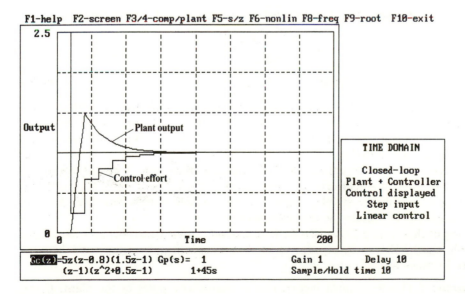

Figure 9.17(a) Step response for Example 9.6

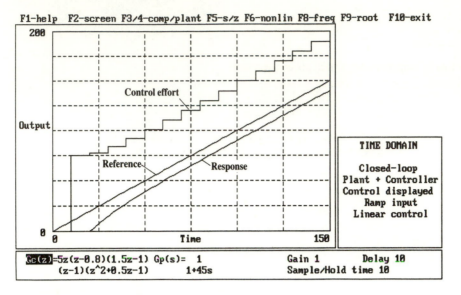

F1-help F2-screen F3/4-comp/plant F5-s/z F6-nonlin F8-freq F9-root F10-exit

Figure 9.17(b) Ramp response for Example 9.6

Stability Constraints

The stability of the closed-loop system is determined by the roots of the characteristic equation

$$1 + G_c(z)G_p(z) = 0 \qquad (9.38)$$

Expressing the controller and plant as separate numerator and denominator

$$1 + \frac{N_c(z)N_p(z)}{D_c(z)D_p(z)} = 0$$

or
$$D_c(z)D_p(z) + N_c(z)N_p(z) = 0 \qquad (9.39)$$

Now the direct design formula calls for a controller which cancels plant dynamics. That is, a controller with zeros to cancel plant poles and poles to cancel plant zeros. Cancelling a plant pole with a controller zero means that $D_p(z)$ and $N_c(z)$ have a common factor. Examination of Eq. (9.39) shows that this common factor is also a factor of the closed-loop characteristic equation. Thus cancelling a plant pole does not remove it from the closed-loop characteristic equation. Problems will occur when the plant has a pole which is outside the stability region. If the unstable pole is cancelled then the closed-loop system will retain the unstable mode.

Turning back to Eq. (9.32) we can see that the undesirable controller zero cancelling an unstable plant pole can be avoided by making the plant pole cancel with the term $(1 - F(z))$. In other words:

Constraint 1 $(1 - F(z))$ must incorporate as zeros, any poles of the plant, $G_p(z)$, which lie outside the unit circle.

Precisely the same arguments apply to plant zeros which are cancelled by controller poles. The cancelled pole/zero remains a factor of the closed-loop characteristic equation. When the plant has a zero which is outside the unit circle and it is cancelled by a controller pole, the instability will again be retained by the closed-loop system.

The cancellation of zeros outside the unit circle can be avoided if $F(z)$ contains the same zero so that the factor does not appear in the direct design equation for the controller.

Constraint 2 $F(z)$ must incorporate as zeros, any zeros of the plant, $G_p(z)$, which lie outside the unit circle.

Another way of looking at the above stability problems is from the root-locus view point. The root locus shows how the system open-loop poles are moved to desirable closed-loop locations as the controller gain is increased. Now consider an almost cancelling pole–zero pair. The pair will give rise to a root locus branch which passes directly from the pole to the zero. The closer the cancellation the shorter the branch. If the pair is located outside the region of stability then the branch will be entirely in the unstable region. Thus no matter what gain we employ the unstable pole will not be able to be moved into the stable region. Only when the unstable pole is not cancelled can it be dragged to a desirable location as the loop gain is increased.

Example 9.7 Direct design for an open-loop unstable system Design a stable controller for the discrete time system given by

$$G_p(z) = \frac{0.1z}{z^2 - 2z + 0.95}$$

The controller should give the system steady-state reference following, and closed-loop poles at $z = 0.4 \pm j0.3$.

SOLUTION The plant has two poles, at $z = 0.776$ and 1.224.

The pole which is outside the unit circle represents an unstable mode and must therefore be a factor of $(1 - F(z))$. The design approach used here is to derive a suitable function for $(1 - F(z))$.

The closed-loop characteristic equation will be

$$(z - 0.4 - j0.3)(z - 0.4 + j0.3) = 0$$

or
$$z^2 - 0.8z + 0.25 = 0$$

This polynomial is the denominator of $F(z)$ and will also be the denominator of $(1 - F(z))$. Because $F(z)$ must be proper, $(1 - F(z))$ must have the same number of zeros as poles (i.e. two).

Let
$$1 - F(z) = \frac{(z - 1.224)(b_0 + b_1 z)}{z^2 - 0.8z + 0.25}$$

Since $F(1) = 1$ then $(1 - F(1)) = 0$ so that

$$0 = \frac{(1 - 1.224)(b_0 + b_1)}{1 - 0.8 + 0.25}$$

or
$$b_1 = -b_0$$

giving
$$1 - F(z) = \frac{(z - 1.224)b_0(1 - z)}{z^2 - 0.8z + 0.25}$$

Therefore
$$F(z) = 1 - \frac{b_0(z - 1.224)(1 - z)}{z^2 - 0.8z + 0.25}$$

or
$$F(z) = \frac{(1 + b_0)z^2 - (0.8 + b_0 + 1.224b_0)z + (0.25 + 1.224b_0)}{z^2 - 0.8z + 0.25}$$

Finally, because of the causality constraint, $F(z)$ must have a pole/zero excess of 1. The coefficient of z^2 must therefore be zero and

$$b_0 = -1$$

Thus
$$F(z) = \frac{1.424z - 0.974}{z^2 - 0.8z + 0.25}$$

and
$$1 - F(z) = \frac{(z - 1.224)(z - 1)}{z^2 - 0.8z + 0.25}$$

The controller can now be found from the direct design equation

$$G_c(z) = \frac{(z - 1.224)(z - 0.776)}{0.1z} \frac{(1.424z - 0.974)}{(z - 1.224)(z - 1)}$$

$$G_c(z) = \frac{10(z - 0.776)(1.424z - 0.974)}{z(z - 1)}$$

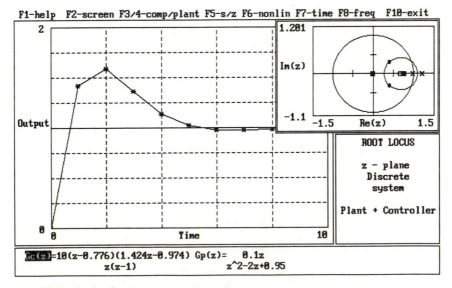

Figure 9.18   Design for the open-loop unstable system of Example 9.7

Figure 9.18 shows the simulated step response obtained with this controller together with the root locus diagram. The open-loop unstable pole has been pulled inside the unit circle by the action of the controller to give a stable closed-loop system.

The Ringing Pole Problem

It is common for sampled-data systems to have z-plane zeros which lie on the negative real axis. This is particularly the case when fractional transport delays are present. The stability constraint tells us how to deal with such zeros which lie outside the unit circle, but we will see that even zeros which are inside the stability region can cause problems if cancelled by controller poles. Such cancelling poles, if close to the unit circle, will contribute an oscillatory mode to the system characteristic equation and as such are called *ringing poles*.

Example 9.8 Direct design with ringing pole The plant of Example 9.5 had first-order dynamics with time constant 45 seconds plus a transport delay of 10 seconds. The controller design will now be repeated for a transport delay of only 5 seconds. The closed-loop system should again have first-order dynamics with a time constant of less than 15 seconds and accurate steady-state reference following. The sample interval of 10 seconds should also be retained.

SOLUTION The sampled data process has a fractional delay equal to half the sample interval and so the system pulse transfer function must be found using modified z-transforms. The discrete process can be modelled as

$$G_p(z) = \frac{0.105(z + 0.895)}{z(z - 0.8)}$$

All the plant poles and zeros are within the unit circle and so the closed-loop system can be chosen with a pole at $z = 0.5$ (as for Example 9.5). Because of the extra plant zero introduced by the fractional transport delay a pole/zero excess of only 1 is required and so a suitable closed-loop transfer function is

$$F(z) = \frac{0.5}{z - 0.5}$$

The numerator coefficient of 0.5 satisfies the steady-state requirement.

Therefore
$$1 - F(z) = \frac{z - 0.5 - 0.5}{z - 0.5} = \frac{z - 1}{z - 0.5}$$

and direct design formula gives the controller as

$$G_c(z) = \frac{z - 0.8}{0.105(z + 0.895)} \frac{0.5}{(z - 1)}$$

$$= \frac{4.76z(z - 0.8)}{(z - 1)(z + 0.895)}$$

Figure 9.19(a) shows the step response of the closed-loop discrete time system which

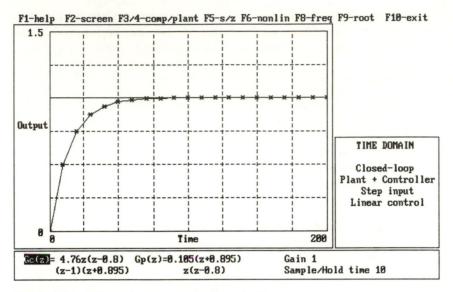

Figure 9.19(a) Discrete time response for Example 9.8

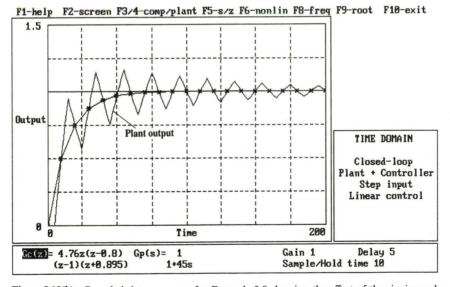

Figure 9.19(b) Sampled-data response for Example 9.8 showing the effect of the ringing pole

shows exactly the response as specified by $F(z)$. However, Fig. 9.19(b) shows the response of the sampled-data system. The controller has a ringing pole at $z = -0.895$ which although stable, gives rise to an undesirable oscillatory control effort. Such control action would in practice quickly wear out any actuator or control valve. Furthermore the continuous time system shows this oscillatory response between samples and the control system is clearly unsatisfactory.

The design will now be repeated but this time not allowing the controller to cancel the problem plant zero and hence avoiding the ringing pole. In other words the offending plant zero is treated as though it were unstable.

In order to avoid the ringing pole appearing in the controller transfer function, the closed-loop system must incorporate the problem plant zero, i.e.

$$F(z) = \frac{b_0(z + 0.895)}{z(z - 0.5)}$$

The steady-state requirement that $F(1) = 1$ gives b_0 as 0.264. Therefore

$$F(z) = \frac{0.264(z + 0.895)}{z(z - 0.5)}$$

$$1 - F(z) = \frac{z(z - 0.5) - 0.264(z + 0.895)}{z - 0.5} = \frac{(z - 1)(z + 0.236)}{z - 0.5}$$

so that now

$$G_c(z) = \frac{z - 0.8}{0.105(z + 0.895)} \frac{0.264(z + 0.895)}{(z - 1)(z + 0.236)}$$

$$= \frac{2.514z(z - 0.8)}{(z - 1)(z + 0.236)}$$

Figure 9.20 shows the sampled-data system response with this alternative controller. The oscillatory control action has completely disappeared and the

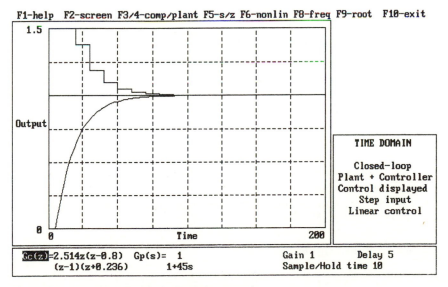

F1-help F2-screen F3/4-comp/plant F5-s/z F6-nonlin F8-freq F9-root F10-exit

TIME DOMAIN

Closed-loop
Plant + Controller
Control displayed
Step input
Linear control

Gc(z)=2.514z(z-0.8)   Gp(s)= 1
       (z-1)(z+0.236)   1+45s

Gain 1   Delay 5
Sample/Hold time 10

Figure 9.20 Example 9.8 with ringing pole removed

intersample response is smooth. The slight penalty which is paid for this is that $F(z)$ has an additional zero/pole pair causing the closed-loop response to have the same fractional delay as the plant.

Dahlin's Method

A popular direct design method developed by Dahlin aims at achieving a closed-loop system with first-order dynamics together with a transport delay. In other words

$$F(z) = \frac{(1-a)z}{z^n(z-a)}; \qquad a = e^{-T/\tau} \tag{9.40}$$

where n is equal to the integer transport delay of the plant. Dahlin's algorithm is similar to the direct design method developed above except that the stability constraints are not applied. Instead any unstable or ringing controller poles are replaced by a pole at the origin with equivalent steady-state gain (i.e. the term $(z + \alpha)$ is replaced by $(1 + \alpha)z$). Dahlin's algorithm is simple to apply and can give acceptable performance. Unfortunately, the shifting of these problem poles is a misrepresentation and can cause the closed-loop response to be far from first order. Dahlin's method is therefore not generally recommended. A better approach is to acknowledge the existence of any problem zeros and deal with them correctly by incorporating them into the closed-loop design.

Example 9.9 Dahlin's method with ringing pole removal The system in Example 9.8 included a zero which led to a ringing controller pole. Here we will apply Dahlin's method to the controller. The controller which gave the closed-loop system a simple first-order response was

$$G_c(z) = \frac{4.76z(z - 0.8)}{(z - 1)(z + 0.895)}$$

Replacing the ringing pole, $(z + 0.895)$, by $1.895z$ gives

$$G_c(z) = \frac{4.76(z - 0.8)}{1.895(z - 1)}$$

$$= \frac{2.513(z - 0.8)}{z - 1}$$

Figure 9.21 shows the system step response using Dahlin's controller. There is no evidence of the ringing observed in Fig. 9.19(b) but the system shows a slight overshoot. Overall this simple controller is probably quite adequate for most purposes.

Exercise A process with the transfer function

$$G_p(s) = \frac{1.24 \, e^{-0.3s}}{(1 + 0.377s)(1 + 0.132s)}$$

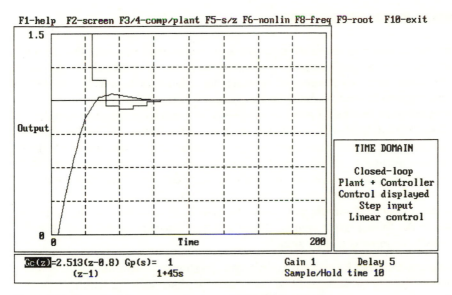

Figure 9.21 Example 9.9 response of system with Dahlin's controller

is to be digitally controlled using a sample period of 0.15 second. Design controllers that give the closed-loop system, steady-state set-point following and a first-order characteristic with a time constant of 0.15 second using the following methods:

(a) direct design without removal of any ringing pole;
(b) Dahlin's method of removing any ringing pole;
(c) direct design with removal of ringing pole by incorporating of the offending plant zero in the closed-loop transfer function.

Summary

Direct design methods give good performance when the closed-loop objective is realistic. When an accurate model of the process is available the target closed-loop system can be chosen to provide fast and accurate response. The practical realities of nonlinearity and uncertain or time-varying dynamics can mean that controllers designed to achieve ambitious goals can perform badly and may even be unstable. There is always a compromise to be made between performance and robustness.

The direct design approach can lead to high order controllers, particularly when the plant includes transport delay and fast sample rates are used. Dahlin's method of shifting poles to the origin of the z-plane can be used to simplify control laws but the method must be used with care. The frequency domain methods covered in Section 9.4, which are based on analogue designs, have the advantage that the order of the controller is independent of the order of the plant and so generally simpler control laws will result. Of course this also limits the achievable performance.

9.6 SAMPLE RATE SELECTION

The selection of sample rate or sampling period is of crucial importance to the eventual performance and cost of any computer control system and is often one of the most important decisions made by the control engineer. Sample rate selection is also one of the first decisions which must be made in the design of digital controllers.

Any sampled-data feedback system is effectively in open loop between samples so it may be thought that the faster we sample the process the better will be the control. This assumption is incorrect because rapid sample rates can lead to numerical problems with calculation accuracy, stability and robustness problems, and problems with quantization and measurement noise. In addition, particularly when a computer is controlling many loops, system cost will increase with sampling rate.

The correct choice of sample rate must be based upon achieving the required performance of the final control system. There are two distinct categories of discrete controller which we must consider, namely:

1. Discrete controllers which are based upon approximation of continuous control laws. For example digitally implemented PID controllers or phase lead/lag compensators.
2. Discrete controllers which are correctly designed using discrete design methods without approximation.

Controllers which are based upon approximation of continuous designs will in general require faster sample rates than those which are designed without approximation.

General Considerations

Shannon's sampling theorem gives a theoretical limit on sample rate for band-limited signals and is often used as a starting point for sample rate selection for sampled-data control systems. The designer should have a good idea of the required bandwidth of the closed-loop system. In the absence of noise, this bandwidth will also correspond to the frequency range of the controlled variable which we need to sample. Thus using the ideas developed from Shannon, we can say that the sampling frequency, $f_s = 1/T$, should be of the order of 5 to 20 times the desired closed-loop bandwidth, f_b.

$$f_s \geqslant 5 \text{ to } 20 \, f_b \qquad (9.41)$$

This desired closed-loop bandwidth can of course be related to a desired closed-loop time constant or rise time, etc.

Controllers Based on Analogue Designs

We have seen (Section 9.3) that the zero-order hold introduced by the DAC in a sampled-data control system can be approximated as a delay of half the sample period. Considered from the frequency domain point of view the hold introduces an extra phase lag of $\omega T/2$ into the plant dynamics.

Assuming that a given analogue compensator or controller gives acceptable performance then, for sample times which are small compared with the plant response time, any discrete approximation of the controller will have a degraded phase margin of approximately $\omega_c T/2$ radian, where ω_c is the gain-crossover frequency (i.e. the frequency at which the gain of the open-loop system is unity. Limiting this degradation to not more than 5 degrees gives

$$\frac{\omega_c T}{2} \leqslant \frac{5\pi}{180}$$

Expressing the crossover frequency in Hz ($f_c = \omega_c/2\pi$) gives

$$f_s \geqslant 36 f_c \tag{9.42}$$

Directly Designed Digital Controllers

Open-loop time constant considerations When a continuous system is sampled, the equivalent discrete time system will have poles in the z-plane which correspond to the system s-plane poles via the mapping

$$z = e^{sT}$$

Thus a plant with a simple lag of time constant, τ, will have an s-plane pole at $s = -1/\tau$ and the sampled system will have a z-plane pole at $z = e^{-T/\tau}$. Now as T becomes small compared to the time constant, the z-plane pole tends to move toward $z = +1$. The discrete time plant model used to design a digital controller will, for small sample times, have poles bunched in the region of $z = +1$. Any small inaccuracy in the model coefficients will correspond to a large difference in plant dynamics. In order that the sensitivity of a digital controller to parameter inaccuracy be acceptable, the system poles must not be allowed to bunch too close to the $z = +1$ point.

For the dominant pole to be no closer than $z = 0.98$ then

$$e^{-T/\tau} \leqslant 0.98$$

or

$$T \geqslant -\tau \ln(0.98)$$

$$f_s \leqslant \frac{50}{\tau} \tag{9.43}$$

Transport delay considerations A pure transport delay when part of a sampled-data system produces one or more poles at the origin of the z-plane, each corresponding to a delay of one sample. Thus a small sample time relative to any system transport delay will introduce many poles at the z-plane origin. The complexity of any controller which is to take account of these poles will thus increase as the ratio of sample time to delay time shrinks. For example, a sample period of half the system transport delay, T_D, gives two extra z-plane poles. It is reasonable to limit these additional poles to between 2 and 4, i.e.

$$f_s \leqslant \frac{2}{T_D} \quad \text{to} \quad \frac{4}{T_D} \tag{9.44}$$

All of the above criteria are somewhat heuristic and can only give a guide to selecting a suitable sample rate. For some plants the above rules may give contradictory results.

Anti-aliasing Filters

We have seen that aliasing will occur when sampling signals which contain components at frequencies above the Nyquist frequency. Aliasing can cause problems in sampled-data control systems if the sampled signal includes any components above half the sample frequency. Normally the sample rate of a digital controller is chosen so that the frequencies associated with the system dynamics are well below the Nyquist frequency. However, the measured value may also contain high frequency components caused by measurement noise, pickup and interference. Aliasing causes such high frequency inputs to appear as low frequencies to the controller. The controller will then attempt to apply control effort to control or regulate the aliased input. Because, in reality, the cause of the input is a high frequency signal the much slower control effort cannot hope to have any effect. The result is a wildly fluctuating control effort which, far from regulating the process, actually creates a disturbance in the plant.

It is important in any sampled-data system to guard against components above the Nyquist frequency entering the sampler. An *anti-aliasing filter* is a filter placed on the input to the ADC which (ideally) rejects all signal components above the Nyquist frequency and passes all those below. Anti-aliasing filters must of course be analogue filters since any digital filter would itself need to sample the signal. The ideal low-pass characteristic is in practice impossible to achieve and real anti-aliasing filters will have an attenuation which increases gradually with frequency. Thus some components at frequencies above the cutoff will get through and some components below the cutoff will be attenuated and may be distorted by phase shift (Fig. 9.22).

The main problem for the control engineer is to decide on the cutoff frequency and order of the filter. At first sight it is tempting to place the filter cutoff frequency well above the control system bandwidth so that the filter has little effect on system performance. The sample frequency must then be chosen well above the filter cutoff frequency. Such an approach can lead to very high sample rates.

An alternative approach is to decide on the control system sample rate based on system performance criteria and then choose the anti-aliasing filter order and cutoff frequency to give the required attenuation at the Nyquist frequency. The control system design should then be based on the plant dynamics and anti-aliasing filter combined.

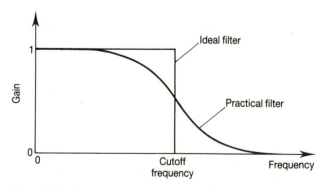

Figure 9.22 Ideal and practical anti-aliasing filter characteristics

Quantization Effects

In the development of discrete controllers considered so far we have neglected any effects due to digital implementation. The assumption was that the control law can be implemented exactly as though the discrete time signals were analogue. In practice of course all the values manipulated by a digital computer system will be digital and will be subject to quantization. Generally, the major cause for concern is quantization error introduced by analogue to digital conversion.

Quantization error in analogue to digital conversion is caused by the finite resolution available with a given ADC. Quantization introduces a nonlinearity into the control loop (Fig. 9.23) which can cause inaccurate control performance, and even limit cycle behaviour. For example, changes in the measured value which are less than the ADC resolution will go undetected by the controller.

Since the sign and magnitude of any roundoff caused by quantization depends on the exact value of the converter input, quantization errors are in general unpredictable and emerge as a random perturbation of the true value. Quantization errors can thus be thought of as a random noise which is added to the converted signal. When the converter input is changing by less than one quantization level per sample, the converted value will occasionally jump from one digital value to the next (Fig. 9.24). If the converted value is the measured value in a control loop it can appear as if the plant output suddenly changes in steps of size equal to the ADC resolution. The effect of such quantization noise can be very significant particularly when the control law includes phase lead or derivative compensation.

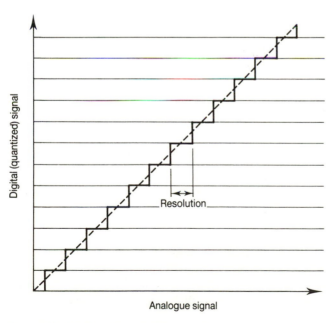

Figure 9.23   Nonlinearity caused by quantization

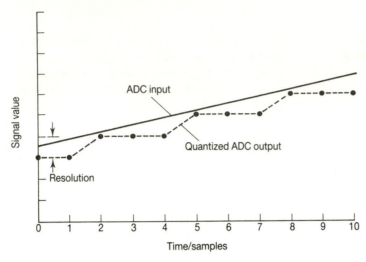

Figure 9.24 Slowly changing signals appear to be stepped due to quantization

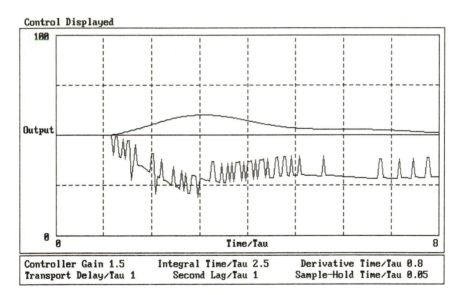

Figure 9.25 Digital three-term controller showing derivative kicks due to ADC quantization (8-bit converter)

Example 9.10 ADC quantization noise in a discrete controller Figure 9.25 shows the response of a system under digital three-term control. The control effort shows quite large kicks appearing on top of the average control action. These sharp pulses of control action are quantization noise which is amplified by the derivative term of the controller.

The derivative contribution of the digital control law used in Fig. 9.25 is

$$u_d = K \frac{T_d}{T} [e(i) - e(i-1)]$$

When the measured value is changing slowly, ADC quantization will cause the error to occasionally change by one quantization step;

$$q = \frac{\text{converter range}}{2^n}$$

where n = number of bits in converter.

Thus the control effort will suddenly change by

$$\Delta_u = \frac{KT_d q}{T} \tag{9.45}$$

For the case shown in Fig. 9.25, the converter is an 8-bit device, and with the controller parameters as shown

$$\Delta_u = \frac{1.5 \times 0.8 \times 100}{0.05 \times 2^8} \% = 9.4\%$$

These derivative kicks do not in themselves degrade the control performance since the average derivative action is correct. The plant acts as a low-pass filter and smooths the effect of the control effort. However, derivative kick can cause excessive actuator wear and should be minimized.

Equation (9.45) shows that for this controller the derivative kicks can be reduced by using a higher resolution ADC or by reducing the sample rate.

9.7 CASE STUDY

Figure 9.26 shows an industrial heat exchanger which is designed to heat a product stream from a steam supply. The product outlet temperature is monitored by a platinum resistance thermometer with signal conditioning. The product flow rate is monitored by a turbine flow meter with a pulse output, the frequency being proportional to the flow rate. The steam supply and product flow are both controlled by pneumatically actuated control valves fitted with current to pressure converters. The control system is required to maintain the product at a constant temperature of 55°C. The flow rate is regularly required to be altered within the range 1.5 to 3.0 litre/min to cater for downstream process requirements. This case study will examine the design and implementation of digital controllers for the product temperature and flow rate. Feedforward compensation will also be examined.

Temperature Control Loop

Figure 9.27 shows the reaction curve for the product temperature obtained by applying a step change to the steam control valve with the flow rate held constant at the design condition (2.1 litre/min). The measured temperature shows considerable fluctuations or

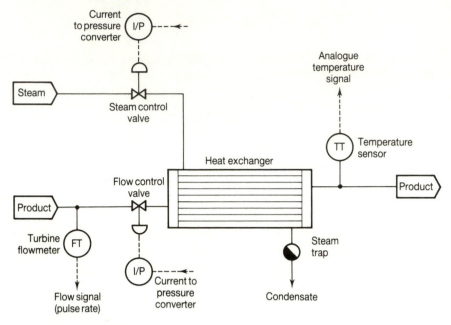

Figure 9.26 Industrial heat exchanger used in case study

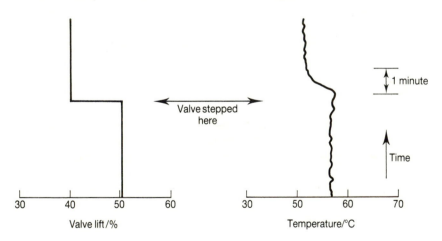

Figure 9.27 Open-loop response of product temperature with a step applied to the steam control valve

process noise. This is probably due to poor mixing of the heated product as it leaves the heat exchanger. In addition, disturbances in the steam quality and product inlet temperature cause the temperature to drift significantly. Analysis of this and other responses indicates that the process can be approximately modelled as a double lag with time constant 15 seconds and a transport delay also of 15 seconds. The transfer function relating the temperature rise to the percentage control valve lift is

$$G_1(s) = \frac{0.54\,e^{-15s}}{(1+15s)^2} \qquad °C/\% \tag{9.46}$$

The transport delay is very dependent upon the product flow rate and can vary between 10 and 20 seconds over the range of flow rates expected.

Accurate steady-state regulation of the product temperature is essential but because the reference temperature will be constant, there is no velocity error requirement. The expected changes in transport delay suggest that to achieve a robust control loop we should not aim for an excessively fast closed-loop performance. Aiming for a closed-loop response with a 50% faster rise time than in open-loop should provide a reasonable compromise between performance and robustness. Trying to achieve a first-order response for this clearly second-order system would require considerably more control effort with little benefit. A closed-loop characteristic equation consisting of a double lag with a time constant of about 10 seconds will therefore be chosen.

Selection of sample rate can be based on the criteria in Section 9.6. The -3 dB bandwidth of a double lag of time constant 10 seconds is about 0.01 Hz. The sample rate should be at least 5 times the closed-loop bandwidth requiring a sample period of not more than 20 seconds.

The transport delay at normal operating conditions is 15 seconds and thus choosing a sample time of 7.5 seconds would give the plant pulse transfer function only two excess z-plane poles. All other criteria of Section 9.6 are satisfied with this sample rate and so $T = 7.5$ s will be used. The resulting discrete time model of the plant is

$$G_l(z) = \frac{0.049(z+0.72)}{z^2(z-0.607)^2} \qquad °C/\% \tag{9.47}$$

For a double lag of time constant 10 seconds the closed-loop characteristic will need a pair of poles at

$$z = e^{-T/\tau} = e^{-7.5/10} = 0.47 \approx 0.5$$

The discrete plant, Eq. (9.47), has a zero which is inside the unit circle. If the controller is designed to cancel this zero the resulting control action may be oscillatory. Furthermore, the zero location is dependent upon the transport delay and so it will move as the delay changes. The chosen target closed-loop system will therefore retain this zero to avoid cancellation by the controller. With a steady-state gain of unity and pole-zero excess of three, the design for the closed-loop transfer function is:

$$F_1(z) = \frac{0.145(z+0.72)}{z^2(z-0.5)^2}$$

The control law is then obtained from the direct design formula (Eq. (9.32)) as

$$G_{c1}(z) = \frac{3z^2(z-0.607)^2}{(z-1)(z^3+0.25z+0.105)} \qquad \%/°C \tag{9.48}$$

Aliasing should not be a problem with this loop since the signal conditioning unit for the resistance thermometer incorporates sufficient smoothing to filter out unwanted

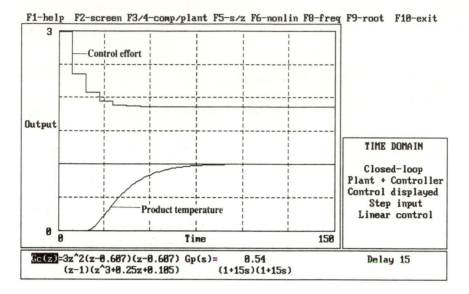

F1-help F2-screen F3/4-comp/plant F5-s/z F6-nonlin F8-freq F9-root F10-exit

Figure 9.28 Simulated closed-loop step response of sampled-data temperature control loop

frequencies. Precautions should however always be taken to avoid interference or mains pickup entering the sampler.

The choice of converter will be governed by resolution requirements. A 10-bit ADC covering the temperature range 0 to 100°C would give an acceptable resolution of better than 0.1°C. Quantization noise when amplified by the controller instantaneous gain of 3 will produce steps in the control effort of 0.3%. In view of the accuracy with which the steam valve can be positioned an 8-bit DAC is quite acceptable with an output resolution of 0.4%.

Figure 9.28 shows the simulated step response of the temperature control loop at design conditions. The response is acceptable with no adverse inter-sample ripple or violent control action. Figure 9.29 shows that the simulated system is quite robust to changes in plant transport delay between 10 and 20 seconds. Figure 9.30 shows some responses for the actual plant obtained using the controller of Eq. (9.48). Trace (a) shows the response to a step change in reference temperature at design conditions. Trace (b) shows the regulation performance of the control system when a step increase of product flow rate is applied. The product temperature drops suddenly but the feedback controller applies a correcting action which soon brings the product back to the reference value. Trace (c) shows the effect of a decrease in flow rate down to the minimum value of 1.5 litre/min. The overshoot and resulting slow oscillation are caused by the increase in plant transport delay at this low flowrate.

Flow Control Loop

The open-loop step response for the product flow is shown in Fig. 9.31. The flow measurement is rather noisy but the basic dynamics relating flow to percentage valve

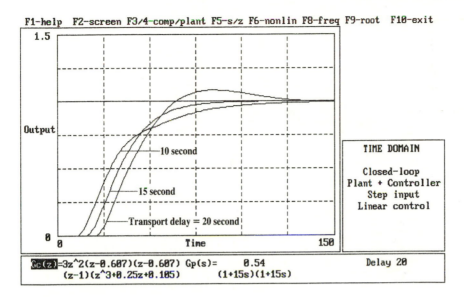

F1-help F2-screen F3/4-comp/plant F5-s/z F6-nonlin F8-freq F9-root F10-exit

TIME DOMAIN

Closed-loop
Plant + Controller
Step input
Linear control

Gc(z)=3z^2(z-0.607)(z-0.607) Gp(s)= 0.54   Delay 20
        (z-1)(z^3+0.25z+0.105)   (1+15s)(1+15s)

Figure 9.29 Effect of transport delay changes on the simulated system step response

lift can be modelled by the first-order transfer function

$$G_2(s) = \frac{0.04}{1 + 0.4s} \qquad \text{litre/min\%} \qquad (9.49)$$

The flow dynamics are very fast in comparison with the temperature loop and further there is no need for a fast closed-loop response. This is an unusual situation since most control problems involve speeding up the system dynamics. The control action in this case is to trim the flowrate rather than achieve a certain dynamic response. The main problem concerns the noisy signal obtained from the flow meter. Choosing slow closed-loop dynamics will reduce the problem of noise amplification in the controller.

Since the flow rate measurement noise is of quite a high frequency, problems due to aliasing could occur. The signal from the turbine flow meter is a pulse rate and so an anti-aliasing filter cannot be used. In this case the sampling process is not carried out by an ADC but by a counter/timer counting pulses over a given time interval. Such a pulse-counting mechanism gives the average pulse rate over the time interval and thus averages out fluctuations at higher frequencies. Any signal components which change rapidly compared to the counting period will be filtered by the averaging and their effect on the measured value reduced. There is an advantage in sampling this loop slowly in that the counting period can be made longer to filter out the high frequency noise.

Another factor is that of resolution: the pulse count will be an integer quantity (i.e. rounded to the next whole number) and so quantization error will exist. The longer the counting period, the larger the value of the count and the less significant will be the quantization. The turbine flow meter produces a frequency of 87 Hz at 1 litre/min. A counting interval of 1 second will therefore result in a sensitivity of 87 per litre/min

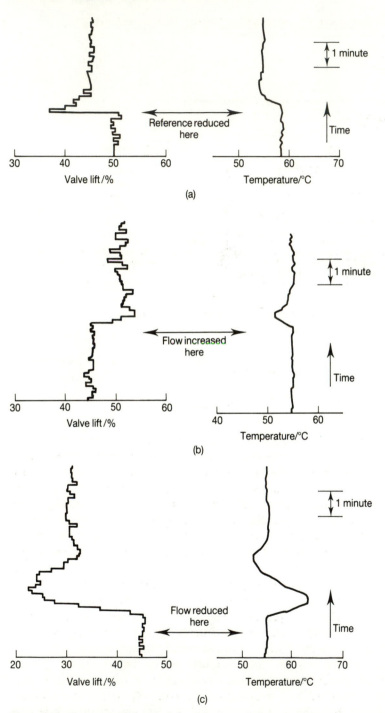

Figure 9.30 Response of actual plant with direct digital temperature control. (a) Step change in reference temperature at design flow rate. (b) Step increase in flow rate from 2.1 to 2.4 litre/min. (c) Step decrease in flow rate from 2.1 to 1.5 litre/min

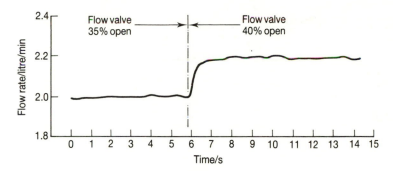

Figure 9.31 Product flow rate open-loop response

which gives acceptable resolution and noise rejection. Taking this measurement gain into account the steady-state gain of the process is $0.04 \times 87 = 3.48$ per % valve lift.

Using a slow sample rate with such fast open-loop dynamics produces some interesting results. With a sample interval of 1 second the pulse transfer function relating the count value to valve lift is

$$G_2(z) = \frac{3.2}{z - 0.08} \quad \text{per \%} \tag{9.50(a)}$$

or
$$G_2(z) \approx \frac{3.2}{(1 - 0.08)z} = \frac{3.5}{z} \quad \text{per \%} \tag{9.50(b)}$$

Because the sample interval is greater than the open-loop settling time, the open-loop pulse transfer function can be approximated as a pure delay of one sample. Now, this makes the controller particularly simple. For example, aiming for a first-order closed-loop response with

$$F_2(z) = \frac{(1 - a)}{z - a}$$

results in the controller

$$G_{c2}(z) = \frac{z}{3.5} \frac{(1 - a)}{z - 1} \quad \% \tag{9.51}$$

i.e., pure integral action with a gain of $(1 - a)/3.5\%$. By changing the controller gain the closed-loop time constant can be altered as desired.

Figure 9.32 shows the step response of the actual flow control loop with a sample time of 1 second and a gain of 0.25% (corresponding to approximately deadbeat closed-loop response). This control loop is very robust to changes in flow dynamics since the sample rate is so slow.

Feedforward

Since one of the major disturbances to the temperature control loop is the change in product flow rate, compensation by feedforward of the flow signal to the steam valve

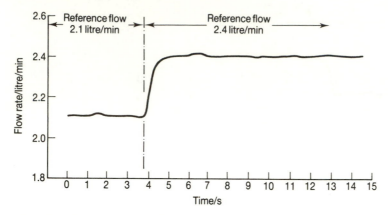

Figure 9.32 Flow control system closed-loop step response

should give significant improvement in the regulation of product temperature. In addition, since the flow signal is readily available, this scheme is simple and cheap to implement.

Figure 9.33 shows the effect of a small increase in product flow rate on the product temperature. The dynamics can be modelled by the transfer function

$$G_3(s) = \frac{-18\,e^{-15s}}{1+15s} \qquad °C/\text{litre/min} \tag{9.52}$$

Not surprisingly, the transport delay is the same as that for $G_1(s)$. Figure 9.34 shows the block diagram for feedforward compensation. To achieve perfect compensation, a change in the measured flow rate should produce a response at point 'A' (via G_{ff} and G_1) which is equal but opposite to the disturbance produced at point 'B' (via G_3). This requires that

$$G_{ff}G_1 = -G_3$$

The discrete feedforward compensator should therefore have the transfer function

$$G_{ff}(z) = \frac{-G_3(z)}{G_1(z)}$$

Using a sample period of 7.5 seconds, $G_3(z)$ becomes

$$G_3(z) = \frac{7.08}{z^2(z-0.607)} \qquad °C/\text{litre/min}$$

and so
$$G_{ff}(z) = \frac{z^2(z-0.607)^2}{0.049(z+0.72)}\frac{7.08}{z^2(z-0.607)}$$

$$= \frac{145(z-0.607)}{(z+0.72)} \qquad \%/\text{litre/min}$$

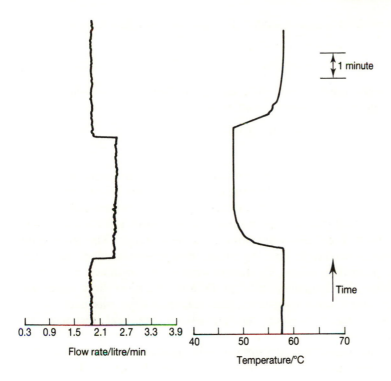

Figure 9.33 Open-loop response of product temperature to a step change in flow rate

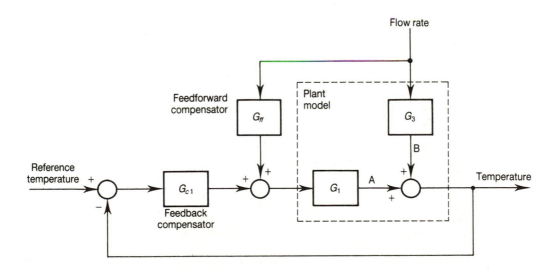

Figure 9.34 Block diagram of temperature control with feedforward

Taking the flowmeter/pulse counter sensitivity of 87 per litre/min into account

$$G_{ff} = \frac{1.67(z - 0.607)}{(z + 0.72)} \quad \%$$

Finally, the oscillatory pole at $z = -0.72$ can be shifted to the origin and the equivalent steady-state gain incorporated, resulting in

$$G_{ff}(z) = \frac{0.97(z - 0.607)}{z} \quad \%$$

It must be remembered that the feedforward compensation is dynamic in that it is **changes** in flow rate which are fed forward to the steam valve. One way to implement this is as a velocity algorithm. The difference in flow rate between the current and previous samples is applied to the feedforward compensator and the result is then used as an increment which is added to the steam valve position at each sample. The differencing and subsequent summation cancel to give the correct compensation effect. However, the resulting velocity algorithm allows the feedforward to be switched in without introducing a disturbance (i.e. bumpless transfer). Because the noisy flow signal was found to give violent steam valve excursions, the reference flow was used as the feedforward signal. Since the flow loop has an almost deadbeat response, the consequences of this are minimal.

Figure 9.35 shows the regulation performance with feedforward incorporated as described above. The compensation gives almost exact cancellation of the effect of flow rate changes at design conditions but the compensation is less good at other operating conditions. Even so, the regulation performance is considerably improved by the feedforward at all operating points.

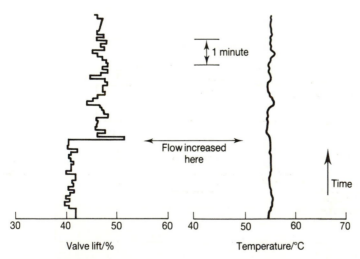

Figure 9.35 Improved regulation performance with feedforward incorporated. Step increase in flow rate from 2.1 to 2.4 litre/min

Exercise Use CODAS-II to compare the performance and robustness of the above temperature controller (Eq. (9.48)) with that of a discrete three-term controller (Eq. (9.24)) with a sample time of 1 second. The controller settings can be found by using the Ziegler–Nichols continuous cycling method (Section 7.7).

PROBLEMS

9.1 A position control system can be modelled by the open-loop transfer function

$$G(s) = \frac{1}{s(1+s)}$$

Proportional feedback is applied by a computer sampling the output at a rate of 1 Hz. Using CODAS-II:

(a) Obtain the pulse transfer function of the system as seen by the control computer.

(b) Examine the root locus of the system as the controller gain, K, is varied. Find the controller gain which gives marginal stability. Calculate the gain which gives the system a gain margin of 2. From the resulting closed-loop poles estimate the system damped natural frequency and the 5% settling time.

(c) Examine the system open-loop Nyquist diagram and confirm the gain margin is as desired. Determine the closed-loop resonant frequency and peak magnification.

(d) Simulate the closed-loop step response of the sampled-data system (i.e. with the plant defined as $G(s)$ but with the sample hold time set). Verify the settling time and damped natural frequency as estimated in (b). Using a user-defined input apply a cosine input at the closed-loop resonant frequency and confirm the system peak magnification. Make sure that the initial transient has died away before making the measurement.

9.2 A continuous time second-order system designed to pass only frequencies in a given band (a band-pass filter) has a transfer function

$$G(s) = \frac{s}{s^2 + s + 25}$$

Using Tustin's rule with prewarping determine a digital filter which has the same transmission at the filter natural frequency. The sample frequency can be taken as 5 Hz. Confirm your design using CODAS-II and determine the 3 dB bandwidth of the two filters, the 3 dB bandwidth being here defined as the frequency range over which the filter gain is within 3 dB of the maximum.

9.3 A process with the transfer function

$$G_p(s) = \frac{1}{s(1+s)^2}$$

is controlled by a proportional controller.

Examine the open-loop frequency response of the continuous system and determine its phase margin and gain-crossover frequency. Using a sample time of 0.5 second obtain a discrete transfer function for the process as a sampled-data system. Again determine the phase margin and gain-crossover frequency. Compare the degradation in phase margin with that predicted by the approximation of a zero-order hold by a transport delay of $T/2$.

9.4 A data logging system samples a signal, x, from a strain gauge every 10 ms and calculates the moving average, y, over the last two samples:

$$y(i) = \frac{x(i) + x(i-1)}{2}$$

Show that any mains pickup at 50 Hz is totally rejected by this averaging technique.

9.5 Using Tustin's rule derive the discrete time equivalent of the modified (proper) PID controller

$$G_c(s) = K\left[1 + \frac{1}{sT_r} + \frac{sT_d}{(1 + \alpha sT_d)}\right]$$

Show that when $\alpha = (T/2T_d)$ this discrete controller is identical to Eq. (9.23).

9.6 The position servo which formed the basis for many design examples in Chapter 5 had an open-loop transfer function

$$G_p(s) = \frac{640}{s(s+4)(s+16)}$$

This problem examines the design of digital compensators for this system when sampled at a frequency of 5 Hz. The design objective is to achieve a phase margin of 45°.

Using Tustin's rule with prewarping of the design frequency:

(a) Design a phase lag compensator which achieves the design objective with no reduction in system Bode gain.

(b) Design a phase lead compensator to achieve the design objective. Note that the plant numerator coefficient will have to be reduced from 640 to 300.

9.7 The phase lead compensator

$$G_c(s) = \frac{1 + a\tau s}{1 + \tau s}; \qquad a > 1$$

gives maximum phase lead at a frequency

$$\omega_m = \frac{1}{\sqrt{a\tau}}$$

Show that the discrete approximation obtained by using Tustin's rule with prewarping of the maximum phase lead frequency gives the pulse transfer function

$$G_c(z) = \frac{(b + \sqrt{a})z + (b - \sqrt{a})}{(b + 1/\sqrt{a})z + (b - 1/\sqrt{a})}$$

where

$$b = \tan\left(\frac{T}{2\sqrt{a\tau}}\right)$$

Examine the Bode diagram of the discrete compensator with $a = 10$ and $\tau = 1$ second with a sample frequency of ten times the maximum phase lead frequency ($T = 2\pi\sqrt{a\tau}/10$) and compare it with that of the continuous compensator. Repeat for a sample frequency of only three times the maximum phase lead frequency.

9.8 Design a stable digital controller for the process with transfer function

$$G_p(s) = \frac{e^{-0.5s}}{s(1+s)}$$

The sample rate may be taken as 1 Hz. The closed-loop system should exhibit first-order dynamics with a pole at $z = 0.5$.

Note that the process transfer function incorporates a free integrator and so the closed-loop system should have a steady-state gain of unity.

9.9 A standard first-order system with transfer function

$$G_p(s) = \frac{1}{1 + \tau s}$$

is controlled by the proportional plus integral controller

$$G_c(z) = \frac{K(z - b)}{z - 1}$$

(a) Show that deadbeat control is achieved with the following controller parameters:

$$K = \frac{1 + \beta}{1 - \beta} \qquad \text{and} \qquad b = \frac{\beta}{1 + \beta} \qquad \text{where } \beta = e^{-T/\tau}$$

(b) Determine the minimum sample interval which can be used if the controller gain, K, is not to exceed 10. Simulate this situation using CODAS-II and verify your answer.

9.10 Design a stable and realizable controller for the discrete system

$$G_p(z) = \frac{0.2z + 0.3}{z^2 - 1.8z + 0.8}$$

The design should give the closed-loop system a characteristic equation

$$z^2 - 0.8z + 0.32 = 0$$

Show that the controller can be approximated by the difference equation

$$u(i) = e(i) - 0.8e(i - 1)$$

9.11 The dynamics relating the position of a frictionless mass to the applied force are described by the transfer function

$$G_p(s) = \frac{1}{ms^2}$$

Obtain a digital control law, in terms of the mass, m, and the sample time, T, which gives stable, deadbeat control of the mass position. Note that the closed-loop transfer function should have unity steady-state gain, and zero velocity error since the system is naturally Type-2.

9.12 A process can be modelled as a pure integrator with a Bode gain of 0.1 per second together with a transport delay of 20 seconds. Design a discrete controller which gives the closed-loop system first-order dynamics with a time constant of 20 seconds. The sample interval may be taken as 10 seconds.

Examine the robustness of your control system by decreasing the process transport delay by 25%.

Repeat the exercise designing this time for a closed-loop time constant of 5 seconds.

9.13 Design a stable discrete compensator for the system in problem 9.6. The design should give the closed-loop system poles at $z = 0.0 \pm j0.4$. Compare the Bode gain of your design with that of the original system.

9.14 A digital compensator with transfer function

$$\frac{U}{E}(z) = \frac{22z^2 - 25z + 6}{z(z - 1)}$$

is to be implemented on an industrial control computer by sampling the measured value using an ADC. Obtain a difference equation for the change in control effort and hence determine the required ADC resolution such that the control effort kicks caused by quantization noise are less than 5% of the controller range.

NONLINEARITIES IN CONTROL SYSTEMS

10.1 INTRODUCTION

Nonlinearities exist for a number of reasons. We have already seen in the case study of the ship steering system in Chapter 6 that the rudder angle was constrained to certain limits. This type of behaviour is true of any real control system because there is always a finite limit to the amount of control effort that can be developed or applied. For example, motors can only produce so much torque, valve movements and sizes are limited, the heating capacity of a steam heater is constrained by its area and the pressure and capacity of the boiler. When dealing with the normal behaviour of the control system these limitations do not matter because the apparatus or plant is working comfortably within its design limits and linear models and methods of analysis will predict behaviour adequately. However if there is a sudden change in demand the equipment may hit a constraint and so the results will be significantly different from those in the linear operating region.

There are many examples where nonlinearities affect the behaviour of the system within the normal design range and long before there is any saturation or limiting. For example in a mechanical system a 'hard' spring may be present, i.e. one whose stiffness increases with deflection. In vehicle systems drag forces are proportional to the square of the wind speed. In process control systems the flow through a restriction is proportional to the square root of the pressure drop. All these examples will not allow linear methods to be adopted directly. However, provided the changes in the operating conditions are **small**, linear models can be derived which are valid for the chosen operating conditions. Thus apart from deriving these 'small signal' models no special techniques are required.

There are, however, classes of nonlinearity which either have been introduced deliberately or may be present because of the type of system where linear models and methods are wholly inappropriate. For example the behaviour of a thermostatically controlled central heating system cannot be adequately modelled or designed using linear techniques. The central heating boiler comes on when the temperature is below the thermostat setting and goes off when the temperature rises above the thermostat setting. Here a nonlinear controller has been deliberately introduced because of cost factors and because there is a very low requirement for the performance of the control system.

Certain systems exhibit nonlinearities which are significant even for small signal amplitudes. The effects of backlash in gear trains, hysteresis and stiction are all examples of phenomena which are present in many position control systems, and whose effect is **more** pronounced at small signal amplitudes than with larger changes.

10.2 PERTURBATION ANALYSIS

The basic concepts behind this technique were introduced in Chapter 2, but the essential points are repeated here for completeness.

When a system includes nonlinearities with no sharp discontinuities (i.e. 'smooth' nonlinearities) and the operating conditions only change by small amounts, the nonlinearity can be treated as a simple gain whose value is equal to the slope of the nonlinear characterisrtic. This technique involves determining the small signal sensitivity of nonlinear elements and so is known as 'small signal' or 'perturbation' analysis. Once a small signal model is obtained, block diagrams, Laplace transforms and all the normal techniques of linear analysis can be applied. It must always be remembered that the model has limited validity and results should not be extrapolated beyond reasonable bounds.

The method is particularly useful for process control systems where smooth nonlinearities are common and where the primary function of the loop is to maintain the process variable at a constant value and so, by design, the changes in operating conditions are small.

Consider the nonlinearity in Fig. 10.1. The output, z, is related to the input, x, by some function, $f(x)$ i.e.

$$z = f(x) \tag{10.1}$$

The nominal (quiescent) operating condition is $\bar{x}$, $\bar{z}$ and δx and δz represent small changes or perturbations from the quiescent conditions. Positive values of δx and δz represent increases in x and z from their respective quiescent values and similarly negative

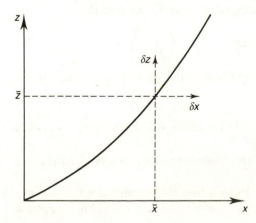

Figure 10.1 Typical 'smooth' nonlinearity

values of δx and δz represent decreases.

$$\delta z = \frac{\partial f}{\partial x}\bigg|_{x=\bar{x}} \delta x = k_x \delta x \tag{10.2}$$

The term k_x represents the sensitivity of the nonlinear characteristic with respect to small changes in the input variable, x, about the quiescent value, $\bar{x}$. The sensitivity can be determined by experiment, by measuring the slope of the characteristics from data sheets, or by partial differentiation if an analytical expression is available of the function.

Quite often the output of an element in a control system depends on two or more changing parameters. For example the head developed by a centrifugal pump depends on the pump speed and the flowrate of fluid through the pump, the flow through a control valve depends on the valve lift and the pressure drop across it. The relationship can be generalized as

$$z = f(x, y) \tag{10.3}$$

By applying the concept of 'total differential' one obtains:

$$\delta z = \frac{\partial f}{\partial x}\bigg|_{x=\bar{x}} \delta x + \frac{\partial f}{\partial y}\bigg|_{y=\bar{y}} \delta y$$

$$\delta z = k_x \, \delta x + k_y \, \delta y \tag{10.4}$$

Again the values of k_x and k_y can be determined empirically or by differentiation.

Example 10.1 Obtain a small signal model relating changes in z to changes in x and y for the equation:

$$z = 6x^3 \sqrt{y}$$

The normal operating values are $\bar{x} = 5$ and $\bar{y} = 4$.

Find the change in z predicted by the small signal model when x increases by 0.2 and y decreases by 0.3. Compare the result obtained using small signal analysis with the result using the nonlinear equation.

SOLUTION Differentiating partially with respect to x and y we obtain:

$$\frac{\partial f}{\partial x} = 6(3x^2 \sqrt{y}) \quad \text{and} \quad \frac{\partial f}{\partial y} = 6\left(\frac{x^3}{2\sqrt{y}}\right)$$

Applying the total differential rule and evaluating the partial derivatives at the quiescent conditions, we obtain:

$$\delta z = 6\left(3\bar{x}^2 \sqrt{\bar{y}}\,\delta x + \frac{1}{2}\frac{\bar{x}^3}{\sqrt{\bar{y}}}\,\delta y\right) = 990\delta x + 112.5\delta y \tag{10.5}$$

Thus for $\delta x = +0.2$ and $\delta y = -0.3$, the small signal model predicts a change in z (δz) of 146.25.

The exact result can be obtained by substituting the actual values of x and y into the nonlinear equation, i.e. when $x = 5.2$ and $y = 3.7$, z is 1622.8. The value of z at the quiescent conditions ($\bar{z}$) is 1500, hence the actual change in z is 122.8.

There is of course a discrepancy between the small signal result and the exact 'large signal' value. The significance of this discrepancy can only be gauged by considering it relative to the quiescent value of z. In relative terms the difference between the two results is insignificant.

It is quite instructive to write the small signal model in a non-dimensional or 'per-unit' manner by dividing both sides of Eq. (10.5) by $\bar{z}$, namely:

$$\frac{\delta z}{\bar{z}} = 3\frac{\delta x}{\bar{x}} + \frac{1}{2}\frac{\delta y}{\bar{y}}$$

This shows very clearly that the fractional change in z is three times the fractional change in x and half the fractional change in y. In the above example the fractional change in x was 0.04 and -0.075 in y. Hence the predicted fractional change in z is 0.0825. The actual fractional change in z is 0.082 (122.8/1500). This non-dimensional treatment serves to show how good the small signal model is and offers a clearer insight than models which use absolute values and changes.

10.3 PERTURBATION ANALYSIS APPLIED TO PROCESS PLANT

Perturbation analysis is particularly suitable for process control systems for the reasons given earlier. Thus this technique will be mainly directed to simple but typical examples of process control systems.

Control Valves

Control valves are commonly encountered elements in process plant and the equations that describe their flow behaviour are nonlinear. Other nonlinear effects may exist because of the valve characteristic and the equipment surrounding the valve. A short treatment is given here but the subject is very important and texts are available which are devoted almost exclusively to this topic. In addition, control valve manufacturers are a very useful source of information on this subject.

Figure 10.2 shows a simplified schematic diagram of a control valve with the graphic symbol that is used to represent it in process flow diagrams. The control valve modulates the flow of a fluid by introducing a variable area aperture into the pipeline. Only liquid service is considered here as the treatment of gaseous flow is rather more complex.

The volumetric flow rate, Q, of a particular liquid through a valve is proportional to the pressure drop across it, ΔP.

$$Q = k_v\sqrt{\Delta P} \tag{10.6}$$

In the case of a control valve, the valve coefficient, k_v, is a function of the valve opening or lift, h. In order to avoid dimensionality, the lift, h, is defined as a fractional lift, i.e. when h is 1 the control valve is fully open, and when h is 0 the valve is shut. Similarly rather than deal with a dimensional equation for the valve coefficient, the

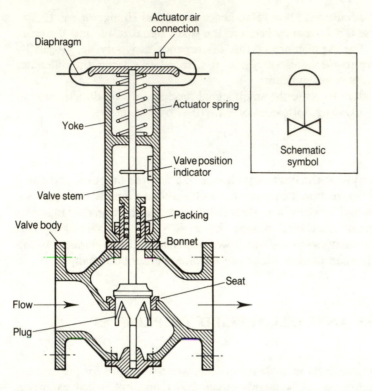

Figure 10.2 Single-ported globe valve with pneumatic actuator and its schematic symbol

equation for the valve coefficient can be written as

$$\frac{k_v}{k_{vo}} = f(h) \qquad (10.7)$$

k_{vo} is the valve coefficient when the control valve is fully open. One would think that k_v should be zero when a valve is completely closed ($h = 0$). In fact with most valves there is a small but finite flow ('leakage flow') when the valve is shut. That is, when h is 0

$$k_v = Lk_{vo} \qquad (10.8)$$

'L' is termed the leakage coefficient.

When the control valves are sized, the valve coefficient 'Cv' is used. The Cv of a valve is the number of US gallons per minute of water that flow through the valve when the pressure drop across it is one pound-force per square inch. The Cv factor is an internationally recognized way of defining control valve sizes. When valves are specified, the Cv factor refers to the flow through the **fully open** valve. k_v is of course proportional to Cv but the two are equal only if the dimensions of the variables in Eq. (10.6) correspond to the dimensions of the variables used to define the Cv factor.

Equation (10.6) can be written in a wholly non-dimensional way:

$$\frac{Q}{Q_o} = \frac{k_v}{k_{vo}} \sqrt{\frac{\Delta P}{\Delta P_0}} \tag{10.9}$$

The graph of Eq. (10.7) which shows how the fractional valve coefficient depends on valve lift is known as the *valve characteristic*. There are two commonly encountered types of control valve characteristics. One is the 'linear characteristic' and the other is the 'equal percentage' characteristic. Neglecting leakage, the valve characteristic of a linear valve is

$$\frac{k_v}{k_{vo}} = h \tag{10.10}$$

An equal percentage characteritic means that the fractional change in k_v is proportional to the change in lift, i.e.

$$\frac{dk_v}{k_v} \propto dh$$

Upon integrating we obtain

$$k_v = k_1 \, e^{k_2 h}$$

Substituting the boundary conditions for $h = 0$ and $h = 1$, the arbitrary constants k_1 and k_2 can be eliminated giving the equation:

$$\frac{k_v}{k_{vo}} = L^{(1-h)} \tag{10.11}$$

Figure 10.3 shows a graph of both types of characteristic. It should be realized that the behaviour of a valve as it nears total shut off is erratic, and Eqs (10.10) and (10.11) derived above are idealized. However above a certain opening they are good representations of the valve behaviour as a function of lift, and the leakage coefficient,

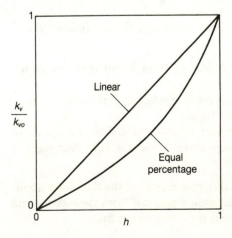

Figure 10.3 Linear and equal percentage control valve characteristics

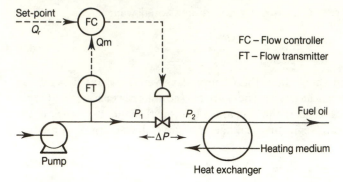

Figure 10.4 Flow control loop

L, should not be interpreted literally, but rather it should be considered as an empirical parameter to curve fit the actual characteristic.

Other valve characteristics are encountered depending on the valve type and application. For example with butterfly valves the Cv factor changes quickly at first and then remains more or less constant for the rest of the valve opening.

Example 10.2 Flow control system Figure 10.4 shows a typical process control system where the flow of fuel oil is controlled by a proportional controller which modulates the opening of a control valve. The fuel flows through a heat exchanger where it is heated so that its viscosity is suitable for firing in fuel oil burners. The fuel oil flow is measured by means of an orifice plate and an associated differential pressure transmitter (dp cell).

This sort of system is quite common in process plant. One of the problems is that the measurement system is nonlinear, i.e. the signal produced by the orifice plate/dp cell combination is proportional to the square of the flow rate. Another complicating factor is that if a general purpose controller is used, the set-point scale is simply calibrated linearly from 0–100%. Thus the set-point demand is simply a certain percentage of the measurement span.

Plant data The following data is available on the plant and the instrumentation:

(a) The design fuel oil flow rate is 8 m³/h.
(b) The control valve is linear with negligible leakage and its lift is 0.7 at design conditions.
(c) The measurement range of the orifice plate/dp cell arrangement is 0 to 10 m³/h.
(d) The pump delivers a constant head of 2.5 barg.
(e) The pressure drop across the heat exchanger is proportional to the square of the flow rate. At design conditions the pressure drop across the heat exchanger is 1 bar.

Requirements Obtain a non-dimensional small signal model of the flow loop about the design conditions. Hence find the controller gain which will limit the steady-state error to 5% of the change in flow rate demand (i.e. set-point change).

SOLUTION The first step is to write down the large signal equations of the components and then to perturb them about their respective design points and finally to combine the small signal models into an overall block diagram.

The controller The controller is a simple proportional controller and acts directly on the valve. The equation is

$$h = K(Q_r - Q_m) + \bar{h}$$

Q_r is the set-point and represents the required flow rate as a fraction of the measurement span. Q_m is the measured flow rate, again expressed as a fraction or percentage of the measurement span. K is the controller gain. The above equation is already non-dimensional and so the small signal non-dimensional equation is simply

$$\delta h = K(\delta Q_r - \delta Q_m)$$

The measurement system The pressure drop across an orifice plate is proportional to the square of the flow rate. We know from the plant data that the range of the dp cell is 0 to 10 m³/h. Hence it produces 100% output when the flow rate is 10 m³/h, i.e. at a flow rate of $1.25\bar{Q}$, where $\bar{Q}$ is the design fuel oil flow rate of 8 m³/h.

Thus the signal from the dp cell (Q_m) can conveniently be expressed in a large signal non-dimensional way by

$$Q_m = \left(\frac{Q}{1.25\bar{Q}}\right)^2$$

The small signal non-dimensional equation for the measurement system is obtained by differentiating the previous equation.

$$\delta Q_m = \frac{2}{1.25^2} \frac{\delta Q_m}{\bar{Q}}$$

The control valve The pressure drop across the control valve, ΔP, at design conditions is 1.5 bar ($\bar{P}_1 - \bar{P}_2$). The equation for the control valve can be written directly in a non-dimensional manner, namely

$$\frac{Q}{\bar{Q}} = \frac{h}{\bar{h}} \sqrt{\frac{P_1 - P_2}{\Delta \bar{P}}}$$

The small signal non-dimensional equation for the control valve is

$$\frac{\delta Q}{\bar{Q}} = \frac{\delta h}{\bar{h}} - \frac{1}{2} \frac{\bar{P}_2}{\Delta \bar{P}} \frac{\delta P_2}{\bar{P}_2}$$

The heat exchanger Finally the equation for the heat exchanger back pressure, P_2, can be written as

$$\frac{P_2}{\bar{P}_2} = \left(\frac{Q}{\bar{Q}}\right)^2$$

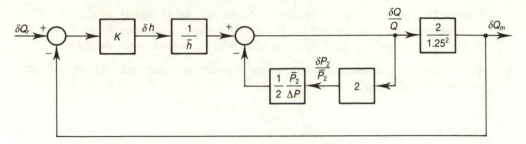

Figure 10.5 Small-signal block diagram of a flow control loop

which in turn can be expressed in a small signal manner as:

$$\frac{\delta P_2}{\bar{P}_2} = 2 \frac{\delta Q}{\bar{Q}}$$

The block diagram The four small signal equations can be combined to form a block diagram as shown in Fig. 10.5. The open loop gain is thus:

$$\frac{K}{\bar{h}} \frac{2}{1.25^2} \frac{1}{(1 + \bar{P}_2/\bar{\varDelta P})} = \frac{K}{0.7} \frac{2}{1.25^2} \frac{1}{(1 + 1/1.25)} = 1.097K$$

The requirement for steady-state errors of less than 5% is equivalent to having a steady-state loop gain which is greater than 19, (i.e. the positional error constant must exceed 19). Hence at quiescent conditions, a controller gain of approximately 17.3 lift/fractional error in flow will meet the specification.

10.4 DYNAMIC SYSTEM WITH NONLINEARITIES

When there is nonlinearity present in a dynamic system, it is often possible to divorce the nonlinear and dynamic effects. All the nonlinear effects can be lumped into one nonlinear block with no dynamics, i.e. the input and output relationship of the block is just the static characteristic of the nonlinearity. Equally, the dynamics of the system can be lumped into another block which represents the linear differential equations governing the dynamic behaviour of the system.

This idea can be presented as a 'block diagram' as shown in Fig. 10.6. The block diagram represents the flow of signals through the system as functions of **time**. The relationship between y and x is nonlinear and so Laplace transforms cannot be used. Moreover, once we are dealing with functions of time, strictly speaking, the input–output relationship of the linear dynamic element must be expressed as a transfer operator, $G(D)$, rather than the transfer function, $G(s)$.

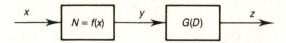

Figure 10.6 Block diagram of a nonlinear dynamic system

Simulating Nonlinearities in CODAS-II

CODAS-II includes an expression evaluator. The available functions are summarized in Appendix D.

The expression evaluator may be used simply as a calculator for obtaining the value of a mathematical expression, e.g. log(pi/4.2). It can also be used in the time domain to produce special input signals, e.g. sin(a*t + b), but in the nonlinear domain, the available functions can be used to synthesize almost any nonlinearity.

Nonlinearities are expressed as

$$y = f(x)$$

where 'y' (the dependent variable) can be considered to be the output of the nonlinearity and 'x' (the independent variable) can be considered as the input to the nonlinearity.

Smooth nonlinearities

Smooth nonlinearities can be modelled by expressing the nonlinearity as a polynomial or by using combinations of functions. For example

$$y = a + b*x + c*x^2 + d*x^3$$

where a, b, c and d can be parameters, numerical values or other expressions. As a further example an equal percentage valve characteristic can be expressed as

$$y = a^{(1-x)} \tag{10.12}$$

where a is the fractional leakage and x is the valve lift. The problem with this last example is that the independent variable, x, representing the valve lift is unconstrained and so valve lifts greater than 1 are possible. We shall see shortly that by using a piecewise linear function this problem can be solved.

Exercise Simulate using CODAS-II a 'hard-spring' non-linearity of the form

$$y = 0.5x + 0.07x^3$$

Use the cursor to establish the value of x (the deflection) when y (the force) is 1.

Piecewise Linear Functions

In addition to the standard mathematical functions listed in Appendix D, the expression evaluator in CODAS-II includes several 'piecewise' linear functions, i.e. functions which are linear over a certain region and then change slope suddenly, or functions which have a step discontinuity.

Figure 10.7 summarizes the most important of the piecewise linear functions. Two of the functions take a single operand, namely the 'relay' function and the 'sgn' function. Only the relay characteristic has been drawn in Fig. 10.7. The relay function can only take on one of two values (± 1). Its definition is

$$\text{relay}(x) = \begin{cases} +1 & x \geqslant 0 \\ -1 & x < 0 \end{cases}$$

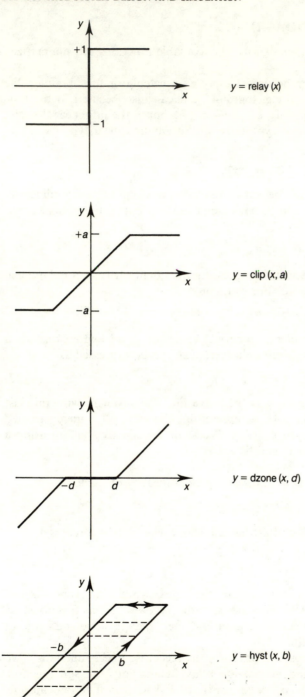

Figure 10.7 Piecewise linear functions in CODAS-II

The characteristic of the sgn function is very similar to the relay function but there is a subtle difference in that the sgn function can adopt one of three values (± 1 and 0). Its definition is

$$\text{sgn(x)} = \begin{cases} +1 & x > 0 \\ 0 & x = 0 \\ -1 & x < 0 \end{cases}$$

We shall see later how the sgn function can be applied. The relay function clearly mimics the behaviour of an ideal relay.

If the relay to be modelled switches from 0 to $+a$, the function

$$y = a*(\text{relay(x)} + 1)/2 \tag{10.13}$$

will model that situation.

There are three other standard piecewise linear functions that are particularly useful for simulating common nonlinearities. These functions require two operands, one is the independent variable, x, and the other is a constant. For example 'clip(x, a)' is used to simulate a saturation characteristic which limits or clips at $\pm a$. This function was used in the ship steering case study of Chapter 6 (Fig. 6.20) to simulate the limiting of the rudder angle. The deadzone function ('dzone') is illustrated in Fig. 10.7. Deadzone is encountered in certain types of amplifier where there is no output until the input reaches a certain threshold.

The function 'hyst' is a little different from the others in that it is not single-valued (i.e. the output, y, can take on many values for the **same** value of the input, x). 'hyst(x, b)' is used to simulate elements with hysteresis or backlash. The amount of hysteresis is governed by the value, b. This type of nonlinearity occurs when there is slack in the meshing of gears, when static or coulomb friction is present in position control systems and magnetic systems because of the effects of residual magnetism. Hysteresis is often introduced deliberately into relays to reduce chatter or the frequency of switching. It must be understood that there is no constraint on the x and y values and there is an infinite set of hysteresis curves. The horizontal region of the characteristic is entered as soon as the direction of x changes.

Exercise Simulate the four standard piecewise linear functions shown in Fig. 10.7 using CODAS-II. In the case of hysteresis use the cursor to show its multivalued nature.

The standard functions can be modified to produce other nonlinearities which are slightly different. For example, suppose that you want to simulate a unipolar amplifier of gain 'k' that limits at 'M' volts as shown in Fig. 10.8. This involves changing the slope of the standard function translating it to the right and raising it vertically. The resulting function is

$$y = \text{clip}(kx - M/2, M/2) + M/2 \tag{10.14}$$

In CODAS-II you will have to use numerical values for k and M or two of the parameters.

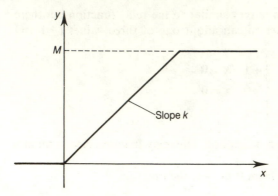

Figure 10.8 'Unipolar' saturation characteristic

Cascading Nonlinearities

One very powerful feature of the expression evaluator in CODAS is the ability to use functions of functions. This feature can be employed to create complex nonlinearities in a straightforward manner.

Earlier we saw (Eq. (10.12)) that the simulation of the equal percentage valve characteristic was not quite right because x (representing valve lift) could go above 1. This blemish can be corrected by first clipping x before passing it to the actual nonlinear characteristic, i.e. Eq. (10.12) is modified to read

$$y = a\char`^(1 - \mathrm{clip}(x,1)) \tag{10.15}$$

This equation is still not a hundred percent correct because x can go negative. From a practical point of view it does not matter greatly because all it means is that the leakage flow simulated in the model can go down to zero. Using the ideas in Eq. (10.14), the characteristic can be modelled more accurately if desired.

Exercise Use CODAS-II to plot the characteristic of an equal percentage control valve with a 5% leakage factor with and without clipping. At what value of lift is the valve flow coefficient 60% of maximum?

Suppose we want to simulate an amplifier that exhibits both deadzone and saturation as shown in Fig. 10.9(a). This can be simulated in CODAS-II by cascading (concatenating) two nonlinearities as shown in Fig. 10.9(b). The required function for CODAS-II is

$$y = \mathrm{clip}(c*\mathrm{dzone}(x,d),a)$$

where the parameters 'c', 'd', and 'a' represent the gain, deadzone and saturation limits respectively of the amplifier.

If the amplifier exhibits hysteresis as well (as shown in Fig. 10.9(c)) this can be included by introducing a hysteresis block as shown in Fig. 10.9(d). The CODAS-II function is

$$y = \mathrm{clip}(c*\mathrm{dzone}(\mathrm{hyst}(x,b),d),a)$$

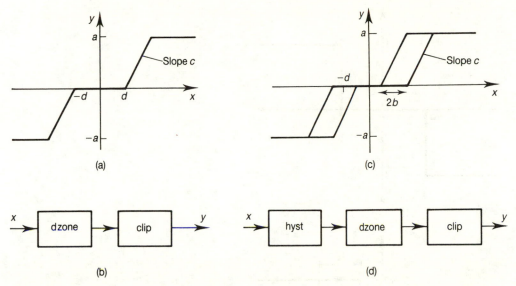

Figure 10.9 Production of complex nonlinearities by cascading. (a) Amplifier with deadzone and saturation. (b) Concatenation of two nonlinearities to produce overall characteristic. (c) Amplifier with hysteresis, deadzone and saturation. (d) Cascading of three nonlinearities

Great care must be taken over the order in which these functions operate. Quite different results are obtained if the order is changed.

One final example of the production of nonlinearities by cascading is a relay which exhibits deadzone. To produce this characteristic the relay function cannot be used because it only takes on two values. Here is where the 'sgn' function is useful.

The function

$$y = a*sgn(dzone(x,d))$$

produces the desired characteristic. If the relay exhibits hysteresis too, the function describing the more complex characteristic is

$$y = a*sgn(dzone(hyst(x,b),d))$$

Exercise
Use CODAS-II to simulate a relay with deadzone and a relay with deadzone and hysteresis as shown in Figs 10.10(a) and 10.10(b). Can you simulate a relay with no deadzone, but which exhibits hysteresis?

10.5 SIMULATING CLOSED-LOOP CONTROL WITH NONLINEARITIES

When nonlinearities are present in a feedback control system its behaviour can be altered significantly. In the following sections we will deal primarily with nonlinearities which are not smooth, i.e. they are represented by piecewise linear functions. Systems which

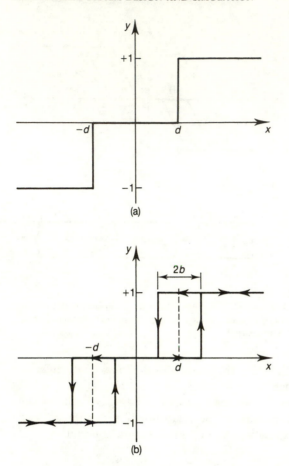

Figure 10.10 (a) Relay with deadzone. (b) Relay with deadzone and hysteresis

incorporate relays or where hysteresis occurs, which are subject to coulomb friction, etc., cannot be tackled using perturbation methods and their responses are quite different from linear systems. In order to illustrate some of the aspects of this class of control systems two examples will be considered.

In CODAS-II, the nonlinear block N, can appear either in the forward path or the feedback path of the control system as shown in Fig. 10.11. When it is in the forward path, the input, x, to the nonlinearity is the error, e, or the output of the controller (compensator), and the output of the nonlinear element is the control effort, u, applied to the plant. When the nonlinear element is in the feedback path, its input is the output of the plant, c, and its output, y, is subtracted from the reference value, r.

There is a third possibility where both the compensator and the nonlinearity appear in the feedback path. This third case is also shown in Fig. 10.11 for completeness.

As was discussed in Section 10.4, strictly speaking transfer operators should be used rather than transfer functions when dealing with nonlinear systems. In CODAS-II, however, you simply enter linear dynamics as transfer functions (i.e. functions of 's' for continuous systems or 'z' for discrete time systems).

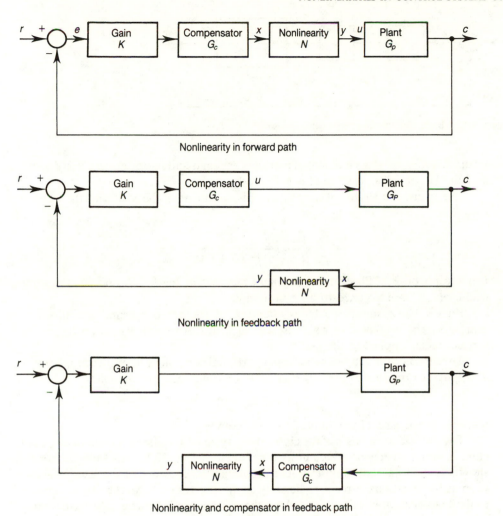

Figure 10.11 Block diagram of CODAS-II simulation with nonlinear element, N

Example 10.3 Position servo with deadzone The object of this example is to examine the behaviour of a servo system when the servo amplifier exhibits deadzone, i.e. it does not produce any output when the input is less than a certain value. This effect is encountered in certain types of power amplifier.

Consider the problem of a servo system positioning a rotational load, which was initially covered in Chapter 2 (Example 2.6). The differential equation relating the angular position of the load, θ, to the motor torque, T, is

$$T = J\frac{d^2\theta}{dt^2} + C\frac{d\theta}{dt}$$

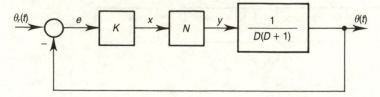

Figure 10.12 Block diagram of a servo with deadzone

where the effective mass moment of inertia of the load and motor armature is J kgm^2 and the viscous drag coefficient is C Ns/rad.

The transfer operator of the angular displacement of the load and the applied torque is

$$\frac{\theta}{T}(D) = \frac{1}{D(JD + C)}$$

For the purpose of illustration, the mass moment of inertia, J, and the damping coefficient, C, are assumed to have unit value.

Figure 10.12 shows the block diagram of the servo system. The block N represents the nonlinearity which is that of a deadzone and its output, y, is the torque that is applied to the load.

The servo system employs simple proportional control, i.e. the correcting torque, T, applied to the load is proportional to the misalignment of the load:

$$T = K(\theta_r - \theta)$$

where K is the gain of the amplifier/motor combination.

The unit step responses of the closed-loop system for different values of amplifier gain, K, for a deadzone of 0.1 units are shown in Fig. 10.13. All three responses show a certain amount of offset notwithstanding the presence of the integrator. As soon as the positional error is sufficiently small (less than $0.1/K$) the torque applied to the load is zero. In this regime where there is no applied torque, the load simply 'coasts' and it is the external viscous drag alone that decelerates the load. If the load stops moving within the error band $\pm 0.1/K$ then it will just remain at that position wherever it happens to be: hence the offset.

Exercise Simulate the above system with the amplifier gain set at unity. Obtain unit step responses for the linear system with initial position of the load at 0, $+0.8$ and -0.8. Repeat with the deadzone nonlinearity included. Why is the offset different in each case?

Coulomb Friction

Before looking at the next example, the effect of Coulomb friction (dry or static friction) will be considered briefly. Coulomb friction is experienced when there is insufficient lubrication in sliding contact or where there is unintentional rubbing. The effect of static friction is to apply a constant retarding load or torque that always opposes motion as

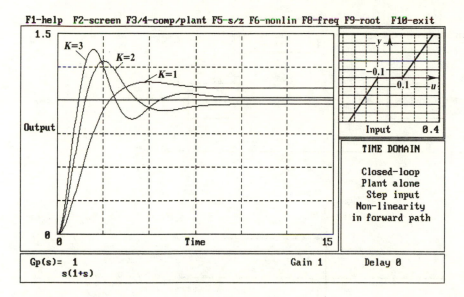

Figure 10.13 Step responses of servo with deadzone

shown in Fig. 10.14(b). Its characteristic is like that of a relay, but it has to be simulated by a nonlinearity in the feedback path as shown in Fig. 10.14(a).

Unfortunately it is not possible to simulate this particular topology using CODAS-II. However the effect of Coulomb friction is similar to that of deadzone discussed above, in that offsets arise when the load stops within a certain band, and the sinusoidal behaviour is identical to one with deadzone in the forward path. The transient behaviour of a system with coulomb friction is different, however, from that of one with deadzone as there is always an extra retarding torque or force present which will give the system greater damping.

Example 10.4 Thermostatic control of temperature Thermostatic control of temperature is quite common in domestic appliances and in simple industrial processes. A thermostat is just a relay that switches a heater on when the temperature is below the set temperature and switches it off if the temperature rises above the set temperature. It is obvious that if the temperature starts to change as soon as the heater comes on or off, the relay will be switching on and off very rapidly. This is termed chatter. In order to reduce chatter some hysteresis is often included in the relay. Clearly, introducing hysteresis will reduce the amount of chatter, but increase the fluctuations in temperature.

Consider the following example of the thermostatic control of an electrically heated hot water tank. In the steady state, the water temperature rises by 8°C above ambient for every amp of heater current. The thermostat has 2°C hysteresis and when it is energized the heater current is 10 amps. The dynamic response of the heater can be modelled by a dominant lag of 10 minutes and a second lag of 1 minute.

Given that the ambient temperature is 20°C we shall use CODAS-II to simulate the response of the system from ambient with a temperature set-point of 50°C.

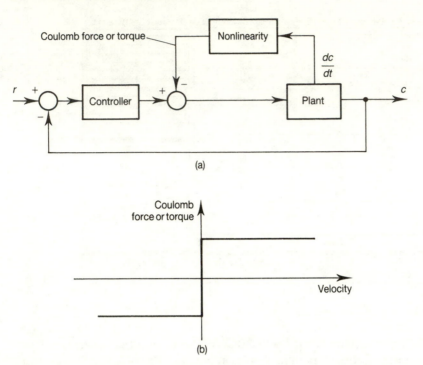

(a)

(b)

Figure 10.14 Coulomb friction and its simulation. (a) Block diagram of system with Coulomb friction. (b) Coulomb friction characteristics

A straightforward linear simulation would simply model the system as having a gain of 8°C/A with two lags. With a nonlinear system a little more care must be taken. The most direct approach is to model the thermostat as producing an output temperature of 20°C (ambient) when the temperarture error is less than −2°C (i.e. actual temperature is more than 2°C above the set temperature) and as producing an output temperature of 100°C (80°C above ambient) when the temperature error is greater than 2°C.

Following the earlier discussion on cascaded nonlinearities, the equation for the thermostat then becomes

$$y = 40*(\text{relay}(\text{hyst}(x,2)) + 1) + 20$$

which can be thought of as a variable element of 80°C and a steady component of 20°C.

The plant dynamics are simply

$$G(s) = \frac{1}{(1 + 600s)(1 + 60s)}$$

The initial condition of the system is 20°C. The results of the simulation are shown in Fig. 10.15. The interesting feature about the response is that it displays a clear and sustained oscillation which is **not** sinusoidal. This oscillation is termed a *limit cycle*.

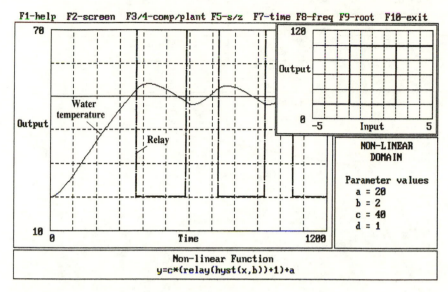

F1-help F2-screen F3/4-comp/plant F5-s/z F7-time F8-freq F9-root F10-exit

Figure 10.15 Transient response of a thermostatically controlled water heater

10.6 STABILITY OF NONLINEAR SYSTEMS

When a linear system is on the point of instability (marginal stability), theory states that it will oscillate sinusoidally at the phase-crossover frequency. In practice it takes a determined effort to generate a pure stable sinewave from a real system. Again in theory, if the gain is increased slightly beyond the critical gain, the output of the system grows indefinitely. In practice a system may be unstable but nonlinearities will always come into effect which will limit the size of the oscillation. Another way of thinking about it is that as the oscillations become larger the perturbation model is no longer valid and the effective gain of the system drops as the sizes of the excursions grow.

To illustrate these ideas consider the system whose block diagram is shown in Fig. 10.16. This system consists of a PI controller and a second-order Type-0 plant. The output of the controller is subject to soft saturation, which means that the sensitivity of the system to changes in control effort drops as the working point increases. A smooth nonlinearity is really required to simulate this situation but to simplify matters a piecewise linear function has been used as shown in Fig. 10.16, where the slope of the nonlinearity changes to a much lower value when the operating point goes above 1.75. Figure 10.17 shows how this system responds to a demand signal that initially is 0.5 units, then after 40 seconds increases to 3.5, is maintaned at that value for a further 40 seconds and then reduced to the original value of 0.5.

At first the system operates in a region where the system gain is relatively large and so it behaves in an unstable manner and the output grows in an oscillatory fashion. When the excursions increase beyond a certain point the effective gain becomes lower and the oscillations reach a steady amplitude, i.e. they exhibit a *stable limit cycle*. When the demand level is changed to 3.5, the operating point is now in a region where the small signal gain is low, and the response is stable. Hence the output of the system

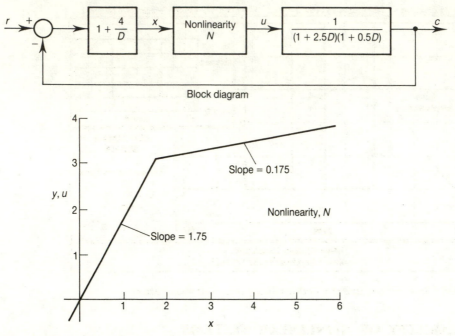

Figure 10.16 System with soft saturation

settles down with no error to the demand value of 3.5. Finally the original demand value is restored and the initial oscillations are large, but they gradually reduce until there is a stable limit cycle once more.

Thus a limit cycle exists when the effective open-loop gain of the system is unity and where the overall open-loop phase shift is 180°. In the above example, the nonlinearity did not introduce any phase shift and so the frequency of oscillations corresponds to the phase-crossover frequency of the system without the nonlinearity. The phase-crossover frequency of the linear part of the system is 0.224 Hz and the open-loop gain at this frequency is about 0.65. The simulation shows that the frequency of the limit cycle is about 0.22 Hz which is in good agreement.

Following the previous argument, for the system to have an overall loop gain of unity, the effective gain of the nonlinearity needs to be about $1/0.65$ (1.5) to produce a stable limit cycle. It can be seen by examining the characteristic shown in Fig. 10.16 that the overall 'gain' of the nonlinearity drops rapidly once the peak of the input signal reaches the corner of the characteristic at $x = 1.75$, $y = 3.2$. This will limit the amplitude of the controller output oscillations to about 2.7 (i.e. $3.2 - 0.5$). Because of the frequency response characteristics of the plant this amplitude is further reduced, explaining in general terms the amplitude of the observed limit cycle.

Describing Functions

The underlying ideas of the above example can be used to obtain an equivalent gain of a nonlinearity in a more generalized and rigorous manner. The reader may have

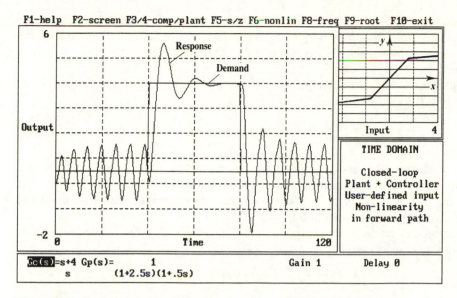

Figure 10.17 Response of system with soft saturation to different levels of demand

observed that the limit cycle oscillation is approximately sinusoidal in nature, rather more so in Fig. 10.17 than with the thermostatically controlled water heater of Fig. 10.15. The reason for the sinusoidal appearance of the limit cycle can be understood by considering the frequency components in the control signal and the frequency response of the system.

If a limit cycle exists then the output of the controller is periodic. The frequency of a periodic signal is termed the *fundamental frequency*. Any periodic signal is made up of a fundamental component (sinusoid) at the fundamental frequency and a set of harmonic components at frequencies which are integer multiples of the fundamental frequency. For most signals the higher frequency harmonics tend to have lower amplitude than the low frequency components. The control effort is applied to the plant and so the set of harmonics constituting the control signal are filtered or shaped by the frequency response characteristics of the plant. Generally speaking, the system under control can be considered to be a low-pass filter, i.e. high frequencies are attenuated. Indeed any real system tends to be low pass as discussed in Chapter 4.

The reason, therefore, for the sinusoidal appearance of the limit cycle, even in the case of the thermostat where the control effort is a square wave, is that the system under control filters out the high frequency harmonics leaving the fundamental component as the preponderant contribution in the output signal. The preponderance of the fundamental component in the output signal is the basis of the method known as the 'describing function' method, which allows stability analysis of nonlinear systems and the prediction of the frequency and amplitude of limit cycles.

The assumption of the method is that the input to the nonlinearity is sinusoidal and only the fundamental component in the output need be considered. In this way the nonlinearity can be treated as a complex gain, termed the *describing function*, **N**. The describing function is analogous to the frequency response of a linear dynamic system as described in Section 4.2. The main difference is that the complex gain, $G(j\omega)$, associated with linear dynamic system is frequency dependent and independent of amplitude, whereas the complex gain, $\mathbf{N}$, is amplitude dependent but generally frequency independent.

Consider a nonlinearity subjected to a sinusoidal input signal, $x(t)$, of amplitude X and period T, as shown in Fig. 10.18(a),

$$x(t) = X \cos(\omega t) = X \cos\left(\frac{2\pi t}{T}\right) \tag{10.16}$$

When a sinusoidal signal is applied to a nonlinearity, the output signal is still periodic in T, but its shape has been distorted and so it will contain harmonics. The shape of the output waveform, $y(t)$, can be obtained by 'projecting' $x(t)$ on to the nonlinear characteristic as illustrated in Fig. 10.18(b).

Neglecting any steady or dc component, the output signal, $y(t)$, can be represented as a harmonic (Fourier) series

$$y(t) = Y_1 \cos(\omega t + \theta_1) + Y_2 \cos(2\omega t + \theta_2) + \ldots$$

$$= \sum_k Y_k \cos(k\omega t + \theta_k) \tag{10.17}$$

Y_1 and θ_1 represent the amplitude and phase of the fundamental component in the output signal, Y_2 and θ_2 represent the amplitude and phase of the second harmonic, etc.

Because the signal, $x(t)$, is periodic, it is convenient to express it in terms of an angle α, where

$$\alpha = \frac{2\pi}{T} t$$

i.e.

$$x(\alpha) = X \cos(\alpha)$$

Similarly the output signal can be expressed in terms of the angle, α,

$$y(\alpha) = \sum_k Y_k \cos(k\alpha + \theta_k)$$

The output signal can conveniently be considered as a series of harmonic vectors $\mathbf{Y}_1$, $\mathbf{Y}_2$ etc., each of which is rotating at a frequency which is an integer multiple, k, of the excitation frequency ω. The describing function, $\mathbf{N}$, is the vector ratio of the fundamental of the output, $\mathbf{Y}_1$, to the input vector, $\mathbf{X}$ (as illustrated in Fig. 10.18(c))

$$\mathbf{N} = \frac{\mathbf{Y}_1}{\mathbf{X}} = \frac{Y_1}{X} e^{j\theta_1} \tag{10.18}$$

The describing function will depend on the amplitude, X, of the excitation signal. Generally the vector notation is not used and the describing function is therefore simply written as $N(X)$.

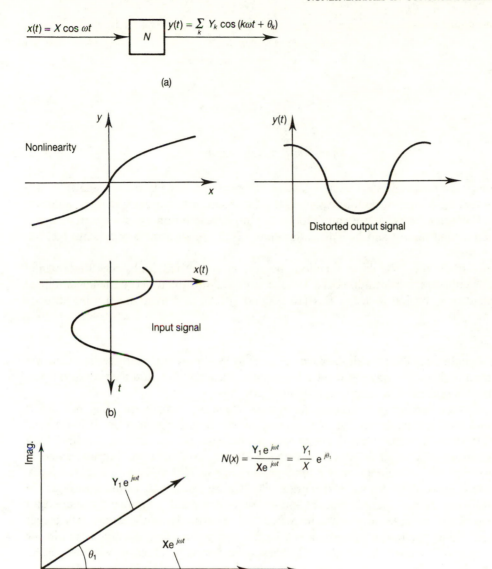

Figure 10.18 (a) Input/output signals of a nonlinearity. (b) Waveform distortion through a nonlinearity. (c) Vector representation of describing function

Calculating the Describing Function of a Nonlinear Element

The harmonic components that constitute the output of the nonlinearity can be calculated using Fourier analysis. The amplitude of the kth harmonic is given by

$$Y_k = (a_k^2 + b_k^2)^{1/2} \tag{10.19}$$

and its phase is

$$\theta_k = \tan^{-1}(-b_k/a_k) \tag{10.20}$$

where

$$a_k = \frac{1}{\pi} \int_{-\pi}^{\pi} y(\alpha) \cos(k\alpha) \, d\alpha \tag{10.21}$$

$$b_k = \frac{1}{\pi} \int_{-\pi}^{\pi} y(\alpha) \sin(k\alpha) \, d\alpha \tag{10.22}$$

Although the range of integration is taken as $-\pi$ to $+\pi$, any interval of length 2π will do as well. a_k, b_k are the familiar terms in the Cartesian or trigonometric representation of the Fourier series. a_k is the amplitude of the in-phase component in the output signal (i.e. the cosine terms) and b_k is the amplitude of the quadrature components (i.e. the sine terms).

As far as the describing function method is concerned, only the fundamental components need to be evaluated, i.e. $k = 1$. The describing function can then be obtained by calculating Y_1 and θ_1 using Eqs (10.19) and (10.20) and then substituting directly into Eq. (10.18).

Example 10.5 Relay with deadzone Figure 10.19 shows the characteristic and the input and output waveforms of a relay with a deadzone, d, and maximum output, M, when excited by a sinusoid of amplitude X.

Before applying the theory developed above it is worth spending just a few minutes thinking about the likely nature of the describing function. It is obvious that when X is less than d, the output of the relay is zero and so the describing function, $N(X)$, will be zero too. When X slightly exceeds d the output waveform will contain a small amount of fundamental and so $N(X)$ increases slightly. As X gets bigger the output waveform gradually becomes more like a square wave as the deadzone becomes less significant and so the fundamental component in the output waveform increases and $N(X)$ increases too. Eventually as the input gets bigger and bigger, the output waveform will hardly change and so the fundamental in the output will be essentially constant and hence the describing function will start getting smaller. The nonlinearity does not introduce any phase shift and so the describing function will be real.

By inspecting the characteristic of the deadzone nonlinearity and the input/output waveforms in Fig. 10.19, it can be seen that no phase shift is introduced when the signal passes through the nonlinearity. Therefore the harmonic components in the output signal are in phase with the input signal and thus the b_k terms in the Fourier expansion are zero.

The nonlinearity is an odd function (i.e. $f(x) = -f(-x)$) and so the output waveform retains the half wave rotation symmetry of the excitation signal, i.e. $y(\alpha) = -y(\alpha \pm \pi)$. Furthermore, as an excitation signal with even symmetry has been used (i.e. a cosine wave where $x(\alpha) = x(-\alpha)$), the output signal has even symmetry too. Thus the output waveform exhibits two symmetries and hence there

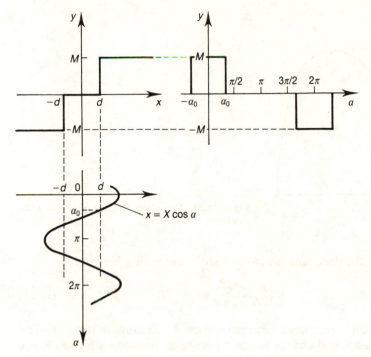

Figure 10.19 Input and output waveforms of a relay with deadzone

is only need to integrate over a quarter of a period. Thus

$$Y_1 = \frac{4M}{\pi} \int_0^{\alpha_0} \cos(\alpha) \, d\alpha = \frac{4M}{\pi} \sin(\alpha_0)$$

where α_0 is $\cos^{-1}(d/X)$. The describing function, $N(X)$, is thus

$$N(X) = \frac{4M}{\pi X} \sin(\alpha_0) \tag{10.23}$$

The above relationship for $N(X)$ shows that it depends on the excitation amplitude and, via the angle α_0, on the size of the deadzone, d. As the angle, α_0, is a function of the ratio of the deadzone to the excitation amplitude, it is more convenient to express $N(X)$ in terms if the ratio d/X. Multiplying both sides of Eq. (10.23) by d and rearranging we obtain

$$\frac{N(X)}{M/d} = \frac{4}{\pi} \frac{d}{X} \sin(\alpha_0) \tag{10.24}$$

Figure 10.20 shows the graph of $N(X)/(M/d)$ as a function of d/X. The describing function reaches a peak when d/X is about 0.7. It can be proved that the peak actually occurs when d/X is $1/\sqrt{2}$, and that the peak value of $N(X)d/M$ is $2/\pi$.

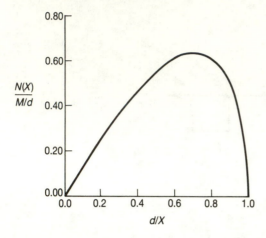

Figure 10.20 Describing function of a relay with deadzone

In an ideal relay the deadzone is zero and so its describing function is

$$N(X) = \frac{4M}{\pi X} \tag{10.25}$$

The describing function is very large for small values of excitation in the case of an ideal relay as an infinitesimal change in the input signal produces a finite change in the output. As the amplitude of the relay output is always the same, at higher amplitudes of input signal the effective gain (describing function) drops. Unlike the relay with deadzone, the describing function of an ideal relay is a monotonic function. The describing function of a relay with deadzone approximates to the ideal relay when $X \gg d$ (i.e. $\alpha_0 \to 0$).

Describing Function and Stability of Closed-loop Systems

Consider a simple feedback system which includes a nonlinearity in the forward path as shown in Fig. 10.21. In this system the linear dynamics have been lumped into a single transfer operator $G(D)$. The open-loop frequency response function is obtained by replacing the D operator by $j\omega$ in the plant transfer operator and replacing the nonlinearity by its describing function, $N(E)$. ($N(E)$ is applicable here because it is assumed that the input to the nonlinearity is the error, $e(t)$.)

It was argued earlier in this section that a limit cycle existed for that system when the effective open-loop gain was unity at the frequency where the system exhibited 180° phase shift. Using the ideas of describing function we can write this condition as:

$$N(E)G(j\omega) = -1$$

or

$$G(j\omega) = \frac{-1}{N(E)}$$

The Nyquist stability criterion for linear systems required that the $G(j\omega)$ locus did not

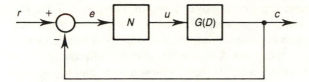

Figure 10.21 Feedback system with a nonlinearity

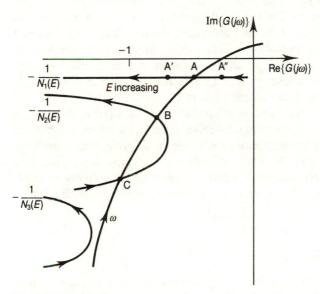

Figure 10.22 Stability assessment using describing function contours in the $G(j\omega)$ plane

encircle the -1 point. The above equation can be viewed as an extension of this concept. The right hand side of the above equation can be considered as a 'wandering' -1 point, i.e. the $-1/N(E)$ can be drawn as a contour in the complex $G(j\omega)$ plane.

Drawing an analogy with the stability criterion for linear systems and ignoring any approximations in the theory of describing functions one can state that if $-1/N(E)$ lies inside the $G(j\omega)$ locus, the closed-loop system is unstable. If $-1/N(E)$ lies outside the $G(j\omega)$ contour the system is stable and a limit cycle exists if the $G(j\omega)$ locus intersects the $-1/N(E)$ locus.

However we need to consider the situation in a little more detail before drawing any firm conclusions.

Consider the locus $-1/N_1(E)$ in Fig. 10.22. The describing function contour crosses the frequency response locus $G(j\omega)$ at the point A. The arrows indicate the direction of increasing amplitude of error, E. Suppose an oscillation exists whose amplitude places it at the point A' on the describing function contour. A' represents a point where

$$|G(j\omega)| < 1/|N_1(E)|$$

i.e. the effective '-1' point is not enclosed by the $G(j\omega)$ locus. The effective loop gain at this point is less than 1 ($|G(j\omega)|\,|N_1(E)| < 1$) and so the amplitude of the oscillation will tend to decrease. Thus the operating point moves towards the point A on the describing function contour. By a similar argument, at point A″, where the amplitude of the oscillation is smaller, the loop gain is greater than 1 and so there will be a tendency for oscillations to grow so moving to the point A again. Thus there will always be a tendency to 'fall' back to the point A. Hence a stable limit cycle will exist and the system will operate at point A. The amplitude and frequency of the limit cycle can be determined from the point of intersection.

The locus $-1/N_2(E)$ shows two points of intersection. Following the previous arguments, point B represents a stable limit cycle. If however, a condition exists where the oscillation is very small so that the operating point on the describing function contour lies to the left of C, the oscillations will decrease in amplitude. Hence point C does not represent any stable limit cycle. In fact the system will be stable for small disturbances and will only develop a limit cycle if an excitation of sufficient amplitude is introduced to push the operating point to the right of C.

The third locus, $-1/N_3(E)$, is the most problematic. In simple terms there is no crossing with the $G(j\omega)$ locus and so the system is ostensibly stable. However no absolute statement can be made regarding the existence of a stable limit cycle. The describing function method is approximate and the effect of higher-order harmonics is ignored. All that can be said about the last situation is that a limit cycle is unlikely to exist. Clearly if the two loci come very close there is less certainty about the prediction of stability.

Stability of Systems with Relays

The $-1/N(X)$ contour for an ideal relay is a straight line that starts at the origin of the $G(j\omega)$ plane for small values of X and goes to $-\infty$ as X increases. Hence any system with a frequency response curve that crosses the negative real axis will exhibit a limit cycle if there is a relay in the loop.

The describing function of a relay with deadzone was developed in Example 10.5. The $-1/N(X)$ locus of this nonlinear element also lies along the negative real axis but starts at $-\infty$ for small values of X, moves towards the origin as X increases and turns back on itself when d/X is $1/\sqrt{2}$. Thus the presence of deadzone in a relay can act as a stabilizing factor and prevent cycling.

> **Example 10.6 Relay with hysteresis** The relay considered above introduced no phase shift between its input and output, hence the associated describing function was real. When elements have hysteresis a phase shift is introduced between the input and output signals and so the describing function associated with such elements is complex.
>
> A relay with hysteresis, b, was encountered in Example 10.4 which considered the thermostatic control of a hot water tank. The input and output waveforms of a relay with hysteresis are shown in Fig. 10.23 and a similar set of graphs are shown in Fig. 10.24 which were obtained using CODAS-II. The output waveform is simply a square wave which lags the applied signal by an angle, β, where

$$\beta = \sin^{-1}(b/X)$$

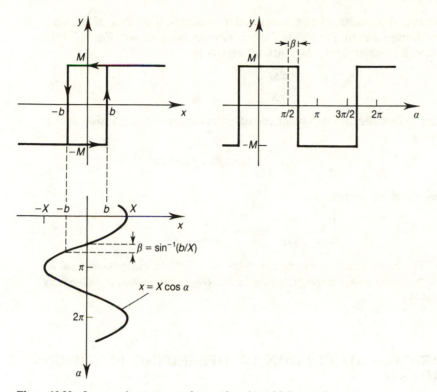

Figure 10.23 Input and output waveforms of a relay with hysteresis

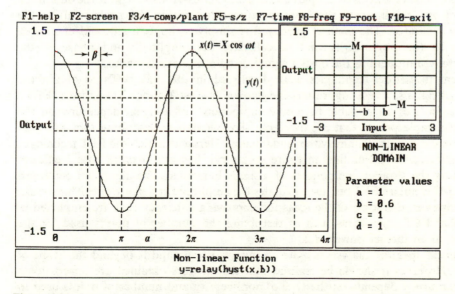

Figure 10.24 Input and output waveforms of a relay with hysteresis using CODAS-II

Neglecting the lag, the output of this nonlinearity is identical to that of an ideal relay. The describing function for an ideal relay was obtained above (Eq. (10.25)), thus the describing function of a relay with hysteresis is

$$N = \frac{4M}{\pi X} e^{-j\beta} \tag{10.26}$$

The form of the $-1/N$ contour can be obtained by inverting the expression in Eq. (10.26) as follows

$$-\frac{1}{N} = -\frac{\pi X}{4M} e^{+j\beta} = -\frac{\pi X}{4M} (\cos(\beta) + j\sin(\beta))$$

By substituting for β we obtain

$$-\frac{1}{N} = -\frac{\pi X}{4M} \cos(\beta) - j\frac{b\pi}{4M} \tag{10.27}$$

Equation (10.27) shows that the imaginary part of $-1/N$ is independent of X. Hence the $-1/N$ contour is a horizontal line in the $G(j\omega)$ plane whose imaginary value is $-b\pi/(4M)$.

10.7 NUMERICAL CALCULATION OF DESCRIBING FUNCTIONS USING CODAS-II

The analytical methods described above require a good knowledge of calculus and of Fourier methods. Analytical techniques are limited to simple nonlinearities whose output waveforms can be described by simple equations. CODAS-II will calculate the describing function of any nonlinearity that can be represented within the package using the expression evaluator. The method employed is simply to apply a sinusoid of a given amplitude to the nonlinearity and to calculate the amplitude and phase of the fundamental component in the output signal using a numerical method.

Efficient numerical techniques were developed in the mid 1960s to obtain the spectrum of signals. These algorithms are known as Fast Fourier Transforms (FFT). An FFT calculates all possible harmonic components of a signal depending on the number of sample points that are available. To calculate a describing function only one harmonic is required, namely the fundamental. Hence although the FFT techniques are very efficient for calculating an entire spectrum, they are 'overkill' and inefficient for obtaining the Fourier components of a single harmonic. The algorithm employed in CODAS-II is not an FFT but just calculates the amplitude and phase of one harmonic. It does, however, use some of the ideas for improving efficiency that are employed in conventional FFT algorithms such as restricting the number of points used for the calculation to an integer power of 2.

Numerical spectral analysis is quite an involved topic and is beyond the scope of this book. However it should be realized that the answers obtained are approximate and their accuracy depends on the type of nonlinearity and number of points used in the calculation. Errors arise primarily because of aliasing as the output signals of

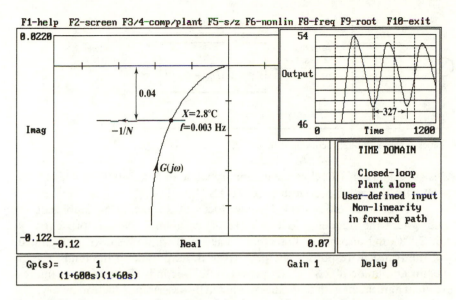

Figure 10.25 Frequency domain analysis of the thermostatic control of temperature

piecewise linear functions are not bandlimited. These errors can be reduced by increasing the number of points used to calculate the fundamental component. The default number of points used in CODAS-II in the calculation for a single value of the describing function is 128, which gives sufficient accuracy for most nonlinearities and is not unduly slow on the average PC. However the number of points per calculation can be increased up to 512.

In order to see how CODAS-II can be used to calculate describing functions and provide stability assessments two examples will be considered.

Example 10.7 Thermostatic temperature control The thermostatic control of temperature was the subject of Example 10.4. The responses drawn in Fig. 10.15 clearly demonstrated that a limit cycle existed. The question is can describing function theory predict the limit cycle, its amplitude and frequency.

Figure 10.25 shows the Nyquist diagram of the system and superimposed on it the $-1/N$ contour. The numerically produced contour is a horizontal line. The theory developed above (Example 10.6) predicts that the imaginary part of the describing function is constant at $-b\pi/4M$. Where M represents the peak output of the relay. In the case of the thermostat the relay effectively produces an output of 100°C when the error is positive and 20° when the error is negative. The theory developed above assumed that the relay output was symmetrically positive and negative. Adding a constant to the output signal of a nonlinearity does not affect the fundamental component and so has no effect on the describing function. It is only the 'alternating' component in the relay output that matters. Thus from the point of view of describing function, the output of the thermostat should be considered as having a quiescent value of 60°C and an alternating value of $\pm40°$C.

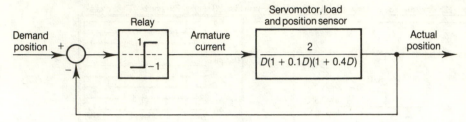

Figure 10.26 Block diagram of a relay-controlled servo

Thus M is 40°C and b is 2°C, thus the imaginary value of N is -0.0393. This figure agrees well with the result produced by CODAS-II.

The $-1/N$ contour crosses the Nyquist locus at an excitation amplitude of 2.8°C and where the frequency is 0.003 Hz. Thus a limit cycle is predicted of amplitude 2.8°C and of period 330 seconds. The small window shows a magnified extract of the transient response. The peak to peak amplitude of the limit cycle is 5.7°C, i.e. an amplitude of 2.85°C. Its period is 327 seconds. Both of these figures are in excellent agreement with the values predicted by describing function theory.

Example 10.8 Relay control of a position servo This last example considers the behaviour of a servo system whose block diagram is given in Fig. 10.26. The relay switches a current of ± 1 amp into the armature of the motor. The transfer function of the motor, load and position dynamics is

$$G(s) = \frac{2}{s(1 + 0.1s)(1 + 0.4s)}$$

Figure 10.27 shows the step response of the closed-loop system to a unit demand of position. There is clearly a limit cycle of amplitude 0.22 and period 1.32 seconds. The small window shows the frequency response of the motor/load combination with the cursor at the position where the locus crosses the describing function contour. Using the cursor, the amplitude of the excitation at that point is 0.206 and the frequency on the Nyquist locus is 5 rad/s which corresponds to a period of about 1.25 seconds. The agreement this time is a little worse, but nevertheless within quite acceptable engineering tolerances.

Exercise Simulate the system in Example 10.8, but modify the nonlinear characteristic so that it incorporates a deadzone, d.

The theory of Example 10.5 showed that the peak value of N is $2M/d\pi$. Hence when there is deadzone present the $-1/N$ contour will not intersect the Nyquist locus provided at the phase-crossover frequency

$$\frac{d\pi}{2M} > |G(j\omega)|$$

Calculate the required value of deadzone. Investigate the stability of the system as the deadzone is increased progressively from this value in steps of about 20%. What are your conclusions regarding using deadzone as a method of stabilizing the system?

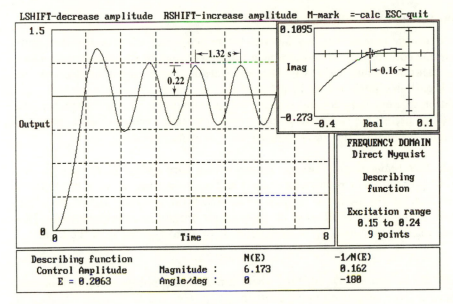

LSHIFT-decrease amplitude   RSHIFT-increase amplitude   M-mark   =-calc ESC-quit

Figure 10.27 Step response and describing function of a relay-controlled servo

Stability Predictions for Systems with Offset

Great care must be taken in using describing functions to predict the behaviour of systems where the operating point of the nonlinearity is 'off centered'. The development of the describing function theory assumed there was no dc bias in the excitation signal to the nonlinear element. In Type-0 systems this is generally not true.

To illustrate the fallacy consider a relay control system similar to the one used in Example 10.8, except instead of the Type-1 plant consider the behaviour of the system with a plant whose transfer function is

$$G(s) = \frac{2}{(1 + 0.1s)(1 + 0.4s)(1 + 0.6s)}$$

This system has very similar gain and frequency characteristics around the phase-crossover region to the one considered in Example 10.8. Introducing a deadzone of 0.15 units in the relay characteristic, according to standard describing function theory, should stabilize the system. However, when the system is simulated, the unit step response clearly shows a stable limit cycle. The reason is that because the system is Type-0, there is an offset and so the mean error is not zero. This means that there is a bias in the excitation signal and the true describing function is much bigger than for signals with zero average value.

Exercise Simulate the above system using CODAS-II when the relay has a deadzone of 0.15 units. Examine its step response, and its response when the set-point is zero but the initial value of the plant output is 1 unit. Comment on the differences in the two responses.

Simulating Systems with Intrinsic Nonlinearities

In many systems the nonlinearity is a function of the output of the system rather than a function of the control effort. For example a transducer may be nonlinear and distort or limit the measured value. CODAS-II has the feature of placing the nonlinearity in the feedback path as shown in Fig. 10.11.

In addition to simulating control systems with nonlinearities in the feedback path, this feature is useful for simulating open-loop dynamic systems with intrinsic nonlinearities. To illustrate the ideas consider the following example.

> **Example 10.9 Simulating a vibration system with a hard spring** 'Hard springs' are springs whose stiffness increases with deflection. They sometimes occur in mechanical systems because the point of support of a structural member moves with the deflection of the spring or because of the inherent properties of some springs. The equation reiating the force, F_k, to deflextion, x, of a hard spring is of the form:

$$F_k = kx + ax^3 \qquad (10.28)$$

Consider the system in Fig. 10.28(a) where a mass is supported on a hard spring and there is some viscous damping present.

The equation of motion is

$$F - C\frac{dx}{dt} - kx - ax^3 = m\frac{d^2x}{dt^2}$$

This equation can be rewritten in operator form as

$$(mD^2 + CD + k)x = F - ax^3 \qquad (10.29)$$

This equation can be recast in the form

$$x = (F - ax^3)G(D) \qquad (10.30)$$

where

$$G(D) = \frac{1}{(mD^2 + CD + k)}$$

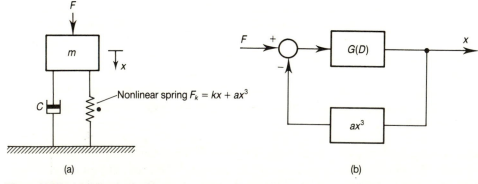

(a) (b)

Figure 10.28 (a) Mass/spring/damper system with nonlinear spring. (b) Block diagram of CODAS-II simulation of a system with a hard spring

Equation (10.30) can be considered to be the equation of a feedback system as shown in Fig. 10.28(b). F is the input to the 'control' system and the term $'-ax^3'$ represents the nonlinear feedback function. This system can now be simulated in CODAS-II by defining the nonlinearity as ax^3, enabling the nonlinearity and placing it in the feedback path.

Exercise The 'jump' phenomenon is a well known characteristic of certain nonlinear systems. The jump effect occurs, for example, during frequency response testing of systems with hard spring nonlinearity. During the frequency response test, the system is excited by a sinewave whose frequency is gradually increased. What the experimenter observes is that the amplitude of oscillation of the system gets bigger as the frequency of the test signal approaches the resonant frequency and then suddenly the amplitude of oscillation becomes smaller.

Use CODAS-II to simulate this effect as follows:
The linear system transfer function is

$$G(s) = \frac{1}{(s^2 + 0.25s + 1)}$$

and the coefficient, a, associated with the nonlinear part of the spring characteristic in Eq. (10.28) is 0.03.

Use a 'swept sine' excitation signal of the form

$$r(t) = \sin((0.95 + t/2000)t)$$

Simulate the system as described above; use 1000 points and observe the response of the nonlinear system over a time interval of 500 seconds. You should see the amplitude of oscillation grow and then suddenly decrease.

Contrast the behaviour when a linear spring is used $(a=0)$. Don't forget with a linear system the resonant frequency will be lower because the effective stiffness is less.

Example 10.10 Simulating a pendulum The equation of motion of a simple pendulum of mass, m, and length, l, subject to viscous damping, C, is given by

$$ml\frac{d^2\theta}{dt^2} + Cl\frac{d\theta}{dt} + mg\sin(\theta) = 0 \qquad (10.31)$$

For small angles the term $\sin(\theta)$ on the right hand side can be replaced by θ which leads to a standard second-order linear differential equation. For large angles, however, the nonlinear formulation must be used. Equation (10.31) can be rewritten as

$$D^2\theta + \frac{C}{m}D\theta = 0 - \frac{g}{l}\sin(\theta)$$

This equation can be viewed as that of a non-unity feedback control system with zero set-point as shown in Fig. 10.29.

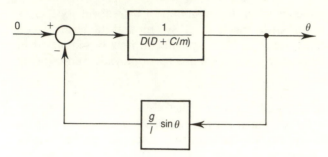

Figure 10.29   Block diagram of CODAS-II simulation of a nonlinear pendulum

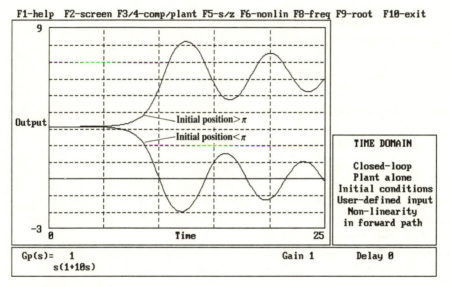

Figure 10.30   Responses of nonlinear pendulum model

The forward path transfer operator is

$$G(D) = \frac{1}{D(D + C/m)}$$

and the nonlinear function in the feedback path is

$$f(\theta) = \frac{g}{l}\sin(\theta)$$

Figure 10.30 shows the response for a pendulum length of 1 m and a damping term C/m of 10 second^{-1}. In this simulation the pendulum is released from initial positions close to vertical (i.e. the initial angle is close to π). The lower trace in Fig. 10.30 shows that the pendulum stays nearly balanced for a while and then falls over and oscillates about the equilibrium position. If the initial angle is slightly greater than vertical, the pendulum falls the other way as shown in the upper trace, i.e. the angle gets bigger and the system oscillates about an angle of 2π and then eventually settles. This equilibrium position of 2π is of course physically the same as in the first trace.

PROBLEMS

10.1 Linearize the following equations about the working points indicated.

(a) $y = kx^3$: $\bar{x} = 4.0$, $\bar{y} = 2.0$

(b) $y = k\,e^x z$: $\bar{x} = 0.8$, $\bar{z} = 2.5$, $\bar{y} = 6.0$

(c) $h = \dfrac{ax^y}{z}$: $\bar{x} = 10.0$, $\bar{y} = 2.0$, $\bar{z} = 4.0$, $\bar{h} = 40.0$.

10.2 The following is the equation for the flow of liquid through a square-edged orifice of width W and length X. The upstream pressure is P_s, and the mass density of the liquid is ρ.

$$Q = C_d W X \left[\frac{2(P_s - P_l)}{\rho} \right]^{1/2}$$

Obtain a small perturbation model for changes in X and changes in P_1 given the following quiescent conditions:

$\bar{W} = 10$ mm $\rho = 0.92 \times 10^3$ kgm^{-3}

$C_d = 0.65$ $\bar{P}_1 = P_s/2$

$P_s = 30$ barg $\bar{X} = 0.05$ mm

10.3 The torque/speed characteristics of a small ac motor are shown in Fig. P10.3. The motor has a mass moment of inertia of 2×10^{-6} kgm^2 and the viscous friction coefficient associated with the motor bearings

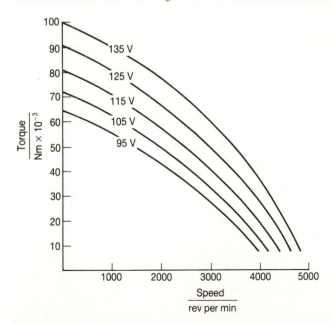

Figure P10.3   Torque/speed characteristics of a 2-phase induction motor

is 10^{-4} Nm/rads^{-1}. The motor is connected by a shaft to a friction load of viscous coefficient of friction 2×10^{-4} Nm/rads^{-1} which is designed to rotate steadily at a speed of 2000 rev/min.

Show that the transfer function relating the motor speed, ω, to the control voltage, V_c is of the form

$$\frac{\omega}{V_c}(s) = \frac{k}{(1 + s\tau)}$$

Furthermore, show using perturbation methods that for the given motor the values of k and τ are approximately 2.3 rads^{-1}/volt and 4.6 ms respectively.

10.4 A steam turbine is used to drive a centrifugal pump. The flow of steam to the turbine is controlled by the position, θ, of a rotary valve. The torque, T, delivered by the turbine expressed in non-dimensional form is given by

$$\frac{T}{\bar{T}} = 1.8\left[\frac{\theta}{\bar{\theta}}\right]^{1/2} - 0.8\left[\frac{\omega}{\bar{\omega}}\right]^{3/2}$$

where ω is the speed of the turbine/pump combination and the bar notation represents quiescent or normal operating conditions. The load torque of the pump is proportional to the square of its speed. Linearize the equation of the turbine/pump and load and hence deduce that a 3.56% change in steam valve position will produce a 1% change in speed.

10.5 Figure P10.5 shows a constant head reservoir feeding a cylindrical tank of uniform cross-sectional area A m^2 via a linear control valve. The outflow, Q_o, is via a fixed valve. It can be assumed that for a given valve opening, the flow rate is proportional to the square root of the head difference across the valve. The normal operating conditions are $\bar{H}$, $\bar{Q}$ and $\bar{h}$ for the level in the tank, the outflow and the valve lift respectively.

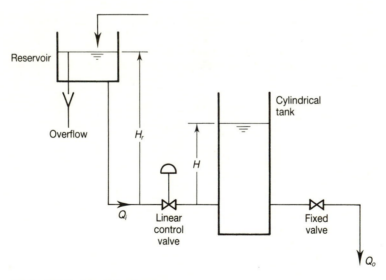

Figure P10.5 Liquid level system

Obtain equations for:

(a) the flow rate, Q_i, through the control valve as a function of lift, h, and head difference $(H_r - H)$;
(b) The flow rate, Q_o, through the fixed valve as a function of H;
(c) The volumetric balance for the tank.

Obtain linearized equations for the system for small changes in control valve lift and hence obtain a linearized block diagram relating changes in level, δH to the change in valve lift δh. Show that the effective time constant, τ, of the system is given by

$$\tau = 2A \frac{\bar{H}}{\bar{Q}}\left(1 - \frac{\bar{H}}{H_r}\right)$$

10.6 A system is controlled by an amplifier that has a gain of 4, and saturates at ± 5 volts as shown in Fig. P10.6. Prove that the describing function of an amplifier of gain, k, saturating at an output voltage of M volts is given by

$$N(X) = \frac{2k}{\pi}\left[\alpha + \frac{\sin 2\alpha}{2}\right]$$

where $\alpha = \sin^{-1}\dfrac{(M/k)}{X}$

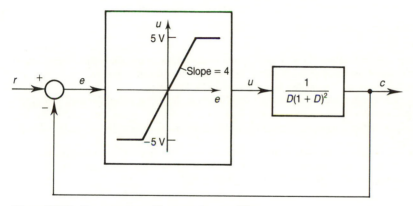

Figure P10.6 Servo system with saturating amplifier

Plot the frequency response of the linear part of the system and superimpose on it the describing function contour of the nonlinearity. Show that a limit cycle must exist. Find its amplitude and frequency and confirm the result by simulation.

10.7 A liquid level control system on a feedstock vessel uses a level switch to operate an inlet and outlet pump. If the level differs by more than ± 10 cm from set-point, liquid is pumped into or out of the vessel at a constant flow rate F m^3 h^{-1} otherwise neither pump operates. The design value of F is 2.5 m^3 h^{-1}.

The effective transfer function of the system relating the inflow, Q, to the level in the vessel, H, is given by

$$\frac{H}{Q}(s) = \frac{k}{s(1 + \tau s)^2}$$

where Q is in m^3 h^{-1}, H is in m, the constant, k, is 1.35 m^{-2} and the time constant, τ, is 6 minutes.

Obtain a block diagram of the system. Simulate the response of the system to a step demand in level of 1 m. Show that a stable limit cycle exists of period 40 minutes and of amplitude 20 cm with the design value of F.

Draw the Nyquist diagram of the linear part of the open-loop transfer function. Using the results for the describing function of a relay with deadzone, deduce the theoretical maximum value of F that will prevent a limit cycle from occurring.

10.8 Figure P10.8 shows the block diagram of an instrument servo for positioning an indicator. There is a $\pm 1°$ backlash present because of the gearing system that is used to position the indicator ($b = 2°$).

The describing function, $N(X)$, of friction controlled backlash is given by

$$\text{Re}(N) = \frac{1}{\pi}\left[\frac{\pi}{2} + \beta + \frac{\sin 2\beta}{2}\right]$$

$$\text{Im}(N) = \frac{\cos^2 \beta}{\pi}$$

where $\beta = \sin^{-1}(1 - b/X)$

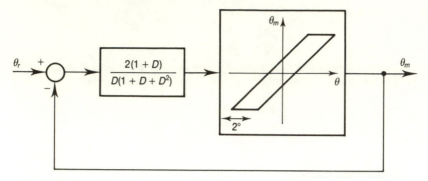

Figure P10.8 Instrument servo with backlash

Plot the describing function contour for a range of values of excitation amplitude from 1.5° to 7° on a Nyquist diagram. Superimpose the frequency response of the linear part of the system.

Hint: put $G(s)$ in the controller transfer function when using CODAS-II.

Which of the two points of intersection represents a stable limit cycle? What is the predicted amplitude and frequency of the limit cycle?

Confirm the frequency domain predictions by examining the response of the system to a 10° step change in the demand position, θ_r.

Hint: CODAS-II cannot simulate the time response of closed-loop systems with the nonlinearity enabled where the plant transfer function is 1. Therefore change $G_p(s)$ to be a very small lag, say of time constant 0.01 seconds. Also use a large number of points to get accurate simulation.

COMPLEX NUMBER MANIPULATION

A1 BASICS

A complex number, z, consists of a real part, x, and an imaginary part, y, i.e.

$$z = x + jy; \qquad j = \sqrt{-1}$$

Alternatively the number can be represented in polar form as a magnitude, r, and an angle, θ, i.e.

$$z = r\, e^{j\theta}$$

The magnitude is also termed the modulus and is written as

$$|z| = r = \sqrt{x^2 + y^2}$$

The angle is also called termed the argument and is written as

$$\angle z = \theta = \tan^{-1} \frac{y}{x}$$

The real and imaginary parts can be expressed in terms of the magnitude and angle

$$x = r \cos \theta; \qquad y = r \sin \theta$$

so that
$$z = r[\cos \theta + j \sin \theta]$$

A2 PRODUCT AND QUOTIENT OF COMPLEX NUMBERS

If z is the product of two complex numbers, z_1 and z_2, then

$$z = z_1 \cdot z_2 = r_1\, e^{j\theta_1} \cdot r_2\, e^{j\theta_2}$$

$$= r_1 \cdot r_2\, e^{j(\theta_1 + \theta_2)}$$

Hence
$$|z| = r_1 \cdot r_2 = |z_1| \cdot |z_2|$$

and
$$\angle z = \theta_1 + \theta_2 = \angle z_1 + \angle z_2$$

Similarly if z is the quotient of two complex numbers, z_1, and z_2, then

$$z = \frac{z_1}{z_2} = \frac{r_1 \, e^{j\theta_1}}{r_2 \, e^{j\theta_2}}$$

$$= \frac{r_1}{r_2} e^{j(\theta_1 - \theta_2)}$$

Hence
$$|z| = \frac{r_1}{r_2} = \frac{|z_1|}{|z_2|}$$

and
$$\angle z = \theta_1 - \theta_2 = \angle z_1 - \angle z_2$$

A3 COMPLEX CONJUGATES

The complex conjugate, $\bar{z}$, of a number, z, has the same real part, but its imaginary part is equal and opposite, i.e.

$$z = x + jy = r \, e^{j\theta}$$

$$\bar{z} = x - jy = r \, e^{-j\theta}$$

The product of a complex conjugate pair is always real:

$$z \cdot \bar{z} = r \, e^{j\theta} \cdot r \, e^{-j\theta} = r^2$$

A4 TRIGONOMETRIC IDENTITIES

Consider the complex number

$$e^{j\theta} = \cos \theta + j \sin \theta$$

and its conjugate

$$e^{-j\theta} = \cos \theta - j \sin \theta$$

Adding the above equations gives

$$\cos \theta = \frac{e^{j\theta} + e^{-j\theta}}{2}$$

and substracting them gives

$$\sin \theta = \frac{e^{j\theta} - e^{-j\theta}}{2j}$$

TABLE OF LAPLACE AND z-TRANSFORMS

	$x(t)$	$X(s)$	$x(i)$	$X(z)$
Impulse	$\delta(t)$†	1	$\delta(i)$‡	1
Unit step	$1(t)$	$\dfrac{1}{s}$	$1(i)$	$\dfrac{z}{z-1}$
Unit ramp	t	$\dfrac{1}{s^2}$	iT	$\dfrac{Tz}{(z-1)^2}$
	t^2	$\dfrac{2}{s^3}$	$(iT)^2$	$\dfrac{T^2 z(z+1)}{(z-1)^3}$
	e^{-at}	$\dfrac{1}{s+a}$	β^i	$\dfrac{z}{z-\beta}$
	$1-e^{-at}$	$\dfrac{a}{s(s+a)}$	$1-\beta^i$	$\dfrac{(1-\beta)z}{(z-1)(z-\beta)}$
	$t\,e^{-at}$	$\dfrac{1}{(s+a)^2}$	$iT\beta^i$	$\dfrac{T\beta z}{(z-\beta)^2}$
	$\sin(\omega t)$	$\dfrac{\omega}{s^2+\omega^2}$	$\sin(i\theta)$	$\dfrac{\sin(\theta)z}{z^2-2\cos(\theta)z+1}$
	$\cos(\omega t)$	$\dfrac{s}{s^2+\omega^2}$	$\cos(i\theta)$	$\dfrac{1-\cos(\theta)z}{z^2-2\cos(\theta)z+1}$
	$e^{-at}\sin(\omega t)$	$\dfrac{\omega}{(s+a)^2+\omega^2}$	$\beta^i\sin(i\theta)$	$\dfrac{\beta\sin(\theta)z}{z^2-2\beta\cos(\theta)z+\beta^2}$
	$e^{-at}\cos(\omega t)$	$\dfrac{s+a}{(s+a)^2+\omega^2}$	$\beta^i\cos(i\theta)$	$\dfrac{1-\beta\cos(\theta)z}{z^2-2\beta\cos(\theta)z+\beta^2}$

$$\beta = e^{-aT}; \qquad \theta = \omega T$$

† Continuous time impulse, $\delta(t) = \begin{cases} \lim\limits_{\tau\to 0} \;\; 1/\tau; & 0 < \tau < t \\ 0\;; & \text{otherwise} \end{cases}$

‡ Discrete time impulse, $\delta(i) = \begin{cases} 1; & i = 0 \\ 0; & i \neq 0 \end{cases}$

TABLE OF SELECTED MODIFIED z-TRANSFORMS

The table lists the modified z-transform for selected continuous time systems with fractional transport delay, δ, when preceeded by a zero-order hold.

$$0 \leqslant \delta < T$$

$$G(z) = \frac{(z-1)}{z} Z_m \left\{ \frac{G(s) \, e^{-s\delta}}{s} \right\}$$

$G(s)$	$G(z)$
$\dfrac{1}{s}$	$\dfrac{(T-\delta)z + \delta}{z(z-1)}$
$\dfrac{1}{s^2}$	$\dfrac{(T-\delta)^2 z^2 + (T^2 + 2T\delta - 2\delta^2)z + \delta^2}{2z(z-1)^2}$
$\dfrac{1}{1+\tau s}$	$\dfrac{(1-\Gamma)z + (\Gamma - \beta)}{z(z-\beta)}$
$\dfrac{1}{s(1+\tau s)}$	$\dfrac{(T-\delta+\tau(\Gamma-1))z^2 + (T-2\tau\Gamma + (\tau - T + \delta)(1+\beta))z + (\tau\Gamma - \beta(\tau+\delta))}{z(z-1)(z-\beta)}$

$$\beta = e^{-T/\tau}; \qquad \Gamma = e^{-(T-\delta)/T}$$

SUMMARY OF CODAS-II COMMANDS

LIST OF CONFIGURATION FILE KEYWORDS

graphpaper   colour of the graph background
textpaper   colour of the text background
axespen   colour of the axes
borderpen   colour of the borders
labelpen   colour of the axes labels
plotpen1   colour of plot 1
plotpen2   colour of plot 2
plotpen3   colour of plot 3
plotpen4   colour of plot 4
textpen   colour of text
font   font type
graphics   adaptor type

COMMANDS AVAILABLE IN CODAS-II

Function Keys

F1   Help
F2   Flip screens
F3   Compensator + plant, $Gc(s)$
F4   Plant only, $Gp(s)$
Alt + F3   Edit compensator
Alt + F4   Edit plant
F5 z-transform transfer function
Alt + F5   Restore continuous transfer function
F6   Non-linear environment (N)
Alt + F6   Edit non-linear function

F7	Time domain environment (*T*)
F8	Frequency domain environment (*F*)
F9	Root locus environment (*R*)
F10	Quit CODAS and exit to MS-DOS
Alt + F10	Temporary exit to MS-DOS

Parameters

Alt + A	Change 'a' parameter
Alt + B	Change 'b' parameter
Alt + C	Change 'c' parameter
Alt + D	Change 'd' parameter

Cursor Control

All cursors

Esc	Cancel cursor
=	Calculator

Time domain environment

@	Enable cursor
Left Shift	Decrease time
Right Shift	Increase time
A	Absolute mode
R	Relative mode

Root locus environment

@	Enable cursor
Left arrow	Move left
Right arrow	Move right
Up arrow	Move up
Down arrow	Move down

Frequency domain environment

@	Enable cursor
Left Shift	Decrease frequency
Right Shift	Increase frequency

Non-linear environment

@	Enable cursor
Left Shift	Decrease input
Right Shift	Increase input

Free-wheeling cursor

.	Enable cursor
Left arrow	Move left
Right arrow	Move right
Up arrow	Move up
Down arrow	Move down
A	Absolute mode
R	Relative mode
M	Mark/line

Annotation cursor

/	Enable cursor
Left arrow	Move left
Right arrow	Move right
Up arrow	Move up
Down arrow	Move down
M	Mark/line
T	Enter text

Key	Meaning	T	F	R	N
B	Base options	×	×	×	×
C	Closed-loop	×	×	×	—
D	Denominator	×	×	×	×
E	Enable nonlinearity	×	×	—	—
F	Frequency range/nonlinear function	—	×	—	×
G	Go, start simulation	×	×	×	×
H	Sample-hold time	×	×	×	×
I	User-defined input/frequency increment	×	×	—	—
J	Number of points (resolution)	×	×	×	×
K	System gain constant	×	×	×	×
L	Load system description, environment	×	×	×	×
M	M-contour	—	×	—	—
N	Numerator	×	×	×	×
O	Open-loop	×	×	×	—
P	Print results to file	×	×	×	×
Q	Quit current branch	—	—	×	—
R	Reference input/excitation range/damping ratio	×	×	×	—
S	Save system description and environment	×	×	×	×
T	Transport delay	×	×	×	×
U	Plot control effort/unit circle	×	—	×	—
V	View (Nyquist/Nichols/Bode)	—	×	—	—
W	Wipe active screen clear	×	×	×	×
X	Plot width	×	×	×	×
Y	Plot height	×	×	×	×
Z	Plot origin/initial conditions	×	×	×	×
=	Calculator	×	×	×	×
Esc	Abort activity, return to command level	×	×	×	×

EXPRESSION EVALUATOR FUNCTIONS

Mathematical Operations

+	Addition
−	Subtraction
*	Multiplication
/	Division
^	Exponentiation

Functions With One Parameter

abs(x)	Absolute value of x
atn(x)	Arctan of x (in radians)
cos(x)	Cosine of x (in radians)
db(x)	Gain x converted to decibels
deg(x)	Angle x in radians converted to degrees
exp(x)	e^x
gauss(x)	Gaussian random variable of variance x

idb(x)	Decibels x converted to gain
ln(x)	Natural logarithm of x
log(x)	Logarithm base 10 of x
rad(x)	Angle x in degrees converted to radians
relay(x)	Relay function ($+1$ for $x > 0$, -1 for $x \leqslant 0$)
sgn(x)	Signum function ($+1$ for $x > 0$, 0 for $x = 0$, -1 for $x < 0$)
sin(x)	Sine of x (in radians)
sqr(x)	Square root of x
tan(x)	Tangent of x (in radians)

Functions With Two Parameters

clip(x,limit)	Saturation
dzone(x,deadzone)	Deadzone
hyst(x,backlash)	Hysteresis
mod(x,y)	x modulus y
pulse(x,duration)	Pulse
quant(x,step-size)	Quantization

SALES ENQUIRIES FOR THE FULL VERSIONS OF CODAS-II AND PCS

Outside North America:

Golten and Verwer Partners
33 Moseley Road
Cheadle Hulme
Cheshire SK8 5HJ
UK

Tel. (061) 485 5435

North American distributors:

Dynamical Systems Inc.
PO Box 35241
Tucson
Arizona 85740
USA

Tel. (602) 292 1962